Urban Sustainability

Editor-in-Chief

Ali Cheshmehzangi, Department of Architecture and Built Environment, University of Nottingham Ningbo China, Ningbo, Zhejiang, China

The Urban Sustainability Book Series is a valuable resource for sustainability and urban-related education and research. It offers an inter-disciplinary platform covering all four areas of practice, policy, education, research, and their nexus. The publications in this series are related to critical areas of sustainability, urban studies, planning, and urban geography.

This book series aims to put together cutting-edge research findings linked to the overarching field of urban sustainability. The scope and nature of the topic are broad and interdisciplinary and bring together various associated disciplines from sustainable development, environmental sciences, urbanism, etc. With many advanced research findings in the field, there is a need to put together various discussions and contributions on specific sustainability fields, covering a good range of topics on sustainable development, sustainable urbanism, and urban sustainability. Despite the broad range of issues, we note the importance of practical and policy-oriented directions, extending the literature and directions and pathways towards achieving urban sustainability.

The series will appeal to urbanists, geographers, planners, engineers, architects, governmental authorities, policymakers, researchers of all levels, and to all of those interested in a wide-ranging overview of urban sustainability and its associated fields. The series includes monographs and edited volumes, covering a range of topics under the urban sustainability topic, which can also be used for teaching materials.

Zaheer Allam
Editor

Sustainable Urban Transitions

Research, Policy and Practice

Springer

Editor
Zaheer Allam
Curtin Mauritius
Charles Telfair Campus
Moka, Mauritius

ISSN 2731-6483 ISSN 2731-6491 (electronic)
Urban Sustainability
ISBN 978-981-99-2694-7 ISBN 978-981-99-2695-4 (eBook)
https://doi.org/10.1007/978-981-99-2695-4

This Springer imprint is published by the registered company Springer Nature Singapore Pte Ltd.
The registered company address is: 152 Beach Road, #21-01/04 Gateway East, Singapore 189721, Singapore

Preface

Cities have been around for millennia and have slowly morphed through time. However, it is until very recently that the urban population boomed, bringing about a new array of challenges, as well as opportunities. Tapping into those opportunities, both the public and private sectors have helped to shape the cities that we know today, and today we experience this change. While the quality of life is of the utmost importance for urban dwellers, is it a key indicator for private developers, or has it far too often been used as a tool for political mileage? As the answers to those questions will surely vary across geographies, there is one common need, that of transitioning to more sustainable, resilient, and safer environments to protect one's culture, home, and life. Those transitions are, however, costly for most urban areas in the global south, demanding new means of financing and problem-solving. Furthermore, while not all solutions can be replicated at face value, there is a need to ponder on what is being done, and what can be done to transform our cities to enable a higher quality of life.

Today, more than ever, we face the need for retrospection and discovery, as we collectively face the combined impacts of the COVID-19 pandemic and the increasing challenges from geopolitical forces. Our cities, of varying sizes, are being impacted, and as they host the global majority, this is resulting in pressures on the quality of life of urban dwellers and on climate change responses and preparedness. This book thus comes at an important time and aims to map out contemporary challenges across geographies and forecast how solutions can be unpacked from both a pragmatic and policy perspective.

The 24 shortlisted chapters featured in this book are grouped under four themes and collectively bring value to the theme of sustainable urban transitions. In the first section entitled 'Sustainable Urban Futures', chapters explore the concepts of One Planet Living, the benefits of using public transit, and how to use social media as a means for sustainability. In the second section entitled 'Mapping Contemporary Urban Challenges', chapters focus on case studies in varying geographies and present empirical assessments and frameworks. The third section, entitled 'Policy Advances in Urban Sustainability', explores broad questions and highlights how urban transitions can be unpacked from a governance perspective, and in doing so showcases case studies. The last section, entitled 'Forecasting Urban Solutions', underlines how urban transformation and urban regeneration can be channeled via various means such as urban mobility, housing, nature-based solutions, and others.

While the questions sought from the authors will surely evolve through time, the excellent submissions provide ground for thought and can help in fueling new models for change.

Moka, Mauritius Zaheer Allam

Contents

Sustainable Urban Futures

The Promise and Potential of One Planet Living 3
Ajaykumar Manivannan and Daina Mazutis

Enjoying the Commute: Wellbeing Benefits of the Public Transport Journey 19
Peter J. Davies, Kathryn McLachlan, Aileen Westbrook, and Waminda Parker

Fostering Sustainable Urban Futures Through Twitter Public Space ... 37
Paola Monachesi

Mapping Contemporary Urban Challenges

An Empirical Approach to Estimating Secondary School Student's Sustainable Mobility Score 57
Isabel Cruz and Luís M. Afonso

The Challenges to Sustainable Urban Future in a Rapidly Changing World: A Case Study of Izmir, Turkey 73
Ebru Ersoy Tonyaloğlu and Engin Nurlu

Nature-Based Solutions for Climate Adaptation in School Environments: An Interdisciplinary Assessment Framework 87
Isabel Ruiz-Mallén, Francesc Baró, Mar Satorras, Funda Atun, Nathalie Blanc, Sarah Bortolamiol, Lidia Casas, Céline Clauzel, Elsa Gallez, Eddy Grand-Meyer, Àgueda Gras-Velázquez, Ivelina Ivanova, Tim Nawrot, Paula Presser, Diana Reckien, and Filka Sekulova

Urban Transition in Monsoon Asian Megacity: Will Paddy Fields in the Rural–Urban Fringe Persist? Insight from Jakarta-Bandung Mega Urban Region, Indonesia 107
Didit Okta Pribadi, Khalid Saifullah, La Ode Syamsul Iman, Muhammad Nurdin, and Andi Syah Putra

Controlling the Valves: Dealing with Conflict in a Transition Initiative for Public–Private Water Governance in Amsterdam 125
E. A. Van Der Heijden, B. C. Van Mierlo, S. J. H. Majoor, and P. J. Beers

Waterfront Revitalization as Opportunity for Sustainability Transitions—The Huangpu River in Shanghai 149
Harry Den Hartog, Fan Ding, Xiaojing Wang, and Jiang Wu

Policy Advances in Urban Sustainability

How 15-min City, Tactical Urbanism, and Superblock Concepts Are Affecting Major Cities in the Post-Covid-19 Era? 163
Luca Maria Francesco Fabris, Federico Camerin, Gerardo Semprebon, and Riccardo Maria Balzarotti

Urban Mobility and Time Geography Within the Smart City Narrative 189
Aditya Dixit and Zaheer Allam

Is Gamification an Enabler of the Sustainable Urban Transition Process? 201
Maria Giovina Pasca, Laura Di Pietro, and Roberta Guglielmetti Mugion

Regional and Project Management in the Context of Shaping Sustainable Urban Futures: Theoretical and Practical Context of the EU Programming Period 2021–2027 in the Czech Republic 219
Martin Pelucha and Jana Kourilova

Transitioning to Circular Plastic Economies in Cities: Conceptual and Policy Implications of a Practice Perspective 239
Olamide Shittu and Christian Nygaard

The Brunswick Spine: An Example of Green TOD in Melbourne, Australia? 255
Rajjan M. Chitrakar, Peter Hogg, and Dillan Eshow

Nightlife and Urban Livability—The Case of Mashhad 275
Kasra Talebian and Muge Riza

Forecasting Urban Solutions

Actualising 'Autonomous Cities' Via Artificial Intelligence for Achieving Netzero Urban Futures 299
Zaheer Allam and Aditya Dixit

Sustainable Urban Mobility Governance: Rethinking the Links Through Movement, Representation and Practice for a Just Transport System 311
Eda Beyazit and Fatih Canitez

Rethinking the Sustainable Housing Concept in the Post-pandemic Era 329
Osama Omar

Transforming Obsolete Spaces into Vital Places: Historic Italian Villages as Laboratories of Sustainability 345
Rossana Galdini, Silvia De Nardis, and Nicoletta Cerreti

Augmenting Mobility Safety in Cities by Increasing Data Pools from Connected Urban Devices 363
Aditya Dixit and Zaheer Allam

***Litoral Besòs*, an Urban Sustainability Transition in the Barcelona Metropolitan Area** 375
Nancy Andrea Ramírez-Agudelo, Joan de Pablo, and Elisabet Roca

Learning from Experience: Reflections on the Delivery of Nature-Based Solutions as Part of a Multi-Partner Collaborative Project 391
Ian Mell, Sarah Clement, Fearghus O'Sullivan, Juliet Staples, Christine Derbyshire, Paul Nolan, Clare Olver, and Stella Shackel

Integrating Indigenous Lifestyle in Net-Zero Energy Buildings. A Case Study of Energy Retrofitting of a Heritage Building in the Southwest of Western Australia 407
Vittoria Strazzeri and Reena Tiwari

About the Editor

Dr. Zaheer Allam holds a Ph.D. in Humanities, a Master of Arts (Res), an MBA, and a Bachelor of Applied Science in Architectural Science from universities in Australia and the United Kingdom. Based in Mauritius, he is the Chairperson of the National Youth Environment Council (NYEC) and a board member of the Mauritius Renewable Energy Agency (MARENA) and works on a number of projects on the theme of Smart Cities and on strategies dwelling in the increasing role of technology in Culture and the Society. Through his engagement in research, Zaheer is an Honorary Fellow at Deakin University and Research Fellow for the Sorbonne Business School and Hiroshima University. Zaheer is also the African Representative of the International Society of Biourbanism (ISB), member of the Advisory Circle of the International Federation of Landscape Architects (IFLA), and a member of a number of other international bodies. He holds a number of awards and commendations, including an elevation to the rank of Officer of the Order of the Star and Key of the Indian Ocean by the President of Mauritius. He is the author of over 115 peer reviewed publications and author of 8 books on the subject of Smart, Sustainable and Future Cities.

Sustainable Urban Futures

The Promise and Potential of One Planet Living

Ajaykumar Manivannan and Daina Mazutis

Abstract This chapter explores the One Planet Living (OPL) framework for sustainable urban development. The OPL Framework is a system of principles and tools for the design, planning and building of new sustainable residential mixed use urban communities, including zero carbon and zero waste objectives. To put progressive climate change mitigation strategies into practice, the OPL framework was first implemented and tested in the BedZed development in England over 25 years ago. Since then, it has been applied to dozens of new construction and development projects worldwide. We begin by introducing the OPL and presenting several case studies of OPL in practice from around the world. We then situate the OPL framework within the broader literature on sustainable urban development to illustrate the promise and potential of this unique approach to construction and development.

1 Introduction

Over the next 40 years, more than 230 billion square meters of new construction will have to be built to accommodate the world's growing population—a number the UN estimates as the equivalent of adding a city the size of Paris to the planet every week [33]. Given the carbon intensity and waste associated with current construction methods, if all new development is executed using current methods, this urban expansion alone will raise global temperatures by 1–2 °C [20]. Urban development at this scale and scope thus comes with great responsibility, not just towards its immediate residents, but also towards future citizens who will inherit the social and environmental consequences of these newly built neighbourhoods. It therefore behoves all stakeholders involved in the built environment sector (i.e., urban planners, landowners, real estate developers, architects, infrastructure engineers, construction companies etc.) to consider the sustainability of every new construction and development project [27].

A. Manivannan · D. Mazutis (✉)
University of Ottawa, 55 Laurier Avenue East, Ottawa, ON K1N 6N5, Canada
e-mail: daina.mazutis@telfer.uottawa.ca

Z. Allam (ed.), *Sustainable Urban Transitions*, Urban Sustainability,
https://doi.org/10.1007/978-981-99-2695-4_1

To address this need, a plethora of frameworks, standards, policies, measurement models, rating systems, assessment tools, technical guidelines and indicators have evolved over the last fifty years to promote sustainable urban development [24–29, 38]. By sustainable urban development (SUD), we mean development that meets the needs of the present without compromising the ability of future generations to meet their own needs by balancing the economic, environmental and social aspects of urban life [1, 28]. Implementing sustainable urban development plans can occur at different scales—from individual buildings (e.g., LEED, BREEAM), to larger communities (e.g., EcoDistricts), cities (e.g., RFSC, EcoCity), countries (e.g., New Urban Agenda) and continents (e.g., ECI), often guided by goals developed by international organizations (e.g. UN SDG 11). Our interest in this chapter, however, is not to compare these different frameworks, but rather to introduce and elaborate on one particular framework, One Planet Living (OPL), and explore how this system of tools and principles has been used to help design, plan and build sustainable residential mixed use urban communities that tackle important climate mitigating objectives such as zero carbon and zero waste goals. While the OPL framework has been successfully implemented around the world, its adoption level remains surprisingly low. This chapter thus aims to highlight the OPL framework's benefits and its differentiating factors from other SUD frameworks.

We begin by describing the OPL framework in more detail and presenting several case studies of OPL in practice from around the world. We then situate the OPL framework within the broader literature on SUDs to illustrate the promise and potential of this unique approach to construction and development. We draw conclusions for both research and practice, as well as highlight important implications for policy in pursuit of more sustainable communities.

2 The One Planet Living (OPL) SUD Framework

One Planet Living is a sustainable urban planning and design framework created by the charity Bioregional and the non-governmental organization World Wildlife Fund (WWF) based on the lessons from building the Beddington Zero Energy Development (BedZED), a large-scale mixed-use development in Sutton, UK [12, 22]. The OPL framework consists of ten principles designed to balance urban development across the three pillars of sustainability—financial, social, and environmental. Its core purpose is to guide real estate developers and other built environment stakeholders to design and build the communities in which we live to operate within the limits of the planet's resources [16, 26].

As can be seen in Table 1, the OPL's ten principles take a holistic approach to sustainable living, that goes beyond just social and environmental sustainability to also include cultural and lifestyle considerations. For example, the OPL includes social goals such as *health & happiness* as well as objectives related to ensuring equity, identity, and heritage (*equity & local economy*, and *culture & community*).

On the environmental pillar, the OPL encourages the construction and development industry to consider the protection of biodiversity, land, and water resources through the goals *land & nature* and *sustainable water*, as well as *zero-carbon energy* and *zero waste*. The OPL is also distinctive in that it includes initiatives that contribute to the overarching goal of living within the earth's planetary boundaries, from encouraging local farming and gardening to cutting down on transportation emissions (*local and sustainable food*), to sourcing more sustainable *materials and products* throughout the supply chain. Due to its system-level focus, OPL's principles and outcomes are interconnected. For example, the reuse of materials in the construction and development process contribute to the project's zero waste goals, while initiatives such as local food sourcing help the local economy, and low-carbon transport and car-free streets (*travel and transport*) improve healthy lifestyles. The ten principles are hence self-reinforcing [4].

Since the completion of BedZED in 2002, the OPL framework and its 10 principles have been adopted worldwide at various scales, from buildings to neighbourhoods, municipalities and entire cities that have aspired to become One Planet Living communities (see Table 2 for a representative sample). These sustainable urban development projects include forward-looking participants in the construction and development industry aiming to incorporate the higher ideals and ambitions for sustainability as set by the OPL principles (e.g. aiming for fully zero carbon vs. just "carbon reduction" goals). Bioregional works with each development to provide flexible, discretionary forms of recognition as a mark of excellence [16].

Table 1 Ten principles of One Planet Living

Health and happiness	Encouraging active, social, meaningful lives to promote good health and wellbeing
Equity and local economy	Creating safe, equitable places to live and work which support local prosperity and international fair trade
Culture and community	Nurturing local identity and heritage, empowering communities and promoting a culture of sustainable living
Land and nature	Protecting and restoring land for the benefit of people and wildlife
Sustainable water	Using water efficiently, protecting local water resources and reducing flooding and drought
Local and sustainable food	Promoting sustainable humane farming and healthy diets high in local, seasonal organic food and vegetable protein
Travel and transport	Reducing the need to travel, encouraging walking, cycling and low carbon transport
Materials and products	Using materials from sustainable sources and promoting products which help people reduce consumption
Zero waste	Reducing consumption, reusing and recycling to achieve zero waste and zero pollution
Zero carbon energy	Making buildings and manufacturing energy efficient and supplying all energy with renewables

Source Bioregional [6]

Table 2 Selected OPL community developments

Case #	Name of development	Location	Developers	Date initiated	Date completed	Description
1	Zibi	Ottawa, Canada	Windmill/ Dream/Theia	2012	Ongoing	Mixed-use sustainable community: 2,700 residential units, 1.5 m sq. ft. of commercial, retail and park space
2	Baker District	Guelph, Canada	Windmill	2018	Ongoing	Mixed-use street development, including new central library
3	BedZED	London, UK	Bioregional/ Peabody Trust	1997	2002	Mixed-use sustainable community comprising 100 homes, office space, a college and community facilities
4	NW Bicester	Bicester, UK	A2Dominion	2010	2017	UK's first Eco-Town with 6,000 homes, plus offices, schools and community facilities
5	One Brighton	Brighton, UK	Bioregional development group and Crest Nicholson	2008	2009	Mixed-use development with 172 apartments, offices, a community space, and a retail
6	White Gum Valley	Perth, Australia	LandCorp	2015	2019	Medium density residential community; 80 homes
7	East Village at Knutsford	Fremantle and Perth, Australia	Development WA	2019	Ongoing	Medium density 36 town houses and 70 apartments on 1.5 hectare site
8	Marrick & Co	Sydney, Australia	Mirvac	2017	Ongoing	Mixed-use community: 225 apartments, community hub, library
9	SALT Torquay	Torquay, Australia	Barwon Water	2018	Ongoing	Medium density residential community

(continued)

Table 2 (continued)

Case #	Name of development	Location	Developers	Date initiated	Date completed	Description
10	Barangaroo	Sydney, Australia	Baulderstone	2009	Ongoing	22 acre multi-use district in the heart of Sydney

3 Case Studies of OPL Communities

Having introduced the OPL SUD very broadly, we now turn to exploring three specific OPL projects in three different countries in more detail: BedZED (UK), Zibi (Canada) & White Gum Valley (Australia). The following mini-case studies introduce each unique development and provide examples of how the OPL principles have been implemented in practice. We also highlight the reputational incentives gained by implementing the OPL framework where applicable.

1. **BedZED (Sutton, UK)**

BedZED is a mixed-use development with 100 homes, office spaces, a college and community facilities in the Hackbridge, London borough of Sutton, UK [3, 22]. It was conceived in 1997 and completed in 2002 by Peabody Trust and Bioregional, using the design of Architect Bill Dunster. BedZED was built as a proof-of-concept for the feasibility of constructing large scale sustainable communities beyond individual sustainable buildings. The plan strived to add only marginal costs for incorporating sustainability features as compared to traditional developments of a similar scope [32]. BedZED was an inspiration to the development of OPL's ten principles and hence, the sustainability goals of this project were slightly different from the string of OPL developments that followed it. The following lists some of the key universal sustainability goals—cultural, environmental and financial—that were achieved at BedZED, while the other two case studies will exhibit examples of all ten OPL principles [3]:

- **Zero carbon energy**: BedZED's energy goals were to attain zero fossil fuel energy supply experimenting with a woodchip-burning combined heat and power (CHP) plant, solar panels and biomass boilers. To reduce energy consumption, BedZED also embedded passive design elements in its structure, including above-standard thermal insulation, double or triple-glazed windows, and a ventilation system that negates the need for electricity for heating, cooling, and fresh, clean air. Between 2012 and 2015, the average energy consumption of BedZED occupants was 27% less, and the gas consumption was 36% less than a development of typical size and type in Sutton [3].
- **Materials and products**: About 50% of BedZED's material weight was sourced within 50 km of its location, and 15% was recycled [3].
- **Sustainable water**: Average water consumption at BedZED is 61% less than the average Sutton town resident.

- **Travel and transport**: BedZED was the first development to offer an in-house car rental service to subscribed members.
- **Culture and community**: On average, BedZED residents know 20 neighbours by name, compared to the neighborhood's average of 8.

This experimental development thus showed over time that building sustainable communities was possible and could have a measurable environmental impact. BedZED benefitted from exceptional leadership from its developers and their appetite for taking extraordinary measures to bring this development into reality. BedZED went on to sweep several awards for its design, including being shortlisted for the prestigious Royal Institute of British Architects (RIBA) Stirling Prize for excellence in architecture [3].

2. **Zibi, Ottawa, Canada**

Zibi, meaning river in Algonquin Anishinaabe language, is a mixed-use development built on 34-acre land across the Ottawa river with 4 million square feet of residential, commercial, retail, and community spaces [26]. Officially launched by Windmill Development Group and Dream Unlimited Corp (now known as Dream) in 2015, Zibi land sits across two cities, Gatineau and Ottawa, from two provinces, Quebec and Ontario. It is built on lands with a complex history. It was once a sacred first nation site, the first European permanent settlement, a hydroelectric power plant facility that operated for over a hundred years, a logging industry that operated a sawmill, and a wood pulp product industry that had operated since the turn of the twentieth century until 2017. Most of this land became an industrial wasteland until it was revived by Windmill for its ambitious sustainable waterfront development.

Zibi's development partners created their first OPL action plan in 2015. Since then, it has regularly published annual updates that get reviewed by Bioregional, tracking the progress of the development as it tries to meet its goals. As a result, Zibi development has made tremendous strides in implementing sustainability features across the ten OPL principles. The following describes the key implementations and examples [38]:

- **Health and happiness**: In an annual survey of residents, 90% of surveyed residents report happiness grade of 4 or higher on a 1 to 5 scale.
- **Equity and local economy**: Zibi's vision attracted like-minded companies to become its tenants, such as Spaces (a co-working space rental company) and Common (a co-living space company). As part of its sustainability goals, the federal government of Canada entered a 15-year lease with Zibi for 158,000 sq. ft of office space.
- **Culture and community**: Due to its historic site, the developers actively engage with First Nations community leaders to ensure that indigenous culture and history are appropriately represented. Zibi also conducts extensive workshops on local food, gardening, and water conservation to educate the Zibi community on sustainability best practices.
- **Land and nature**: Zibi has planned for three parks housing native plant species, including those of cultural value to the Algonquin community and artworks from

indigenous and local artists. Multi-use pathways are also constructed to connect to local green spaces and cycling networks.
- **Sustainable water**: Estimated potable water consumption is below 121 L per day for residents, compared to the provincial average of 271 (Ontario) and 386 (Quebec) L/day/person.
- **Local and sustainable food**: In 2021, it achieved a food-growing garden space to homes ratio of ~0.5.
- **Travel and transport**: The ratio of car parking spaces to homes is currently one, and Zibi projects this ratio to go below one as new residential developments are completed.
- **Materials and products**: It retained and repurposed >50% of the existing buildings.
- **Zero waste**: Zibi diverted >90% of its construction waste from landfill.
- **Zero carbon energy**: Partnering with Hydro Ottawa (City of Ottawa's electricity distribution company) and Kruger Products (paper mill company), Zibi built an innovative district energy system that aims to supply heating and cooling to all of its tenants. The district energy system uses the post-industrial waste heat from nearby Kruger's tissue mill for heating and the Ottawa river for cooling to achieve its goal of zero-carbon energy.

The Zibi development is still under construction and estimated to be completed by 2032. It remains an exemplary case study of the OPL SUD framework's promise and potential. Similar to BedZED, Zibi attracted several awards, including the 2016 Award for Planning Excellence for Neighbourhood by the Canadian Institute of Planners (CIP) [39] and was made possible by outstanding leadership by the developers who leveraged sustainability goals shared by the residents, companies, and local governments to create a development designed to achieve a common objective [26].

3. **White Gum Valley, Australia**

White Gum Valley (WGV) is a 2.3-hectare medium-density residential development with more than 80 homes in the City of Freemantle and the suburbs of Perth, Western Australia. WGV development was launched in 2014 by LandCorp with a diverse group of partners from the government, industry, and university to experiment and build sustainable residential development using the OPL principles. WGV is Western Australia's first OPL community and the second in Australia to be endorsed by Bioregional. The following describes some of the positive results related to the ten OPL principles [5, 10, 36].

- **Health and happiness**: Number of neighbours known by name >18. Percentage of residents that exercise regularly >50%.
- **Equity and local economy**: Number of housing types >5. Percentage of affordable housing ~10%
- **Culture and community**: 22 sustainability events were held in 2020 along with other community events.
- **Land and nature**: Converted a local drainage sump into a park with diverse local flora, known as WGV Sump park.

- **Sustainable water**: Estimated average potable water consumption of 70 L per day per person, 65% less than the baseline of 200 L/day/pax. Sump park is also used here to manage stormwater.
- **Local and sustainable food**: 30% of all street/landscape trees are classified as edible with mostly fruits. All residents have access to food growing gardens, albeit with limited space.
- **Travel and transport**: WGV had an electric vehicle car club and the number of cars owned per home is 1.3, below the city of Perth average.
- **Materials and products**: Use of lightweight frames for 30% of single lots constructions, that can potentially reduce embodied carbon by 30 to 50%.
- **Zero waste**: Waste recycling in construction is 85% and 70% of the household waste is recycled.
- **Zero carbon energy**: 100% of the energy is supplied from renewable energy source, such as from the Solar panels installed in every building.

WGV is also an excellent example of a successful public and private partnership to deliver sustainable solutions for residential developments. WGV was awarded $1 million by the Australian Renewable Energy Agency (ARENA) to research solar battery storage in residential buildings [11] and was termed a 'Living Laboratory' for four years between 2014 and 2019, where Curtin University researchers collaborated with the Cooperative Research Centre for Low Carbon Living (CRCLCL) and other partners to monitor, learn, and develop solutions for efficient use of energy and other resources [9]. The development also collaborated with Cooperative Research Centre for Water Sensitive Cities (CRSWSC), the Department of Water, and Water Corporation to build systems like rainwater harvesting and community bore irrigation supply for gardens to efficiently use water [11].

The WGV development has also won multiple design awards for its sustainability initiatives, such as the three 2016 Planning Institute of Australia awards for excellence (Best Planning Ideas - Small Project (Western Australia), National award for Best Planning Ideas - Small Project, the Planning Minister's Award) and the two 2017 Urban Development Institute of Australia (UDIA, Western Australia) awards for Urban Water Excellence and Sustainable Urban Development [13].

4 OPL and Other Sustainable Urban Development Frameworks

In the last two sections, we broadly introduced the OPL framework and provided some brief case examples for how the OPL principles are being used in practice. We believe that one of the OPL's unique features is the set of universal principles used to plan, design and build communities that can live within the earth's planetary boundaries. In the last two decades, however, the number of available SUD frameworks available to urban planners and developers has increased dramatically with only a handful being globally applied or recognized given the inevitable differences in building contexts and climates [25]. In this section, we briefly compare the OPL to other commonly studied SUD frameworks [2, 15, 29, 31].

The two most globally adopted frameworks are the Leadership in Energy and Environmental Design (LEED) from the United States [34] and the Building Research Establishment Environmental Assessment Method (BREEAM) from the United Kingdom [7], followed by the Deutsche Gesellschaft für Nachhaltiges Bauen (DGNB) system, a German certification system adopted widely in Germany and implemented in 35 other countries [17]. While there are more than 50 other urban sustainability frameworks [25], many others are more closely associated with and adopted by their country of origin, including the Green Star Australia [21], Comprehensive Assessment System for Built Environment Efficiency (CASBEE) from Japan [23], Global Sustainability Assessment System (GSAS) from Qatar [19], and finally, Singapore's Building and Construction Authority (BCA) Greenmark [8]. Because these seven SUD frameworks are the most commonly used and/or commonly studied within the literature [15, 29, 31], we select these as a base for comparison to situate the rather lesser known OPL framework explored in this chapter (see Table 3).

Given that there are many existing reviews of the individual benefits and challenges of each SUD framework (e.g., [25, 31, 37]), we do not explore these in detail here. Rather, we summarize several dimensions of commonality and differences between the frameworks based on overarching elements such as the "type" of SUD framework, either: performance assessment, certification or design and planning [25]. In addition, we also incorporate [16] two basic types of SUDs including standardized SUD frameworks (e.g., rating tools and indices, target setting initiatives, process standards, and indicator guidelines) and more regenerative approaches (focused on holistic system-thinking). We also consider the nature of the SUD framework and whether this is more prescriptive or flexible, the business model of the SUD framework practiced by the parent organization (voluntary or mandatory), the implementation scale of the built environment (building, neighbourhood/district/communities, and cities), the application scope (if the framework is universally applied), the implementation focus (systemic or program), the level of geographic adoption (global, defined here as more than five countries, or national), and the governance structure of the framework's parent organization (market-driven or municipal driven). Finally, we compare these SUD frameworks based on their relative focus on sustainability's social, economic, and environmental pillars through the OPL's ten universal principles.

As can be seen in Table 3, the majority of the most popular SUD frameworks fall under what [25] would designate as a certification system, except for CASBEE, which is defined as a performance assessment framework. However, a stark difference between most other SUD frameworks and OPL, is that the OPL can be classified as a regenerative framework that emphasizes a holistic approach, while the rest of the frameworks follow more standardized approaches [16]. This distinction is important and speaks to a more integrative and forward-thinking perspective on construction and development. That is, while conventional SUD frameworks "focus on technical aspects or fragmented parts of a system, the regenerative worldview aims for thriving whole living systems, emphasising the interdependence and interconnection of their different elements" ([16]: p. 2).

And yet, we know that achieving sustainability development is a complex, complicated and wicked problem, rendering fragmented approaches inadequate [35] While well-defined criteria such as those adopted by prescriptive SUD programs may be more straightforward for evaluation, measurement, and public recognition, the one-size-fits-all approach leaves developers little room to adapt their site plans to deal with these complex problems holistically. The OPL thus stands in stark contrast to other SUD frameworks given its more flexible approach to planning, development and implementation. This flexibility also means that the OPL framework, which is based on a set of universal principles, can be more easily applied across different scales and in different contexts. While some standard frameworks can be adapted, they are not always suitable for local conditions. For example, GSAS is particularly developed for Middle East's environmental conditions [14, 19], and the BCA Green Mark is tailored for tropical climates [8]. This speaks to the advantage of the OPL's focus on the application of universal sustainability principles that allows for flexibility in choosing and applying specific indicators according to local conditions.

One of the most characteristic features of OPL then as compared to other SUD frameworks, is that it is evaluated at the system level [16]. It acknowledges the interconnectedness of its goals and the respective solutions. This contrasts with other SUD frameworks that evaluate its goals at the component (and sub-component) level and represent them in aggregate numbers; the OPL framework's ten principles are broad and general enough that each developer can create their own indicators through which sustainability will be assessed.

None of the other SUD frameworks are comparable in terms of the scope of the ten universal sustainability principles covered. While some frameworks, like LEED, BREEAM, and Green Star, may share 70–80% of the goals with the OPL framework, these goals are unevenly weighted; in reality, they focus on just two or three of the stated goals [29, 30]. For example, LEED weights climate change-related indicators 35% while allotting community and green economy 5% each [34]. In some cases, the weights of these rating tools depend on controllable parameters. For example, indicators that depend on occupants to pursue a goal (e.g., cycling for low-carbon transport) are downrated as they are considered to be an uncontrollable parameter [34]. This is a direct consequence of standardized frameworks' emphasis on measurement. Table 3 provides an overview of how the OPL differs from other SUD frameworks across these dimensions.

Table 3 Comparison of the OPL with other established SUD frameworks

Differentiating Dimensions/ Criterion	OPL	LEED	BREEAM	Green Star	CASBEE	DGNB system	GSAS	BCA Greenmark
Country of origin	UK	US	UK	Australia	Japan	Germany	Qatar	Singapore
Type of SUD framework								
Performance assessment	✗	✗	✗	✗	✓	✗	✗	✗
Certification	✗	✓	✓	✓	✗	✓	✓	✓
Design and planning	✓	✗	✗	✗	✗	✗	✗	✗
Standardized approach	✗	✓	✓	✓	✓	✓	✓	✓
Regenerative approach	✓	✗	✗	✗	✗	✗	✗	✗
Nature of SUD								
Prescriptive	✗	✓	✓	✓	✓	✓	✓	✓
Flexible	✓	✗	✗	✗	✗	✗	✗	✗
Business model								
Voluntary	✓	✓	✓	✓	✓	✓	✓	✗
Mandatory	✗	✗	✗	✗	✗	✗	✓	✓
Scale								
City	✓	✓	✗	✗	✓	✓	✗	✗
Neighbourhood	✓	✓	✓	✓	✓	✓	✓	✓
Building	✓	✓	✓	✓	✓	✓	✓	✓
Application scope								
Universal	✓	✗	✗	✗	✗	✗	✗	✗
Require spin-off	✗	✓	✓	✓	✓	✓	✓	✓
Implementation focus								
System	✓	✗	✗	✗	✗	✗	✗	✗
Program	✗	✓	✓	✓	✓	✓	✓	✓
Relationship between goals/themes								
Even weights	✓	✗	✗	✗	✗	✓	✗	✗
Interconnected	✓	✗	✗	✗	✗	✗	✗	✗
Geographic adoption								
Global	✓	✓	✓	✗	✗	✓	✗	✗
National	✓	✓	✓	✓	✓	✓	✓	✓
Governance structure								

(continued)

Table 3 (continued)

Differentiating Dimensions/ Criterion	OPL	LEED	BREEAM	Green Star	CASBEE	DGNB system	GSAS	BCA Greenmark
Market-driven	✓	✓	✓	✓	✗	✓	✗	✗
Municipal driven	✗	✗	✗	✗	✓	✗	✓	✓
Goals								
Health and happiness	✓	✓	✓	✓	Partly	Partly	✗	✗
Equity and local economy	✓	✓	Partly	✓	Partly	✗	Partly	✗
Culture and community	✓	✓	Partly	✓	Partly	✗	Partly	✗
Land and nature	✓	✓	✓	✓	✓	✓	✓	✓
Sustainable water	✓	✓	✓	✓	✓	✓	✓	✓
Local and sustainable food	✓	Partly	✓	✗	✗	Partly	✗	✗
Travel and transport	✓	Partly	✓	✓	✓	✓	✓	✓
Materials and products	✓	✓	✓	✓	✓	✓	✓	✓
Zero waste	✓	✗	✗	Reduce	Reduce	Reduce	Reduce	Reduce
Zero carbon energy	✓	Reduce	✓	Reduce	Reduce	Reduce	Reduce	Reduce

5 Discussion: The Promise and Potential of OPL

The One Planet Living framework for sustainable urban development is an ambitious, goal-oriented framework. It stands out from other SUD frameworks as it is flexible, universal, embraces systems thinking, acknowledges its interconnectedness, weighs its goals evenly, is all-encompassing, and attempts to uniformly balance sustainability's social, economic, and environmental aspects.

Because it is so flexible, implementing the OPL is an iterative process. Real estate developers and other built environment stakeholders can learn from their peers and their own projects as they monitor progress beyond the initial design and development stage. The decentralized peer review and public evaluation process allows each urban development to create unique solutions to sustainability challenges without needing to conform to strictly prescribed standards. Some OPL developments, like BedZED, have shown that sustainable developments can be built with marginal additional costs to incorporate sustainability features as compared to traditional developments [32]. In general, long-term cost savings are expected to help offset the additional development costs and the sometimes-premium price paid by the occupants. However, this requires exceptional leadership from the construction and development industry to put forth ambitious plans that align with and leverage the shared sustainability goals of social, private, and public bodies. The case studies show that this venture can be enriching with significant reputational returns.

While standardized frameworks like LEED and BREEAM have been globally adopted for thousands of projects, Gerhards and Greenwood [16] argue that these programs benefited from their high-profile parent organizations, providing first mover advantages that have led to greater integration with regulatory requirements and public policy. OPL's flexible and bespoke approach itself can therefore be cited as a challenge for broader adoption and take-up from developers looking for more standardized and cookie-cutter approaches [16].

Meeting the ambitious and lofty goals set in the planning stage of OPL projects can also be challenging as targets can be hard to achieve in practice, especially around zero waste and zero carbon goals. In reality, many of the OPL principles are set for the long term, and incorporate a time frame that is well beyond the development's construction stage. Making progress in practice requires continued monitoring and regular community education on sustainability best practices, including advanced monitoring methods, devices and regular occupant surveys which can come at additional cost. While Bioregional encourages publication of annual progress and review of OPL developments, because of its flexible approach, not all developments have done this consistently. Implementation and monitoring challenges, however, are common to all SUD frameworks and not unique to the OPL framework.

Sustainable urban development in the future will need to be regenerative [18]. Gerhards and Greenwood [16] describe the regenerative approach as flexible, ambitious, holistic, dynamic—considering the relationship between the parts of a system while fostering shared endeavours, communication, engagement, and learning among all actors. The OPL framework aspires to this description of regenerative SUD more than any other more common SUD framework with its holistic framework of universal principles [25]. Existing OPL developments have shown that these goals are worth pursuing and that entire communities can reap economic, social, and environmental benefits for their efforts.

For the construction and development industry to consider ambitious and holistic SUD frameworks such as the OPL, government legislation, regulation, and policies must harmonize at the national to municipal level to champion regenerative approaches to developing the built environment. This could be followed by strong financial incentives (e.g., grants, subsidies, taxation credits) to encourage take up and offset perceived risks of SUDs. In practice, sustainable construction and development is also hindered by a lack of expertise and knowledge within the sector around how to achieve sustainability goals. This could be addressed by adequate training and education in all areas of sustainability. On top of this, visionary and ambitious leaders are required to bring together multiple stakeholders to champion shared goals and break barriers to creating more sustainable urban developments [27].

6 Conclusion

This chapter provided a broad overview the OPL framework for sustainable urban development and its ten universal guiding principles. Through three mini-case studies of OPL communities, we provided examples of how these SUDs have achieved some of the goals related to this program, attracting national recognition for their design plans. While challenges remain for broader adoption of the OPL framework, regenerative approaches show the most promise for the future of sustainable urban development, and the OPL currently serves as its best example in practice.

References

1. Bayulken B, Huisingh D (2015) A literature review of historical trends and emerging theoretical approaches for developing sustainable cities (part 1). J Clean Prod 109:11–24
2. Berardi U (2013) Sustainability assessment of urban communities through rating systems. Environ Dev Sustain 15(6):1573–1591
3. Bioregional (2016) BedZED: the story of a pioneering eco-village. https://www.bioregional.com/resources/bedzed-the-story-of-a-pioneering-eco-village
4. Bioregional (2019) One Planet Living goals and guidance for communities and destinations
5. Bioregional (2021) WGV one planet annual review 2021. https://www.bioregional.com/resources/wgv-one-planet-annual-review-2021
6. Bioregional (2022) One Planet Living. https://www.bioregional.com/one-planet-living
7. BRE Global Limited (2012) BREEAM communities technical manual 2012. https://files.bregroup.com/breeam/technicalmanuals/communitiesmanual/#resources/otherformats/output/10_pdf/20_a4_pdf_screen/sd202_breeam_communities_1.2_screen.pdf
8. Building and Construction Authority (BCA) Singapore (2016) BCA green mark for districts. https://www1.bca.gov.sg/docs/default-source/docs-corp-buildsg/sustainability/gm_district_v2_1.pdf
9. Byrne J, Law A, Hosking R, Breadsell J, Syed M, Babaeff T, Morrison G, Newman P (2019) Mainstreaming low carbon residential precincts—The WGV living laboratory. Cooperative Research Centre for Low Carbon Living, Sydney, Australia
10. Cabanek A, Zingoni de Baro ME, Byrne J, Newman P (2021) Regenerating stormwater infrastructure into biophilic urban assets. Case studies of a Sump Garden and a Sump Park in Western Australia. Sustainability 13(10):5461
11. Cooperative Research Centre for Water Sensitive Cities (CRSWSC) (2017) White Gum Valley case study: a waterwise way of living. https://watersensitivecities.org.au/wp-content/uploads/2017/06/Case_Study_White_Gum_Valley.pdf
12. Desai P (2008) Creating low carbon communities: One Planet Living solutions. Globalizations 5(1):67–71
13. DevelopmentWA (2019) WGV overview. https://developmentwa.com.au/projects/residential/white-gum-valley/overview
14. Ferwati MS, Al Saeed M, Shafaghat A, Keyvanfar A (2019) Qatar sustainability assessment system (QSAS)-neighborhood development (ND) assessment model: coupling green urban planning and green building design. J Build Eng 22:171–180
15. Gerhards J (2021) One Planet Living and the legitimacy of sustainability governance: from standardised information to regenerative systems. PhD thesis University of Westminster School of Social Sciences https://doi.org/10.34737/vq22y
16. Gerhards J, Greenwood D (2021) One Planet Living and the legitimacy of sustainability governance: from standardised information to regenerative systems. J Clean Prod 313:127895

17. German Sustainable Building Council (DGNB) (2022) DGNB systems: planning and certification of sustainable districts. https://www.dgnb-system.de/en/districts/
18. Gibbons LV (2020) Regenerative—The new sustainable? Sustainability 12(13):5483
19. Global Sustainability Assessment System (GSAS) (2019) GSAS 2019 design & build assessment manual. https://gsas.gord.qa/wp-content/uploads/2021/10/GSAS-2019-Design-and-Build-Assessment-Manual-for-Building-Typologies.pdf
20. Georgescu M, Morefield PE, Bierwagen BG, Weaver CP (2014) Urban adaptation can roll back warming of emerging megapolitan regions. Proc Natl Acad Sci 111(8):2909–2914
21. Green Building Council Australia (GBCA) (2016) Green Star Communities National framework. https://gbca-web.s3.amazonaws.com/media/documents/green-star---communitiesframework-final.pdf
22. Hancock T, Desai P, Patrick R (2020) Tools for creating a future of healthy One Planet cities in the Anthropocene. Cities Health 4(2):180–192. https://www.bioregional.com/resources/one-planet-living-goals-and-guidance-for-communities
23. Japan Sustainable Building Consortium (JSBC) (2014) Comprehensive Assessment System for Built Environment Efficiency (CASBEE) for urban development. https://www.ibec.or.jp/CASBEE/english/download/CASBEE-UDe_2014manual.pdf.
24. Joss S, Cowley R, Tomozeiu D (2013) Towards the 'ubiquitous eco-city': an analysis of the internationalisation of eco-city policy and practice. Urban Res Pract 6(1):54–74
25. Joss S, Rydin Y (2018) Prospects for standardising sustainable urban development. In: Bell S, Morse S (eds) Routledge handbook of sustainability indicators. Routledge, pp 364–378
26. Mazutis D, Abolina E (2019) The Five I model of sustainability leadership: lessons from the Zibi One Planet Living sustainable urban development. J Clean Prod 237:117799
27. Mazutis D, Sweet L (2022) The business of accelerating sustainable urban development: a systematic review and synthesis. J Clean Prod 131871
28. Rasoolimanesh SM, Badarulzaman N, Abdullah A, Behrang M (2019) How governance influences the components of sustainable urban development? J Clean Prod 238:117983
29. Sharifi A, Murayama A (2013) A critical review of seven selected neighborhood sustainability assessment tools. Environ Impact Assess Rev 38:73–87
30. Tanguay GA, Rajaonson J, Lefebvre JF, Lanoie P (2010) Measuring the sustainability of cities: an analysis of the use of local indicators. Ecol Ind 10(2):407–418
31. Turcu C (2018) Sustainability indicators and certification schemes for the built environment. In: Bell S, Morse S (eds) Routledge handbook of sustainability indicators. Routledge, pp 156–175
32. Twinn C (2003) BedZED. Arup J 38(1):10–16
33. United Nations Environment Programme & International Energy Agency (UN-EIEA) (2017) Towards a zero-emission, efficient, and resilient buildings and construction sector: global status report 2017. https://www.worldgbc.org/sites/default/files/UNEP%20188_GABC_en%20%28web%29.pdf
34. United States Green Building Council (USGBC) (2013) LEED v4 impact category and point allocation process overview. https://www.usgbc.org/resources/leed-v4-impact-category-and-point-allocation-process-overview
35. Waddock S, Meszoely GM, Waddell S, Dentoni D (2015) The complexity of wicked problems in large scale change. J Organ Change Manag 28(6):993–1012
36. Wiktorowicz J, Babaeff T, Breadsell J, Byrne J, Eggleston J, Newman P (2018) WGV: an Australian urban precinct case study to demonstrate the 1.5 C agenda including multiple SDGs. Urban Plan 3(2):64–81
37. Xia B, Chen Q, Skitmore M, Zuo J, Li M (2015) Comparison of sustainable community rating tools in Australia. J Clean Prod 109:84–91
38. Zibi (2022) Zibi 2021 One Planet report. https://zibi.ca/wp-content/uploads/2022/05/2021-One-Planet-report_ENG-FINAL.pdf
39. Zibi (n.d.) Zibi development website. https://zibi.ca/about/

Enjoying the Commute: Wellbeing Benefits of the Public Transport Journey

Peter J. Davies, Kathryn McLachlan, Aileen Westbrook, and Waminda Parker

Abstract This article presents the argument that the travel experience, not just the infrastructure, is central to how public transport projects are conceptualised, funded and operated. This responds the idea of a sustainable city within which advocates seek the 20-min city in which there is support for a mode switch from the car to public and active transport. The narrow and traditional utilitarian concepts are challenged including cost benefit analysis that asserts that time commuting has little to no value or simply is a 'means to an end', and that of transit orientated development with its focus on infrastructure planning and delivery. In contrast, the experience of travel and the value of transit systems connecting places is fundamental to promoting public transit systems. These modes of transit can be pleasurable and contribute to the user's eudaimonic physical and mental wellbeing. It is proposed that individual and social wellbeing has significant utility, and that evidence-based studies using established measures of subjective wellbeing and social capital, are needed to capture the rich social, communal dimensions and spatial connectivity that public train systems afford. Recommendations for future research, transport policy and urban planning are advanced for addressing the contribution of subjective and social wellbeing in transit studies and these can broaden the consideration of the current focus on physical health benefits in support of liveable cities.

Keywords Wellbeing · Transport evaluation · Liveable cities · Train transport · Place-making

---

This work was supported by a Macquarie University Enterprise Partnership Scheme under grant number 5086920 in partnership with Metro Trains Sydney under Grant number 5086900.

P. J. Davies (✉) · K. McLachlan
School of Natural Sciences, Macquarie University, Sydney, NSW 2109, Australia
e-mail: peter.davies@mq.edu.au

A. Westbrook · W. Parker
Macquarie University, Sydney, Australia

Z. Allam (ed.), *Sustainable Urban Transitions*, Urban Sustainability,
https://doi.org/10.1007/978-981-99-2695-4_2

1 Introduction

The concept of a liveable, sustainable, productive city, has foregrounded the need for a networked and efficient public transport system that also integrates with wider urban design outcomes [51]. Such calls respond to an increasing focus on physical [59] and mental wellbeing, as subjective concepts of liveability [18], that have tended to fall in the shadow of more traditional infrastructure utility-maximisation measures [41]. Liveability and wellbeing are increasingly placed at the top of governmental agenda's as a benchmark indictor of the performance of individuals, communities and cities [3, 14, 15]. However, its value remains hidden when measured through traditional economic indicators such as the Gross Domestic Product (GDP), which fail to reflect important dimensions of citizen welfare, productivity and quality of life [3, 14, 22, 26].[1]

In recent decades there has been worldwide renaissance in public transport, hailed as 'a new mobility paradigm', a strategy to address complex and interrelated urban problems of car dominance, environmental sustainability, climate change, health, social exclusivity and poverty [2, 42]. Yet for public transit systems, their business case is all too often constrained by a narrow economic remit founded on the shorter-term return on investment to support transport funding and policies. This often assumes travel is a disutility, and thus individuals will aim to minimize travel time and cost. This idea is more recently being challenged by social scientists, cultural geographers, and the health professions as they present new methods to offer a compelling case to advocate a shift to car-based transport alternatives [10]. However, the path dependencies that capture public sector transport decision making are yet to fully embrace these new methods [6].

To date, the bulk of research efforts to explore the role of transport in contemporary urban life have focused on measuring subjective wellbeing of commuters (predominantly modalities of car, bus, and active transport) through probing negative aspects of traveling, such as congestion and time saved [12]. However, innovative studies addressing the experiential dimensions of train passengers reveal surprising results, with three-quarters of passengers considering train travel time as time well spent. Train travel, in particular, as a mode of commuting, is associated with a wide variety of benefits [40]. Lyons and Urry [39] propose that because the train has the potential to be a 'mobile office' it is potentially the most 'productive' mode of transport. Other researchers point to the additional exercise gained by those using public transport, achieved through walking or cycling to the station (door-to-station then station-to-door parts of the journey), compared to the car, as a clear positive health

[1] THE GDP "measure only partial economic activities," and is inadequate as a measure of the sustainable development of societies (see Giannetti et al. [26 p. 12]). Alternative and supplements to the GDP have been proposed, to address societal and eco-system wellbeing. These include such as Sustainable Wellbeing Indicators ("greening and socializing"), the Human Development Index, Physical Quality of Life Index (PQLI), Gross National Happiness (GNH) (based upon Buddhist spiritual as opposed to economic activities), Wellbeing Index, Happy Planet Index (see Giannetti et al. [26]). The ideal index needs to incorporate economic, environmental and social determinants of wellbeing, and awaits future research.

outcome [49]. Less well understood are the drivers behind various mental health benefits comparing modes of transport. In particular, walking and cycling are often reported as less stressful than travel by car [24, 36], yet for public transport users, there is less information on how the journey itself impacts mental health impact aside from a few studies that have explored the negative aspects associated with crowding [37], gender and service quality [67] and urgency [20].

A well-designed metro transport system has the potential to impact other wellbeing factors such as health and social problems [25]. These can have significant monetary and social costs but are complex multifactorial problems extending to, for example, cost of road carnage, societal health problems (obesity diabetes, sedentary behaviour), mental health problems (depression), social problems (loneliness, isolation, exclusivity, social inequities) [27]. As city planners and policy makers apply best practice guidelines for developing public transport infrastructure, it is vital they recognize the nexus between transportation, wellbeing and liveability to enable governments to invest in this valuable societal infrastructure [42]. This article appraises the literature, examining the relationship between metro train transport and human wellbeing. In doing so, it draws on the findings of a research collaboration between Macquarie University and Metro Trains Sydney, that explored the question, how do transport systems contribute to commuter wellbeing? This was examined through three theoretical measures of wellbeing: hedonic, eudaimonic and utility, and how when combined these theoretical frameworks can better inform urban planners, transport operators and users, of the benefits of the journey. The article concludes with future policy and research directions for planners and policy makers to support the widening of the scope and value of public transit systems and their value to societal wellbeing.

2 Health and Wellbeing

This section is designed to provide a short overview of the transport literature related to health and wellbeing. It establishes the case for wellbeing to have a greater role in assessing the impact and benefits of transit infrastructure and the need to consider the sub-components of the commute. The evidence points to the need to evaluate not simply the utilitarian aspects of the commute, as most often explored by transit agencies, but also the hedonic and eudaimonic aspects.

Health and well-being are complex concepts that are not easily defined nor quantified. The World Health Organisation (WHO) sees health as a state of complete physical, mental and social well-being and not merely the absence of disease or infirmity [68]. For individual health and wellbeing, this is a state where normal stresses are managed, and an individual can work productively and contribute to community life [57]. Under this definition, there is a deliberate merging of the physical and mental health and wellbeing aspects that are all too frequently separated as part of subjective wellbeing studies. A subjective evaluation of the quality of one's life [14] is critical to understanding individual and collective wellbeing. In this context, the OECD's

Better Life Index [46] considers better health can be achieved through modifying lifestyle choices. It is through this lens that we explore the role of public transport and its capacities to support wellbeing outcomes.

Citizen mobility is a fundamental aspect of urban wellbeing [2]. Access to transport allows movement of people, objects, ideas and information, essential for social connectivity, creativity, economic and political practices [66]. The pivotal role of transport in the wellbeing and resilience of cities and their citizens is declared in Commitment 115 of the UN's New Urban Agenda, Habitat III [63]:

> We will take measures to develop mechanisms and common frameworks at the national, subnational and local levels to evaluate the wider benefits of urban and metropolitan transport schemes, including impacts on the environment, the economy, social cohesion, quality of life, accessibility, road safety, public health and action on climate change [63, p. 29].

In part, this responds to the renaissance in public transport in recent decades as a strategy to address complex and interrelated urban problems. By and large, these problems have emerged as a consequence of the dominance of the car on urban planning which has supported suburbanisation and the associated environmental sustainability, climate change, health, social exclusivity and poverty issues therein [2, 42].

The complexities and multifaceted relationships of the value of the commute and its contribution to wellbeing, therefore, have been the subject of investigation by multiple disciplines, including psychologists, social scientists, and cultural geographers, indicative of its complex, multifaceted relationship [13]. Different theoretical and philosophical frameworks inform the measurement, evaluation and interpretation of subjective wellbeing [22] and have tended to focus on the satisfaction, that lies within a hedonic and utilitarian approach, and lesser on the experience of the commute that would be captured by eudaimonic measures.

In psychology, evaluation of hedonic wellbeing emphasises individual preferences and pleasures of mind and body, versus pain/displeasure [33]. In contrast, the eudaimonic wellbeing tradition is multidimensional and is concerned more broadly with positive psychological development and human flourishing [50]. The body of evidence demonstrating associations between transport and hedonic wellbeing is heterogenous, with modalities of car, bus, cycling, walking being the focus of most studies [6, 13]. Using traveller satisfaction scales, a range of affective experiences have been measured in commuters, including stress, pleasure, boredom [6, 18, 45].

Recently, investigators have begun to inquire more widely about positive experiences and the positive utility of commuting [43, 48]. Several studies have demonstrated that commuting is associated with enjoyment and higher levels of life satisfaction [43]. While economic models predict that reducing commuting time to zero is the most desirable outcome, when passengers are asked, the ideal commute duration is not zero [43, 52]. Commuting, therefore, has potential benefits to wellbeing by serving as a buffer between work and home, providing personal privacy/isolation and 'protected time' [29]. The positive commuting 'utility' can thus be considered to have 3 components: utility for destination, for activities pursued during travel, and for the intrinsic experience of travel [5].

Mobility and specifically transport are recognised as a public health issue [1, 27]. Research demonstrates that transport and physical and mental health are linked in multiple ways. Greater dependence on cars instead of active transport (walking, cycling), contributes to an increasingly sedentary lifestyle, associated with secondary health problems [30, 31]. By definition, public transport involves travelling with others, providing opportunities for social interaction [8] so long as the spatial and design factors facilitate interactions [4]. Handy and Thigpen [29] examined commuter satisfaction (active modes, bus, train, car) in a campus study of UC Davis reporting the perceived commute quality, evaluated using a series of Likert scales regarding levels of commute stress and enjoyment, revealed cyclists and train commuters experienced the highest commute quality. A National Rail Passenger Survey conducted in Britain showed 72% of travellers considered train traveling time to be useful, with evidence for a positive utility of travel time with passengers engaged in reading, working, studying, talking to other passengers, listening to music, window gazing, people watching [40]. Similarly, a 2014 study on Sydney's rail service reported various activities passengers undertook when travelling revealing 57% used their Smartphone, tables or laptop, while the next most common activity was to relax and look at the view (46%) [16]. Although longer commutes are found to be associated with lower positive affect [19], there is growing evidence that commuting has the potential to enhance wellbeing.

Commuting in, and of, itself is not an undesirable activity [45] that seeks an objective of zero time [43, 52], nor is it 'dead time' [28]. Rather, there are wider 'utilities' of the commute that when considered holistically give significant support to public transport [23, 43, 48]. These components suggest that measuring the utility from the 'door-to-door' or 'station-to-station' can miss the nuance of opportunities and positive experiences of the activities and 'places' experienced as part of the wider commuting purpose. Presently, the literature and models examining public transit systems place a greater emphasis on the negative utilitarian aspects (e.g. concerns over transport crowding, timeliness and cleanliness), while discounting positive aspects of the commute [61].

The emerging focus on design and place for people, place-making, seeks to reinvent public spaces as community places. A 'space' becomes a 'place' when culture, meaning, and values in people's lives are rooted within it [7, 62]. The value for a place is embedded in its physical space, economic importance, social relations, and cultural aspects of human experiences [21]. Within the ideas of transit orientated development (TOD), the evaluation of commuting and place-based travel has tended to foreground the urban place, not its relationship to the commute [10, 35]. This is surprising, as these TOD places are the nodes connecting people and place foregrounding the value of the commute, not simply a means to supporting urban planning densification agendas. In this context, the often-heralded triple bottom line agenda to planning, liveability, sustainability and productivity, should be considered as integral part to where and how TOD can support the satisfaction, contribution to wellbeing and broader value of both the place, and how to get there.

A well-designed transport system has the potential to impact wellbeing, health and social outcomes [25], as it links people to places. With increasing city density and mobility, city living is rapidly emerging as being one of the most important health concerns globally [54]. For example, 50% of Australians experience some form of mental illness, with increasing levels of societal stress, anxiety and depression. As city planners and policy makers design and apply best practice guidelines for developing public transport infrastructure, it is vital that they recognise the nexus between transportation, wellbeing and liveability to enable governments to invest in this vital societal, as well as grey, infrastructure [42].

Arguably, there is an urgent need to shift policy frameworks to move beyond narrow economic growth (GDP) and utility measures, towards wellbeing outcomes [12, 38, 55]. There is a strong case for *enhanced wellbeing* to be a primary policy goal of transport [56]. Yet while it is increasingly recognised that mobility conferred by transport systems is fundamental to wellbeing, there are significant gaps in our understanding of the complex relationship between urban transport and wellbeing, and inadequate empirical evidence [6, 11]. In this study we are concerned with emphasising the opportunities to value the commute through multiple perspectives: the transport operator, environmental planner and physiologists. This interdisciplinary focus converges the focal interests of the satisfaction of the infrastructure (operator), the connection of transport and place (planner) and how these contribute to health and wellbeing (health sciences).

3 Research Program: Sydney's North-West Metro System

The re-development of a heavy rail (train) system to Australia's first driverless metro train system in North-West Sydney, provided an opportunity to re-conceptualise the door-to-door commuter experience, to improve community health and wellbeing, and influence sustainability. Metro Trains Sydney (MTS), as operators of the newly commissioned service were interested in exploring ways of reframing city mobility, as a social service and preferred lifestyle choice. This aligned with MTS's sustainability and operational performance criteria and was designed to integrate evidenced-based decision making into the current and future design and operation of the North-West Metro.

'Connection on the Move' was a co-jointly funded, year-long, interdisciplinary research collaboration established between Macquarie University and Metro Trains Sydney (MTS) in December 2018. The research program was based on a whole-of-systems method, as a way of framing the interconnections and interdependencies of the system (the metro) and subsystems (inclusive of the many parts of the commuting journey and experience). The interdisciplinary approach involved academics, professional staff and students exploring impactful, industry relevant and community connected approaches to transform metropolitan transport infrastructure and delivery.

The research aimed to understand the relationship between people's connection with place (the station) and place-making (precinct around the station) and how this may contribute to their wellbeing and travel experience. The research was framed around two broad questions:

- What makes a positive difference to enhancing the safety, health and well-being of individuals, groups, organisations and the community, while commuting on the move? and
- What factors influence cultural change and the creation of a sense of place that are generalizable across urban transit contexts?

3.1 Commuter Survey

This research reports on a commuter survey to inquire into the expectations and needs of commuters at key stations along the North-West metro line. Questions were designed to explore the lenses of utilitarian, hedonic and eudaimonic wellbeing in order to frame an understanding of commuter expectations, needs and challenges, expectations, and needs of commuters at key stations along the North-West metro line.

The survey was undertaken after the commencement of the new metro, to assess the impact of the upgrade (changing from a heavy-rail to a metro system) to focus on the journey experience, inclusive of the journey, station, and place, and commuter needs, expectations and aspirations (Attachment 1). This wider focus on the travel experience aimed to inform an understanding of place-making and its connection to wellbeing, and the types of messaging commuters are interested in, to enhance their journey experience. This included exploring information portals, for example a dedicated travel app, or information kiosk and the content therein, to inform our central questions.

One part of the survey was designed to inform how digital technology could assist the metro operator to sharing information and engage with commuters. Digital technology, including various social media platforms and Applications (Apps) can now offer real time, geospatial data based on customer needs and preferences, yet remain nascent within the public transport realm. Notably, as this was the first metro system in Sydney, the transport operator considered that the commuter/technology interface could represent a point of difference of this system to conventional rail.

This approach provided an opportunity to inform a wider understanding of commuter satisfaction that remains institutionally and contractually grounded to utilitarian and hedonic measures and not inclusive of eudaimonic wellbeing. The potential of this research was to explore ways to enhance the travel experience. This extends to: the broader conceptions of the commute journey, inclusive of the arrival at the station, the station-to-station commute, and the places that may be the destination or the intermediary places); and how well, what and why people enjoyed the station and commute from a hedonic and eudaimonic wellbeing perspective.

4 Results

The survey was distributed face-to-face and on-line with (N = 102) responses. Stations surveyed included North Ryde, Macquarie Park, Macquarie University, Kellyville and Rouse Hill. The first set of questions were framed around time and cost saving of the metro, reflecting a focus on the utility of the metro when compared to other forms of transport. Further questions were specific to the travel experience including, what sort of information commuters wanted access to, and the experience of the metro station itself. This information was designed to inform a future commuter App and information kiosk at the stations (Attachment 1). These questions were oriented around hedonic elements, making the commute more enjoyable, and to gain a wider eudaimonic understanding of commuter experience and expectations.

Responses were sought from adults over the age of 18 years in accordance with the approval of the Macquarie University Ethics Committee (no. 5201938419575). Of the 102 responses, 56% were female, 44% male and 53% were aged between 18–29, 19% were 30–39, 13% 40–49 and the remainder were over 50. English was the preferred language (84%), with Mandarin selected by (6%), other (3%), and Korean and Hindi (2.4%). The respondents were travelling from, and to, suburbs which were spread across the North-West geographical area of Sydney. Primary reasons for travel included: employment (39%), study (34%), resident (19%) and other (8%). Percentage figures for each question have been rounded to the nearest whole number.

From a utilitarian perspective, the commencement of the metro resulted in 84% of respondents reporting a decrease or similar travel time. The cost of travel using the metro also reported similar results with 83% reporting less or same or cost. This result is likely to reflect the evolution of the public transit line from a heavy rail to the more frequent services of the metro system with pricing determined remaining consistent and linked to Sydney's smartcard ticketing system.

Table 1 reports information or services that people wanted to access on the commute. The responses reflect a focus on maximising the utilitarian efficiency of their travel (i.e. timetables), their hedonic wellbeing (what would give them greater pleasure) and eudaimonic insight that broadens their desirability to travel to a specific place.

Tables 2, 3 and 4 report the responses related to questions about what people wanted to know about when arriving at the train station, what they wanted at the station itself, and what might incentivise the travel experience. These questions revealed deeper insights as to the importance of destination, the place, the desirability of the design of the metro station as a place in its own right, and a blending of how the utilitarian, hedonic and eudaimonic elements of wellbeing are critical to the design and operation of public transit systems. Responses to what people want to see at their station (Table 2) reflected a strong preference to how public transport contributes to an experience and life fulfilment, suggesting most response tend to a eudaimonic response. In contrast, to the question what do people want at their

Table 1 Information and services commuters want

Answer	Category	%	Count
Connecting transportation services and times (e.g. connecting bus or ferry)	Utilitarian	36	93
Community activities and events in the local area	Eudaimonic	16	42
Relaxing activities (e.g. mindfulness app, meditative music)	Hedonic	15	39
Articles about the Metro Train services and operations	Utilitarian	9	24
History of the local area and surroundings	Eudaimonic	9	22
Podcasts offering a range of topics	Hedonic	6	16
A selection of online games	Hedonic	4	11
Other (please specify)		4	9
Total		100	256

destinations, suggesting more a tangible outcome or physical features, there was a stronger response towards improving the ambience and stimulation, indicative of a hedonic outcome.

Presently there are limited in-station retail facilities, such as cafés, with commuters having to rely on these services and experiences in the wider precinct. The metro stations, themselves, have limited 'individual' distinguishing character that would otherwise reveal to the commuter a utilitarian sense of identity of the suburb or destination nor specific aesthetic expressions, such as in-station artwork, that may contribute to hedonic and eudaimonic wellbeing. From a place-making perspective, the desire for better links to the station as a means of supporting the experience of arrival to station was notable. This highlights the importance of ensuring state government agencies and local government can collaborate in the master planning process and infrastructure delivery of the station and immediate curtilage (more

Table 2 What people want to see at their station

Answer	%	Count
Range of local restaurants and eateries available	20	67
Local community events (e.g. market days)	16	55
Range of local retail businesses available	14	47
Types of recreational activities within the area (e.g. tennis court)	11	38
Proximity to local parks	10	35
Information about activities within the local community (e.g. community fundraisers)	8	28
Range of other local businesses present	8	27
Local heritage information and sites	8	26
Other (please specify)	3	11
Total	100	334

Table 3 What people want at destination stations

Answer	%	Count
Personalisation of the station (e.g. distinguishable from other stations in a local context)	24	61
Nature features (e.g. more greenery)	23	57
Art feature	21	52
Historical information/photographs of the suburb	18	46
Water feature	12	30
Other (please specify)	2	5
Total	100	251

Table 4 What might incentivise the travel experience

Answer	Category	%	Count
Rewards from employers/university to encourage public transport use	Utilitarian	19	56
Better links (bike/walkways) to/from station	Eudaimonic	19	54
Greater awareness of savings in cost and time	Utilitarian	16	47
Greater awareness of the benefits to the environment	Eudaimonic	14	39
Greater awareness of wellbeing rewards including physical and mental health	Eudaimonic	10	28
Features enabling people to relax and enjoy the train journey and experience (e.g. mindfulness app)	Hedonic	8	24
Learning more about the local areas you are travelling (e.g. stories about place, opportunities to participate in the local events)	Eudaimonic	5	15
Other (please specify)		5	13
Bike loan program	Utilitarian	4	12
Total		100	288

often resting with the state transport and planning agencies) and neighbour surrounds (typically the responsibility of the local council, such as providing the connecting foot and cycle paths).

5 New Framework and Policy Recommendations

The survey findings provide evidence that commuters have multiple expectations with respect to their travel experience. These expectations exceed a primarily utilitarian focus and extend to planning and place [5]. Importantly, the survey reveals that the commuter's subjective experience may commence with hedonic attributes and expectations, but expand into the eudaimonic realm, as the travel experience proceeds. To more fully understand how travel satisfaction, fulfilment and experience is considered it is necessary to recognised the interconnected and interdependent elements that exist across the travel domains. This is inclusive of the door-to-station, station-to-station, and how place, both the station and precinct surrounding, are integral to travel.

For transport planners, those involved in preparing the business case to support transit systems and for those tasked to design of new public transit systems, this insight offers new opportunities. Cost–benefit tools to evaluate transit infrastructure should be expanded to better capture the wider value of travel. In doing so, this must lessen the dominance of the economic ontological approach to one that can more adequately capture subjective wellbeing. This would bring these into line with similar subjective evaluation tools to evaluate city liveability, such as used by the UK Office for National Statistics to measure personal well-being [64]. For the commute itself, the value proposition of public transit systems needs to acknowledge that time spent commuting is valuable for various hedonic and eudaimonic reasons inclusive of the variety of activities that can be undertaken therein. The design of public transit stations and precincts need to depart from a modular efficient and repeatable design to one that can capture and reflect the local identify and character. The surrounding areas and curtilage of transit nodes need to enable and integrate different forms of active transport (e.g. connecting foot and cycle paths) as a means to support physical and mental health outcomes as well as offering aesthetic and functional experiences.

The importance of place was revealed as a notable insight from respondents. This extends beyond the placemaking that is typically applied to the precinct surrounding the station, as usually captured within TOD approach, with a view that stations themselves should have their own identity, be it design, colour, furniture or art. While public transport is central TOD as a way of creating more sustainable, compact, networked eco-cities [10, 35], the experience of the commute would seem to be able to be enriched with the stations themselves the subject of a placemaking design lens. This can overcome the cookie cutter approach to stations and enable them to have a distinct character that reflects for example the local history, the environment or cultural influences. An emerging example is the use of art for Sydney's metro system with aims including to: elevate the customer travel experience, create a benchmark for public art in transport; and raise awareness of and pride in local histories and cultural diversity [44]. The generic utilitarian design may in fact hinder supporting a positive commuter experience and at worse be a barrier to the promotion of mode shift to public transport.

The eudaimonic perspective to the travel experience, therefore, reveals nuances of the commuter journey and how this can be improved. While there are clear and definable utilitarian and hedonic metrics related to travel, such as timeliness and cleanliness of the station and carriages, these comprise a narrower assessment of the commuting experience.

In seeking to understand the value of the commute, a broader interdisciplinary and new ontological approach is required. For transport planners, the current utilitarian focused conception largely forms the basis of transport infrastructure planning: Who will use the service? What are the time savings? What are the congestion savings?

Once built, the focus shifts to the operators who foreground managerial performance such a timeliness and cleanliness. Therefore, as an infrastructure project goes from conception to operation there are different aspects of the commute that have value. These change from predicted outcomes, at the business case stage, to the diversity of reported value once in service. Accordingly, surveys reporting how a current system is valued should form part of a reflection (or a continuous improvement) to the design and operation of new (or modified) systems. In essence, broadening and closing the cost benefit loop that has a focus on satisfaction to capture the expectations of the experience and fulfilment of the journey thus deepen the value proposition.

Figure 1 conceptualises a combined utilitarian, hedonic and eudaimonic approach. In arriving at a wider new ontological position, we suggest the value of public transportation will be increased, thus serving to provide greater support as part of the business case for this investment. This framework is designed for the operational phase of a transit system, with a foundation to support data to reveal the satisfaction, fulfilment and experience of the commute via a stated preference or personal value statement. Within each sphere there is a focus on the utility, hedonic and eudaimonic elements. When combined, data reveals, as was the case in the NW metro study, a deeper insight into the value of the travel for the commuter. While this approach is nascent and lacking robust numeric values that would serve to support an expanded cost–benefit approach, data could be used to improve the operational performance of a system, place and the component parts of the commute (e.g. door to station and station to station). In addition, it can inform the strategic planning and design of new TOD networks and places, emphasising the role of the station and precinct.

6 Conclusions and Policy and Research Directions

Transport infrastructures are key domains of the urban connective tissue. These infrastructures connect people, built environments and nature. Given the increasing mobility of the population, there is a wealth of evidence that transport, in particular, is associated with positive utility and enhanced subjective wellbeing. Rather than a continued emphasis on economic utility models, an interdisciplinary approach is proposed as a means of developing a new narrative of train transport.

It is hypothesised that railway infrastructure can be better utilised through informed design thinking in relation to the broad definitions of health and social wellbeing, by enhancing awareness through digital information provision, and by promoting a sense of place and community belonging. The boundaries of this model would incorporate the first and last mile connecting users to the transit system, the nodes or stations themselves and the connecting linear corridors. New investigative approaches that incorporate the concepts of eudemonic and social wellbeing into transport are necessary using an established range of social and wellbeing indicators, in TOD, and in non-TOD contexts.

The interdisciplinary approach to this research provides a model of enquiry adopting systems thinking and methodological approaches (hedonic and eudaimonic)

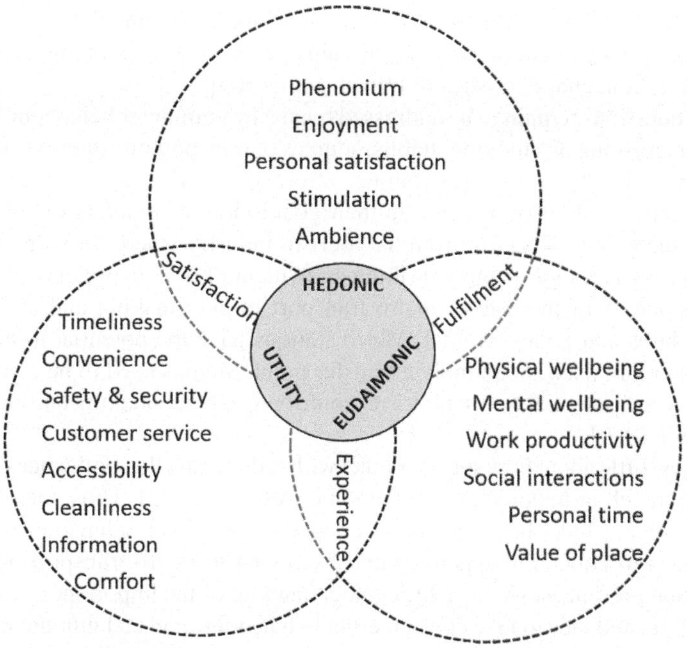

Fig. 1 A combined utilitarian, hedonic and eudaimonic approach for public transport infrastructure

for designing questions/approaches to inform planning and place-making. The findings of this survey can inform an alternative business case for public transit systems, focusing on the relationship between city planning and health (physical and mental) strategies. Commuting on public transport provides an opportunity to improve individual and community wellbeing. This occurs across various components of the door to door journey. Increasingly, social media is informing how people commute. Social media must central to commuter engagement to offer a journey that has utility, is enjoyable and contributes to positive physical and mental health outcomes.

The key policy recommendations and areas for research include:

1. The adoption of social capital models for quantifying the relationship between metro transport and social wellbeing should be examined as it recognises the critical social dimension of public transit systems in addition to economic and environmental aspects [32, 34]. Further, Banister [2] contends that commonly adopted measures of transport accessibility "focus on travel time reduction… and treat time as a commodity rather than a resource" (p. 290). New approach which incorporate the concepts of wellbeing into transport are necessary [2] and these need to examine the relationship between transport and a range of social and wellbeing indicators, in TODs, and in non-TOD contexts [47].
2. New approaches to transport and planning studies are evolving to address qualitative dimensions such as comfort, ambience and sensual stimulation, sensual and perceptive impressions during the journey [52, p. 200]. Such studies need

to address attributes of metro transit stations and surrounds that encourage use of train systems including impact of value of residences, walking proximity to transit station, characteristics of transit station [69].
3. Many questions remain to be addressed regarding commuter behaviour, potential for transforming or 'nudging' habits, sources of transport information, and receptiveness to health-giving opportunities afforded by rail transport [17, 53, 58]. In this regard, the transport sector, in their goal to lure commuters to public transport, much could be learnt from the tourism industry where for many travellers the journey is a significant part of the enjoyment of their experience [60].
4. Investigations of the role of metro transport in placemaking and activation of place, hubs and nodes, is vital. Metro stations have the potential to be a travel destination in itself, not simply a transfer point. Studies need to be conducted to evaluate the potential for metro travel contribute to place-making and community cohesion [8, 9].
5. We know little about how social contact with fellow travellers influences emotions during travel, or hedonic and eudaimonic well-being [13]. Thus, there is a need to investigate links between social networks and travel behaviour, to identify strategies to entice citizens to switch from cars to metro transport. In this era of virtual presences, we have little understanding of the importance of corporeal travelling, and face to face contact, crucial to hedonic and eudaimonic wellbeing, to social networking, given our 'compulsion for proximity' [65].
6. Longitudinal studies are vital to elucidate casual factors linking transport, green infrastructure, health and wellbeing [27, 54]. It is hypothesized that metro travel (as opposed to sedentary car travel) provides incidental opportunities for increased incidental physical activity and social interaction, with the potential to affect significant urban health issues (e.g. loneliness, isolation, depression, cardiovascular disease) [59].

Attachment 1

Interview Questions

1. Demographics: name, age, gender, nationality, postcode, occupation.
2. What modes of transport do you use to access residences, businesses, organisations and services within Macquarie Park?
3. What is your motivation for using each of the identified modes?
4. What expectations do you have of the modes you utilise, e.g. timely, safe, comfortable?
5. What benefits have you found in using these modes of transport?
6. What challenges have you found in using these modes of transport?
7. Do you have any suggestions for improvement?

References

1. Badland H, Whitzman C, Lowe M, Davern M, Aye L, Butterworth I, Giles-Corti B et al (2014) Urban liveability: emerging lessons from Australia for exploring the potential indicators to measure the social determinants of health. Soc Sci Med 111:64–73
2. Banister D (2019) Inequality in transport. Alexandrine Press, Marcham
3. Bache I, Reardon L (2016) The politics and policy of wellbeing: understanding the rise and significance of the new Agenda. Edward Elgar Publishing, London
4. Bertolini L (1999) Spatial development patterns and public transport: the application of an analytical model in the Netherlands. Plan Pract Res 14(2):199–210
5. Cao X, Mokhtarian PL, Handy SL (2009) No particular place to go: an empirical analysis of travel for the sake of travel. Environ Dev 41(2):233–257
6. Chatterjee K,Chng S, Clark B, Davis A, De Vos J, Ettema D, Reardon L et al (2020) Commuting and wellbeing: a critical overview of the literature with implications for policy and future research. Transp Rev 35(3):250–274
7. Cho S, Lambert D, Kim D, Roberts R, Park W (2011) Relationship between value of open space and distance from housing locations within a community. J Geogr Syst 13(4):393–414
8. Currie G, Stanley J (2008) Investigating links between social capital and public transport. Transp Rev 28(4):529–547
9. Currie G, Stanley J, Stanley J (2007) No way to go: transport and social disadvantage in Australian communities. Monash University Press, Clayton, Victoria
10. Curtis C (2015) Public transport-oriented development and network effects. In: Hickman R, Givoni M, Bonilla D (eds) handbook of transport and development. Edward Elgar Publishing, Cheltenham, pp 136–148
11. Delbosc A (2012) The role of well-being in transport policy. Transp Policy 23:5–33
12. De Vos J (2018) Towards happy and healthy travellers: a research agenda. J Transp Health 11:80–85. https://doi.org/10.1016/j.jth.2018.10.009
13. De Vos J, Schwanen T, Van Acker V, Witlox F (2013) Travel and subjective well-being: a focus on findings, methods and future research needs. Transp Rev 33(4):421–442
14. Diener E (2009) Subjective well-being. The science of well-being. In: Diener E (ed) The collected works of Ed Diener. Springer, New York, pp 11–58
15. Dodge R, Daly AP, Huyton J, Sanders LD (2012) The challenge of defining wellbeing. Int J Wellbeing 2(3):222–235
16. Douglas N (2016) Service quality values for Sydney Rail. Published by Douglas Economics
17. Dziekan K (2008) What do people know about their public transport options? Transportation 35:519–538
18. Ettema D, Gärling T, Olsson LE, Friman M (2010) Out-of-home activities, daily travel, and subjective well-being. Transp Res Part A Policy Pract 44(9):723–732. https://doi.org/10.1016/j.tra.2010.07.005
19. Ettema D, Friman M, Garling T, Olsson LE, Fujii S (2012) How in-vehicle activities affect work commuters' satisfaction with public transport. J Transp Geogr 24:215–222
20. Evans GW, Wener RE, Phillips D (2002) The morning rush hour: predictability and commuter stress. Environ Behav 34(4):521–530
21. Friedmann J (1987) Planning in the public domain. Princeton University Press, Princeton, NJ
22. Forgeard MJC, Jayawickreme E, Kern ML, Seligman MEP (2011) Doing the right thing: measuring wellbeing for public policy. Int J Wellbeing 1(1):79–106
23. Gamberini L, Spagnolli A, Miotto A, Ferrari E, Corradi N, Furian S (2013) Passengers' activities during short trips on the London Underground. Transportation 40:251–268
24. Gatersleben B, Uzzell D (2007) Affective appraisals of the daily commute: comparing perceptions of drivers, cyclists, walkers, and users of public transport. Environ Behav 39(3):416–431
25. Geurs KT, Boon W, Van Wee B (2009) Social impacts of transport: literature review and the state of the practice of transport appraisal in the Netherlands and the United Kingdom. Transp Rev 29(1):69–90

26. Giannetti BF, Agostinho F, Almeida CMVB, Huisingh D (2015) A review of limitations of the GDP and alternative indices to monitor human wellbeing and to manage eco-system functionality. J Clean Prod 87:11–25
27. Giles-Corti B, Vernez-Moudon A, Reis R, Turrell G, Dannenberg AL, Badland H, Owen N et al (2016) Urban design, transport and health 1. City planning and population healthy: a global challenge. Lancet 388:2912–2924
28. Gripsrud M, Hjorthol R (2012) Working on the train: from 'dead time' to productive and vital time. Transportation 39:941–956
29. Handy S, Thigpen C (2019) Commute quality and its implications for commute satisfaction: exploring the role of mode, location, and other factors. Travel Behav Soc 16:241–248
30. Hill JO, Peters JC (1998) Environmental contributions to the obesity epidemic. Science 280(5368):1371–1374
31. Hodgson S, Namdeo A, Araujo S, Pless-Mulloli T (2012) Towards an interdisciplinary science of transport and health: a case study on school travel. J Transp Res 21:70–79
32. Jones P, Lucas K (2012) The social consequences of transport decision-making: clarifying concepts, synthesising knowledge and assessing implications. J Transp Geogr 21:4–16
33. Kahnemann D, Diener E, Schwartz N (eds) (1999) Well-being: the foundations of hedonic psychology. Russell Sage Foundation, New York
34. Kamruzzaman M, Wood L, Hine J, Currie G, Giles-Corti B, Turrell G (2014) Patterns of social capital associated with transit-oriented development. J Transp Geogr 35:144–155
35. Kenworthy JR (2006) The eco-city: ten key transport and planning dimensions for sustainable city development. Environ Urban 18(1):67–85
36. Legrain A, Eluru N, El-Geneidy A (2015) Am stressed, must travel: the relationship between mode choice and commuting stress. Transp Res Part F Traffic Psychol Behav 34:141–151
37. Lundberg U (1976) Urban commuting: crowdedness and catecholamine excretion. J Hum Stress 2(3):26–32
38. Lunke EB (2020) Commuters' satisfaction with public transport. J Transp Health 16.https://doi.org/10.1016/j.jth.2020.100842
39. Lyons G, Urry J (2005) Travel time in the information age. Transp Res A Policy Pract 39(2–3):257–276
40. Lyons G, Jain J, Holley D (2007) The use of travel time by rail passengers in Great Britain. Transp Res Part A 41:107–120
41. McFadden D (2001) Disaggregate behavioral travel demand's RUM side–a 30 years retrospective. In: Hensher DA (ed) Travel behavior research. Elsevier, Amsterdam, pp 17–63
42. McLeod S, Scheurer J, Curtis C (2017) Urban public transport: planning principles and emerging practice. J Plan Lit 32(3):223–239
43. Mokhtarian PL, Salomon I, Singer ME (2015) What moves us? An interdisciplinary exploration of reasons for travelling. Transp Rev 35(3):250–274
44. New South Wales Government (2018) Metro art–integrated and sculptural public artwork. Expression if interest information. https://www.create.nsw.gov.au/wp-content/uploads/2018/07/metro-sculptural-public-artwork.pdf
45. Novaco RW, Gonzalez OI (2009) Commuting and well-being. In: Amichai-Hamburger Y (ed) Technology and psychological well-being. Elsevier, Amsterdam, pp 174–205
46. OECD (2019) Better life index. http://www.oecdbetterlifeindex.org/#/11111111111
47. Onyx J, Bullen P (2001) Measuring social capital in five communities. J Appl Behav Sci 36(1):23–42
48. Paez A, Whalen K (2010) Enjoyment of commute: a comparison of different transportation modes. Transp Res A 44(7):537–549
49. Rissel C, Curac N, Greenaway M, Bauman A (2012) Physical activity associated with public transport use: a review and modelling of potential benefits. Int J Environ Res Pub Health 9(7):2454–2478. https://doi.org/10.3390/ijerph9072454
50. Ryff CD (1989) Happiness is everything, or is it? Explorations on the meaning of psychological well-being. J Pers Soc Psychol 57:1069–1081

51. Scott M, Mick Lennon M, Haase D, Kazmierczak A, Clabby G, Beatley T (2016) Nature-based solutions for the contemporary city/Re-naturing the city/Reflections on urban landscapes, ecosystems services and nature-based solutions in cities/Multifunctional green infrastructure and climate change adaptation: brownfield greening as an adaptation strategy for vulnerable communities?/Delivering green infrastructure through planning: Insights from practice in Fingal, Ireland/Planning for biophilic cities: from theory to practice. Plan Theory Pract 17(2):267–300
52. Schiefelbusch M (2010) Rational planning for emotional mobility? The case of public transport development. Plan Theory 9(3):200–222
53. Schwanen T, Banister D, Anable J (2012) Rethinking habits and their role in behaviour change: the case for low-carbon mobility. J Transp Res 24:522–532
54. Shanahan DF, Lin BB, Bush R, Gaston KJ, Dean JH, Barber E, Fuller RA (2015) Toward Improved public health outcomes from urban nature. Am J Pub Health 105(3):470–477
55. Shliselberg R, Givoni M (2017) Motility as policy objective. Transp Rev 38(3):279–297
56. Stanley J, Stanley J (2007) Public transport and social policy goals. Road Transp Res 16:20–30
57. Taylor J (2012) What is wellbeing? https://www.habitsforwellbeing.com/what-is-wellbeing
58. Thaler RH, Sunstein CR (2008) Nudge: improving decisions about health, wealth and happiness. Yale University Press, New Haven
59. Thompson S, Kent J, Lyons C (2014) Building relationships for healthy environments: research leadership and education. Heath Promot J Aust 25:202–208
60. Thompson K, Schofield P (2007) An investigation of the relationship between public transport performance and destination satisfaction. J Transp Geogr 15(2):136–144
61. Transport for New South Wales (Jun 2020) Economic parameter values. https://www.transport.nsw.gov.au/news-and-events/reports-and-publications/tfnsw-economic-parameter-values
62. Tuan Y-F (1977) Space and place: the perspective of experience. University of Minnesota Press, Minneapolis, MN
63. United Nations (2017) The new urban Agenda. http://habitat3.org/the-new-urban-agenda
64. United Kingdom Office of National Statistics (2021) Personal well-being in the UK: April 2020 to March 2021. https://www.ons.gov.uk/peoplepopulationandcommunity/wellbeing/bulletins/measuringnationalwellbeing/april2020tomarch2021
65. Urry J (2001) Transports of delight. Leis Stud 20(4):237–245
66. Urry J (2007) Mobilities. Polity Press, Cambridge
67. Wener R, Evans G, Boately P (2005) Commuting stress: psychophysiological effects of a trip and spillover into the workplace. Transp Res Rec J Transp Res Board 1924:112–117
68. World Health Organisation (2021) Who we are: constitution. https://www.who.int/about/who-we-are/constitution
69. Zhao J, Deng W (2013) Relationship of walk access distance to rapid rail transit stations with personal characteristics and station context. J Plan Dev 139(4):311–321

Fostering Sustainable Urban Futures Through Twitter Public Space

Paola Monachesi

Abstract The smart city is based on digital data that can contribute to a better understanding and efficient management of urban processes. These data, which are produced by devices and citizens through online and social media activities, are usually managed and exploited by big tech companies. We claim that digital sustainability is tightly connected to environmental sustainability. Citizens have lost control over data-driven processes that are happening on a global arena. However, communities, through their online communication, can contribute to an alternative narrative and the creation of public spaces that constitute the premise to build more sustainable cities. In this paper, a social media analysis of the Twitter communication of two grassroots communities (i.e., creative skilled migrants living in Amsterdam and Dutch elderly) is carried out. It is based on a language analysis of geo-tagged data, combined with a social network analysis that show the contribution of these two groups to the smart city debate. Their different voices can give rise to more sustainable cities in which social equity and environmental protection become priority themes. These two communities exploit the available Twitter space, as a contact zone. Even though there might not be a clear political purpose in their narrative, the interaction between the local and global level of communication (mediated by social networks) gives rise to a space that stimulates participation, and has the potential to contribute to a more sustainable urban future.

1 Introduction

The smart city, as eco-efficient city structure [1, 32], is based on digital data that contribute to a better understanding and efficient management of urban processes [16]. These data are produced not only by devices but also by citizens through online and social media activities.

Data collection and monitoring by global technology companies, governments, and knowledge institutions can contribute to more sustainable urban spaces [14, 29,

P. Monachesi (✉)
Utrecht University, Utrecht, The Netherlands
e-mail: P.Monachesi@uu.nl

Z. Allam (ed.), *Sustainable Urban Transitions*, Urban Sustainability,
https://doi.org/10.1007/978-981-99-2695-4_3

30, 52]. However, it remains an open question whether sustainability, understood as environmental protection and social equity, can be really achieved through digitization since economic growth seems the main concern of urban development [36].

Citizens are often not involved or not interested in taking part in the monitoring and exploitation of their digital data. They underestimate the impact of their digital data on environmental and social sustainability, but these data could contribute to make urban processes more efficient and improve social conditions, if properly employed.

In this paper, we claim that digital and environmental, as well as social sustainability are strictly connected. The lack of sustainability in the way we relate to the environment and to digital networks are two aspects of the same problem. They are two symptoms of the same illness: communities have lost control over social and economic data-driven processes that are happening at large scale on a global arena. At all levels, from small local contexts to the state, communities are not being heard in the decisions regarding environmental policies or in their needs to protect their privacy or in their social needs.

The aim of this paper is to analyze whether communities, through their communicative processes, can contribute to more sustainable urban futures. The current smart city discourse relies on themes such as digital digital innovation, economic growth, resource efficiency [28] and concepts such as: 'technology', 'smart grids', 'big data', 'efficiency', 'infrastructure', and 'information' [19, 52]. An alternative storytelling is necessary to give rise to a democratic and progressive planning practice: stories can be a powerful tool in this respect [26, 51].

Communities can contribute to establish an alternative relation between digital and environmental sustainability by exploiting online communication. It should be possible to promote a different narrative to that of big technology companies and private stakeholders that can contribute to the creation of public space [40]. Public space constitutes the premises to build a more sustainable urban future. One in which the interests of government and companies are in balance with the needs of the citizens.

As argued in [40], an unexpected use of media can play an important role in the creation of public space, both in the digital and in the urban dimension. The spatial dimension is relevant in giving rise to communities that can contrast dominant institutional and corporate actors.

In this paper, we present a social media analysis of the Twitter communication of two grassroots communities (i.e., creative skilled migrants living in Amsterdam and Dutch elderly) to show that their voices can have an impact in giving rise to more sustainable cities in which social equity and environmental protection become priority themes. There needs to be an alternative to the discourse promoted by governments and big tech corporations which is often only in the direction of economic growth.

The analysis relies on an innovative methodology proposed in [41] that merges quantitative and qualitative methods. It is based on a language analysis of the geo-tagged data, in combination with a social network analysis, that shows the contribution of these groups to the smart city debate. It looks at urban sustainable development from the vantage point of new media, communication, and language technology [41].

Based on an investigation of the topics of discussion in Twitter, it can be shown that these two apparent different groups share instead important similarities. They both promote an alternative storytelling with respect to the smart city by addressing sustainability in their Twitter communication. This is the case especially for a specific group of creative migrants, that of the architects and designers that are concerned not only with the economic aspects of environmental sustainability, but also with the environment and social justice [42]. Some topics (i.e. circular economy) are also reflected in the communication of the pre-retirement age group of the old adults analyzed. The 67+ group, is especially interesting in this respect because it focusses mainly on environmental sustainability and climate change, by also addressing their political implications.

The analysis relies crucially on an identification of communities on social media since they can play an important role in the creation of public space (and the other way around). Public space is a space where people can discuss, fight for their rights and interests, work for the common good, and gain control over the data they produce. It will be argued that this space is the premise to build a more sustainable way of life.

These two groups exploit Twitter as a digital space of encounter where beliefs and values get confronted and can compete for domination. Even though there might not be a clear political purpose in their narrative, the interaction between the local and global level of communication (mediated by social networks) gives rise to a space that stimulates participation contributing to a more sustainable urban future. Twitter functions thus as a contact zone [50].

This paper is innovative in several respects. At the *methodological* level, it relies on social media data to investigate the narrative of two apparently different actors, that is creative skilled migrants working in the Netherlands and Dutch elderly whose narrative has not yet been analyzed in detail with respect to the smart city. The analysis combines in an innovative way language and social network analysis. The approach presented can complement more traditional works that focus on elicited data, such as focus groups or interviews (i.e. [16, 52]). Social media provides spontaneous and authentic data, that can reflect different voices that would normally not take part in more traditional data collections.

At the *theoretical* level, the paper contributes to the human-centered, smart city literature in which citizens are asked to participate in innovation processes [10, 12, 14, 28]. Both groups analyzed engage in social activities and in projects for the common good, this is the case of the urban designers as well as the elderly that are rather active in promoting local initiatives.

At the *policy* level, we show that a language analysis of social media communication can offer a useful instrument to urban planners and policy makers to work towards more sustainable urban futures in which environmental sustainability [17, 27, 56], social justice [22, 34, 53] and resilience to future shock [19, 28] become the leading themes. There is the need to achieve a balance between corporate driven solutions and those ones arising from the needs and participation of the citizens [3].

2 Smart Cities and Sustainability

The smart city is an urban space shaped by global technology companies, governments, and knowledge institutions that collect and analyze data, regulating urban processes and making them more efficient [20]. It is in this way that data and sustainability are linked. More specifically, the smart city agenda aims to support sustainability, which is often understood as a function of innovation in relation with technology [21]. However, economic growth seems the main concern of urban development since data produced by devices and citizens are exploited by corporations and governments, often through public- private partnership. They create new markets and profit that often do not consider the needs of the citizens and of the environment.

In order to give rise to more sustainable modes of urban development, technology should serve the needs of the citizens and citizens should participate in innovation processes, as co-creators [57]. Data should be conceived as a common good and be open to local companies, social platforms, and organizations, creating long term public value. Citizens should not be mere data providers but decision-makers [8].

Data ownership plays an important role in environmental and social sustainability. Data are valuable and value-creating but are in the hands of big tech corporations that exploit them for their own benefits. In this context, it is imperative to regain control over the data produced by citizens. Data should be conceived as a common good and be open to local companies, social platforms, and organizations. They should be employed to serve the needs of the citizens and of the environment, creating long term public value [4].

The current pandemic due to covid-19 has highlighted the failure of the technocratic smart city model. A model that had already been subject to much criticism because it serves mainly the interests of coorporations and governments, rather than those of the citizens [1, 11, 19, 25, 31, 38, 54, 55].

Smart cities such as London or New York that are driven by data and algorithms failed to use smart technology to carry out contact tracing. They didn't manage to limit the spread of the virus and the high number of deaths. Smart cities proved not to be adequate to support the well- being of their citizens. Amsterdam, which has been much praised for its smart city policies [2, 8], doesn't have sufficient parks and green areas for people to breath fresh air and maintain distancing. Homes are not conceived to have the entire family working and living in the same space 24/7. The smart city is driven by technology, but the recent pandemic made even more clear that the use of technology on which it is based, serves mainly the interests of corporations and governments rather than those of the citizens.

A growing body of research and academic literature has argued for a different approach to the smart city, suggesting an alternative vision that considers the environment and the well-being of citizens as priority. The human-centered smart city literature [2, 10, 12, 28, 38] asks for citizens' awareness and participation in innovation processes and for a view in which data and technology is being used in support of the environment and citizens' needs. This paper is embedded within this research line.

3 Public Space and Contact Zone

Public space is the necessary building stone on which a more sustainable smart city can be built. The well-functioning of public space is a crucial element to foster debate and political action. More specifically, [44] argues for an agonistic space in which subjects with conflicting views and interests can coexist and compete, leading eventually to the creation of new communities and political subjects. Agonistic spaces contribute to the development of a 'counter-hegemony' [, p. 151].45

We witness, however, a progressive privatization of public space by institutional actors and by global capitalism [35] that have taken over public spaces in the physical and digital dimensions [43]. However, (public) physical space is crucial because it structures co-presence and interaction patterns of people that create the conditions that give rise to communities [23]. As argued by Hillier and Hanson [24], social relationships between people can be fostered by spatial solidarity, that is through sharing the same local world and getting together in physical space and by transpatial solidarity trough sharing the same interests that can extend across physical space boundaries.

Since human beings live in a spatial dimension, in addition to a social and temporal one, there is an inevitable relation between spatial and social phenomena that influence each other [9, 23, 37, 53]. Therefore, public space plays a role in giving rise to communities, but at the same time, communities can operate to create public space.

Media are important elements in the creation of (public) space. This is especially the case for mobile media that have contributed to the fusion between the digital and the physical dimensions which is most evident in urban contexts [13, 15]. The privatization of public spaces in the city has run in parallel with the emergence of new digital spaces that favor transpatial solidarity since people can come together based on their interests and goals, as is the case in social media platforms, such as Twitter.

Twitter is particularly relevant in analysing the communication of the groups that are considered in this paper, that is creative skilled migrants and Dutch elderly since they employ it to promote their activities, discuss events and build their network. It is considered a space of encounter where information, beliefs and values are shared, leading to interaction and sometimes conflict: a digital platform that creates mobile meeting places for shared experiences [49]. It is a contact zone [50] which is a space for encounters both voluntary and involuntary, and thus a space for potentiality, for example through unexpected use of media such as the stratagematic actions discussed in [40]. Stratagematic actions are characterized by an unexpected use of the hybrid spaces created by mobile and social media: physical and digital, local, and global.

They produce side effects that could be exploited by communities to refine, define, and eventually reach their goals, if they become aware of their potential.

4 Methodology

Social media platforms can offer massive, dynamic and diverse data that allow for an investigation of the behaviours, opinions and feelings of their users. These are spontaneous and authentic data that allow for the creation of textual corpora based on millions of users. Social media data can complement more traditional data collections, such as those emerging from focus groups, interviews and surveys. They make possible to identify voices that might not easily emerge through traditional approaches since these data emerge from real-world situations, from various people and they are not elicited by the prompt of researchers.

In this paper, an innovative methodology is presented that is based on the creation of a sample of active Twitter users through computational techniques. Their collected data has been analyzed by focusing on a hashtag investigation of the communication topics of the two groups identified (cf. also [6]). An additional innovative aspect of the approach presented is the integration of the language analysis of social media data with a social network analysis to visualize the geographical distribution of the communication patterns of the two groups investigated.

4.1 Data Collection

The data collection carried out has focused on two specific groups that are relevant to analyse in the context of urban sustainability because they might highlight a different discourse from that of global corporations. The two groups are Dutch elderly and creative skilled migrants working in the Netherlands. The identification of these two groups of users, however, can be challenging.

Identifying elderly is rather complex because in Twitter, the age of users is not visible and, in addition, defining age is not a trivial task. We have opted for using chronological age to select users which have been further grouped in three classes, that is users below 55 years old, users between 55 and 67 and above 67 years old. These groups correspond to three life stages that are related to the active working life of the individuals (i.e., below 55), the pre-retirement stage (between 55 and 67) and post retirement (above 67). Users have been identified manually, starting from a profile that was created for the purpose of following organizations, related to ageing, that one could expect to have old adults among its followers. Users have been categorized in the three groups using various tactics, such as profile picture and/or age of birth present in the Twitter name. Data was then extracted automatically from the created sample.

This methodology has been adapted to create a sample of the other group of users under investigation, that is skilled creative migrants working in the Netherlands. Identifying creative migrants is even more challenging than the elderly since they do not characterize themselves as such.

The first step was to identify occupations related to creativity. Three different resources have been employed to single out the relevant creative industries and derive the corresponding occupations. The result was a list of 164 creative professions that were further categorized in 11 sectors, within the creative industries. The professions identified were matched with the profile descriptions of Twitter users. Additional criteria were used, such as migrants being located in The Netherlands. As in the case of the elderly, a profile was created on Twitter in order to follow several organizations related to creative industries that one could expect to have skilled creative migrants among their followers.

The final list of creative migrants was manually verified but a content analysis of the profile description of the users was also carried out to further verify the correctness of the selection. It is based on word frequency and it confirms the validity of the approach. As in the case of the elderly, data was then extracted automatically from the created sample. We refer to [41] for further details on the data collection of skilled creative migrants.

4.2 Analysis

The Twitter data collected from the two groups was analyzed by employing an innovative methodology that combines a language analysis of the (geo-tagged) data with a quantitative investigation of the communication patterns through social network analysis. While the language analysis reveals the interests and activities of the users, the social network analysis provides a geographical visualization of their communicative interactions.

The language analysis proposed relies on computational methods and focusses on the frequency of the hashtags employed by the users to investigate their topics of discussion. Hashtags are a way to add metadata to shared content and highlight in this way the debate that is being carried out on the platform [5, 18]. Hashtags are more informative and thus more interesting to investigate than normal words. A frequency analysis based on hashtags reveals the topics that social media users want to highlight and provide information about what interests them [39, 48].

In the case of the creative migrants, four groups of creatives present in our sample were analysed. The categories are ‘writers’, ‘new media arts’, ‘fine arts’, and ‘architects and designers’. The aim was to investigate whether there are differences in the interests and communication of the various categories of creative skilled migrants, but especially their attitude towards sustainability.

The language analysis carried out can be connected to social network analysis to provide insights on the geographical distribution of the various patterns of communication possible in Twitter. The followers, reply and mention networks of the users were analysed.

In the case of creative migrants, the nature of the relationships between users has been investigated through a quantification and categorization of the users' connections in four dimensions: creative vs. non-creative and local vs. global users. The latter, that is, local and global users, have been calculated based on the geo-location provided. In this way, it is possible to investigate with whom creative skilled migrants interact and where these users are located.

The social network analysis provides insights on whether creative migrants are connected to more non-creative users than creative ones, for example. In this way, one can investigate whether they use the platform for professional or for private use. It is also possible to assess whether creative migrants connect more within the Netherlands or internationally. A visualization of followers, mention and reply networks also allows to analyse the level of engagement of these users within their network. Further details on the analysis of skilled creative migrants can be found in [41] and in[42].

The methodology proposed for the analysis of creative migrants adapts nicely also to the other group considered, that is the elderly, providing evidence for its validity.

More specifically, a frequency analysis of hashtags and words employed by the three age groups identified has been carried out. In the case of the elderly, the methodology has been extended by combining the hashtag frequency analysis with the geo-location information available in Twitter. This approach allows to investigate the spatial distribution of users that is particularly relevant in the case of urban sustainability. It provides an example of how digital sustainability and environmental sustainability are related. The data provided by the users make possible to assess whether certain age groups have an interest for environmental sustainability in connection with the place they live in and can play a role in stimulating an alternative discourse that triggers social action.

The geolocation information available in Twitter profiles, if present, makes possible to assess the distribution of the living places of the various users and to visualize it through maps. In this way, it becomes possible to investigate the interest of users for specific places and to analyze whether there is a correspondence between the places of residence and the locations users mention in their hashtags.

The importance of social network analysis within this methodology should be considered. It provides information on whether the message of the elderly gets broadcasted worldwide. It reveals their communication pattern within and outside of the Netherlands and makes possible to assess their impact in broadcasting their messages to different locations.

4.3 Results

The Twitter analysis of Dutch elderly reveals that location hashtags are used mainly by people living in the place that is being mentioned or close to it. It also shows that elderly above 67 exhibit an extensive use of location hashtags, especially locations within The Netherlands, as can be seen in Table 1. The expectation was that the elderly would mention mainly touristic places because of their available leisure time but this is not the case. A detailed analysis reveals that they use location hashtags to promote the city they live in, showing that they are connected to the local environment. Furthermore, a difference can be noticed in where people live, while big cities (i.e., Amsterdam and Rotterdam) are the domain of young people, elderly live in smaller places, as can be seen in Fig. 1.

The presence in tweets of hashtags related to nature and sustainability has been investigated to have better insight on the behavioural differences among the age groups in this respect. Previous literature has shown that young people don't have a strong interest in nature in comparison with elderly people [7, 47].

The elderly use more nature related hashtags than the other groups. The group 55–67 is the least interested in this topic, as can be seen in Table 2.

However, with respect to sustainability, the elderly are the least interested in sustainable development while the 55–67 group is the most interested one. The elderly use almost only unique terms and it should be noticed that only two terms are used by all three groups which are: 'climate' and 'climate change'.

The most frequent terms used by the under 55 group are those related to 'circular economy', followed by 'the green city' ('groenerstad'). The terms used by the 55–67 groups are very similar, and they also relate to circular economy: 237 of the 314 hashtags used are about it, as can be seen in Fig. 2.

It is interesting to notice, that the elderly belonging to the 67+ group mention instead rather different terms with respect to sustainability, in particular, the term 'circular economy' is not mentioned at all. On the other hand, the terms mentioned are mainly related to climate, climate change and climate actions, including terms related to a political dimension of climate change, as can be seen in Fig. 3.

Furthermore, terms related to the organic movement are being used. It seems thus that the elderly associate sustainable development mainly to the environmental aspects and not to the economic ones, as is the case for the younger groups.

To summarize: the elderly use location hashtags more than the other groups and they do not engage in the current sustainable development discourse based mainly

Table 1 Location hashtags

	Total number hashtags	Locations NL	Locations foreign	Percentage locations (%)	Percentage NL locations (%)
UN55	16076	758	287	6.5	4.7
OV55	7436	339	43	5.1	4.6
OV67	1545	100	16	7.5	6.5

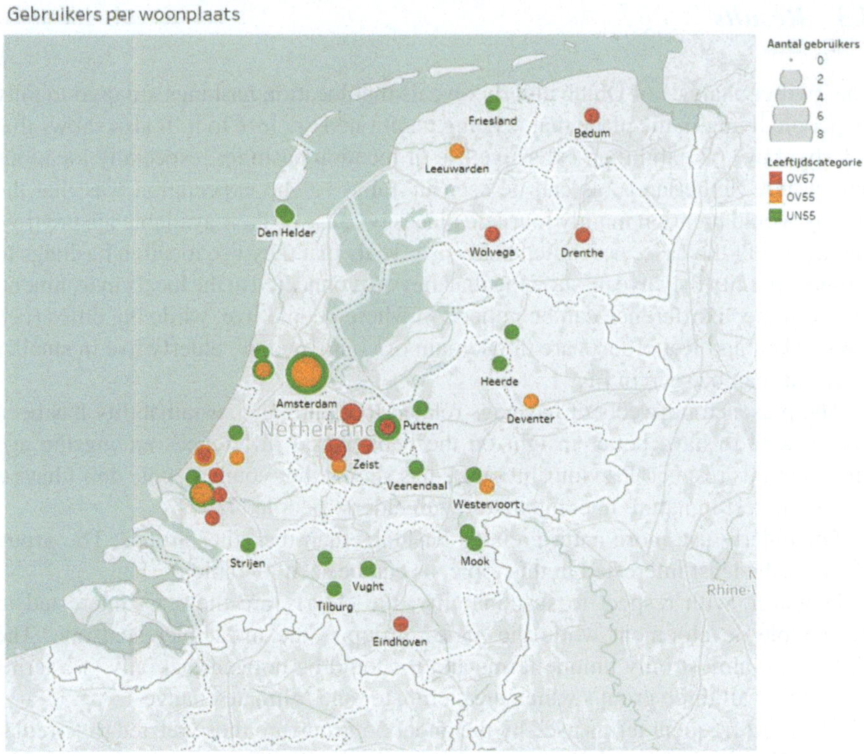

Fig. 1 Living places of all users

Table 2 Nature hashtags

	Total number hashtags	Nature #	Number unique #	Percentage (%)
UN55	16076	243	123	1.5
OV55	7436	30	29	0.4
OV67	1545	27	24	1.7

Fig. 2 Hashtags related to sustainability by 55–67 group

Fig. 3 Hashtags related to sustainability by 67+ group

on economic issues, but instead they have an interest for environmental aspects and for nature. They seem to play a role in broadcasting and alternative discourse on sustainable development in which a bond is established between nature and human beings. In this context, it is relevant to investigate the network of the elderly to assess whether they can have an impact in promoting their messages.

The connections of the group under 55 are located in the centre of the country where are the big cities, that is Amsterdam, Rotterdam and Utrecht. In the case of the international network, connections are mainly located in the east coast of the United States, as well as in Australia and New Zealand which are English speaking countries, a language easily mastered by Dutch users.

In the group 55–67, the core of the network is located in the Randstad where are the big cities Amsterdam, Rotterdam, Den Haag and Utrecht. For the rest, the network is quite well spread across the country. With respect to the international network, the US is a place where many connections are located as well as in Oceania and in South Africa, probably due to language reasons. Users in this group have more connections outside of the Netherlands (i.e. 10,7%) than the younger group of under 55 (i.e. 9,2%).

The network of the 67+ shows relevant hubs in Utrecht and locations close to this city. As for the international network of the elderly, it can be noticed that it is less strong than that of the younger groups since most of the connections of this group are in The Netherlands.

Figure 4 shows the network differences among the three age groups, as expected the young group is the most connected but the elderly do show quite a strong network, as well.

As for the creative migrants, a social media analysis of their communication shows that they contribute to an alternative, sustainable idea of smart city. Co-creation and participation are important issues in the discourse of creative migrants, but they also argue for the importance of datafication of urban space and acknowledge in this way the role of government and corporations in data governance. They combine technology with a participatory and socially driven discourse based on events, initiatives and projects showing social and environmental engagement.

This is especially evident in the case of the group of architects and designers. They are interested not only in technology and data, two of the features of the smart city discourse but they relate them to the common good. They play an important

Fig. 4 Social network analysis of three age groups: from old to young

role in broadcasting grassroots initiatives aimed at social inclusion and the circular economy. An overview of the hashtags used by this group can be seen in Fig. 5, taken from [, p.7].41

Their discourse can contribute to give rise to a more emancipatory and innovative smart city agenda. They can raise awareness within civil society by broadcasting relevant topics and social, local activities through Twitter at global level, even beyond their professional circle, linking Amsterdam to the global networks of power represented by global cities. Further details on this analysis can be found in [41].

A qualitative thematic analysis of the communication of this specific group of creative migrants has provided more details and it has shown that these creative migrants play an important role in broadcasting grassroots initiatives aimed at social inclusion and the circular economy. They address several topics in this respect such as circular economy, alternative forms of energy but also climate change and sustainable

Fig. 5 Hashtags used by the group architects and designers

development. They are also involved in promoting projects that foster social innovation, inclusion, education and they promote makerspaces that trigger inclusion and participation. More details about the thematic analysis and its results can be found in [42].

5 Discussion

In this paper, two apparently different grassroots communities have been analyzed with respect to their social media communication. Similarities emerge from an analysis of the topics of discussion that was based on the frequency of the hashtags used. For example, they both address sustainability in their communication on Twitter. This is the case especially for a specific group of creative migrants, that of the architects and designers. They are concerned with the economic aspects of environmental sustainability and address several topics in this respect such as circular economy and alternative forms of energy like wind, solar, biofuel. Some of these topics can be found also in the communication of the pre-retirement age group of the elderly analyzed. For example, they mention topics like circular economy.

However, the elderly are especially interested in environmental sustainability and nature. Their vision of sustainability is different from older adults (i.e. the age group between 55 and 67) since those focus more on the economic aspects of sustainability.

What emerges from the analysis of the data of the creative migrants is that sustainability is a topic of discussion both in relation to work as well as in connection to their private interests. Architects and designers are concerned especially about economic sustainability, but they are involved in promoting projects that foster social innovation as well as inclusion and education. They also promote makerspaces that trigger inclusion and participation, but they are not very concerned about privacy or data ownership.

Even though we could refer to the elderly and the creative migrants as communities, they do not really profile themselves as such on Twitter, they are individuals that express their views and ideas as individuals and not as a community. They do not join forces with others. They exploit the available space in Twitter that can be seen as a contact zone, as a digital space of encounter where beliefs and values get confronted. However, they do not shape the space they occupy. There is a potential among those users to contribute to a different storytelling, but there isn't a clear political impact in their communication. The elderly and the creative migrants analyzed form groups that share certain characteristics, but they are not a community, yet.

The social network analysis carried out shows that creative migrants are more connected globally, while the elderly are more connected locally. On the other hand, the hashtag analysis reveals connections to local communities in both cases. These communication networks have the effect to create a different storytelling through local participation and these groups can broadcast these activities towards a more human-centred smart city, globally through their international networks. Both groups, manage to exploit the digital and physical spaces that are mediated by their social networks.

It might be that this interaction between the local and the global level mediated by social networks might contribute to the creation of a public space. However, this space was not yet able to foster the creation of a community with a clear political purpose, but it shows potential that it might in the future.

6 Conclusion

The innovative methodology proposed in this paper relies on language and media but can have broad applications in urban studies: the results of the data analysis allow for an understanding of the storytelling behind the smart city discourse. This is important knowledge for urban planners and policy makers as is knowledge based on spontaneous data.

The hashtags employed by the creative migrants in their Twitter communication support social and environmental sustainability, similarly for what concerns the Dutch elderly analysed. The groups investigated are interesting because they provide a good example of the way digital and environmental sustainability interact. The data produced by these groups can promote an alternative human-centred smart city discourse. Citizens should become more aware of the relevance of the data they produce and make sure that they are being used to support their needs and not only economic growth.

The analysis of the communication of the two groups has revealed that they employ Twitter to broadcast local initiatives at a global level, creating an interaction between the physical urban dimension and the digital one through the social media platform. They give rise to an alternative discourse that has the potential to bring change if strategically exploited.

More specifically, elderly can play an important role in reshaping the discourse on sustainable development and redefine the relation between the human and the natural world, a necessary condition for a new perspective on sustainable urban futures. Evidence comes from Japan that is active in supporting the creation of communities, giving the opportunity to retired people to carry out caring activities in relation to agriculture, food and community support [46]. Japan is at the forefront in this respect since it precedes other countries in acknowledging ageing as an urban issue.

In this vision, elderly can play a very concrete role towards the creation of a more human-centred smart city. The elderly can act as innovators since they have time and a lifetime experience at their disposal. This is especially the case for Dutch old adults since they are at the forefront with respect to volunteering activities and social media use. It is thus important to give rise to initiatives to stimulate an alternative debate on sustainable development driven by volunteering activities where older adults can play a central role, for example through social media communication.

Environmental sustainability has received attention in the last years thanks to the active involvement of the younger generations through initiatives such as Fridays for Future, but the elderly have not been much involved. They could instead play an important role since this paper has argued that interests of the younger age group and of the 67+ group with respect to nature is comparable. Social media could be exploited by both groups that could join forces to give rise to an active ageless community. The political interests of the 67+ group that emerged from the hashtag analysis carried out could be beneficial for the younger generation and could help in creating a stronger impact at political level and contribute to re-shaping the environmental sustainability discourse.

The current pandemic has shown the vulnerability of cities and challenged the smart city agenda. One wonders whether the sustainability discourse should be promoted in relation to the city or whether one should go beyond it. A recent exhibition by Rem Koolhaas at the Guggenheim Museum of New York, has drawn attention on what he calls the countryside. He defines with this term everything which is not city, that amounts to 98% of the earth surface. Koolhaas claims that rural areas are changing as consequence of the changes that occur in urban areas [33]. Transformations outside of the city are just as radical as those that occur in relation to the urban, an aspect often neglected. However, the countryside that is represented in the exhibition is subordinated to the needs of the city, without a clear identity, it accepts gentrification phenomena passively. The polarization that we have witnessed in recent political elections, in the US, but not only, shows that the non-urban spaces have their own identity that usually sharply conflicts with the urban ones. Polarization is often triggered by this subordinate role.

The Covid-19 pandemic has made the borders between the rural and the urban less sharp, it has brought many people to leave the city and to experiment new ways of living and new possibilities. There is the need to acknowledge the big transformations that have characterized the non-urban in the last years, exploit their potential and maybe conceive new ways of merging the rural and the urban in the future, as a way to support environmental and social sustainability and the existence of the human species.

References

1. Allen A, Lampis A, Swilling M (2016) Untamed urbanism. Routledge advances in regional economics, science and policy. Routledge, New York
2. Angelidou M (2014) Smart city policies: a spatial approach. Cities 41:S3–S11
3. Angelidou M (2015) Smart cities, a conjuncture of four forces. Cities 47:95–106
4. Bria F (2018) Our data is valuable. Here's how we can take that value back. The Guardian. https://www.theguardian.com/commentisfree/2018/apr/05/data-valuable-citizens-silicon-valley-barcelona. Accessed 15 Jan 2019
5. Bruns A, Burgess JE (2011). The use of Twitter hashtags in the formation of ad hoc publics. In: Proceedings of the 6th European consortium for political research (ECPR) general conference 2011, pp 1–9
6. Bruns A, Burgess JE (2012) Researching news discussions on Twitter. New methodologies. Journal Stud 13(5–6):801–814

7. Buijs A, Custers M, Langers F (2007) Natuur door andere ogen bekeken. Wageningen, Wageningen UR
8. Calzada I (2018) (Smart) citizens from data providers to decision-makers? The case study of Barcelona. Sustainability 10(9):1–25
9. Castells M (1996) The rise of the network society. Blackwell Publishers, Oxford, UK
10. Castelnovo W, Misuraca G, Savoldelli A (2015) Smart cities governance: the need for a holistic approach to assessing urban participatory policy making. Soc Sci Comput Rev
11. Cugurullo F (2018) Exposing smart cities and eco-cities: Frankenstein urbanism and the sustainability challenges of the experimental city. Environ Plan Econ Space 50(1):73–92
12. Dameri RP (22 Oct 2013). Searching for smart city definition: a comprehensive proposal. Int J Comput Technol 11(5):2544–2551
13. de Souza e Silva A, Firth J (2010) Locative mobile social networks: mapping communication and location in urban space. Mobilities 5(4):485–506
14. de Waal M, Dignum M (2017) The citizen in the smart city. How the smart city could transform citizenship. It-Inf Technol 59(6):1–11
15. Farman J (2012) Mobile interface theory: embodied space and locative media. Routledge, New York, NY
16. Fernandez-Anez V, Fernández-Güell JM, Giffinger R (2018) Smart city implementation and discourses: an integrated conceptual model. The case of Vienna. Cities 78:4–16
17. Gargiulo Morelli V, Weijnen M, Van Bueren E, Wenzler I, Salvati L, De Reuver M (2013) Towards intelligently-sustainable cities? J Land Use Mob Environ 1:73–86
18. Golder SA, Huberman BA (2006) Usage patterns of collaborative tagging systems. J Inf Sci 32(2):198–208
19. Hajer M (2016) On being smart about cities. Seven considerations for a new urban planning and design. In: Allen A, Lampis A, Swilling M (eds) Untamed urbanism. Routledge advances in regional economics, science and policy. Routledge, New York, pp 50–63
20. Hancke GP, de Carvalho e Silva B, Hancke GP Jr (2013) The role of advanced sensing in smart cities. Sensors 13(1):393–425
21. Haarstad H (2017) Constructing the sustainable city: examining the role of sustainability in the 'smart city' discourse. J Environ Planning Policy Manag 19(4):423–437
22. Harvey D (2008) The right to the city. New Left Rev 53:18
23. Hillier B (1996) Space is the machine: a configurational theory of architecture. Cambridge University Press, Cambridge, UK
24. Hillier B, Hanson J (1984) The social logic of space. Cambridge University Press, Cambridge
25. Hollands R (2015) Critical interventions into the corporate smart city. Camb J Reg Econ Soc 8(1):61–77
26. van Hulst M (2012) Storytelling, a model of and a model for planning. Plan Theory 11(3):299–318
27. de Jong M, Joss S, Schraven D, Zhan C, Weijnen M (2015) Sustainable–smart– resilient–low carbon–cco–knowledge cities; making sense of a multitude of concepts promoting sustainable urbanization. J Clean Prod 109:25–38
28. Joss S, Cook M, Dayot Y (2017) Smart cities: towards a new citizenship regime? A discourse analysis of the British Smart City standard. J Urban Technol 24(4):29–49
29. Kitchin R (2014) Big data, new epistemologies and paradigm shifts. Big Data Soc 1(1):1–12
30. Kitchin R (2014) The data revolution: big data, open data, data infrastructures and their consequences. Sage, London
31. Kitchin R (2014) The real time city. Big data and smart urbanism. GeoJournal 79:1–14
32. Kitchin R (2015) Making sense of smart cities: addressing present shortcomings. Camb J Reg Econ Soc 8(1):131–136
33. Koolhaas R (2020) Countryside. A report. Taschen
34. Lefebvre H (1996) The right to the city. In: Kofman E, Lebas E (eds) Writings on cities. Blackwell, Oxford
35. Lefebvre H (2003) The urban revolution (R. Bononno, Trans.). University of Minnesota Press, Minneapolis, MN

36. Martin CJ, Evans J, Karvonen A (2018) Smart and sustainable? Five tensions in the visions and practices of the smart-sustainable city in Europe and North America. Technol Forecast Soc Chang 133(4):269–278
37. Massey D (2005) For space. Sage, Thousand Oaks, CA
38. McFarlane C, Söderström O (2017) On alternative smart cities. City 21(3–4):312–328
39. Monachesi P, de Leeuw T (2018) Analyzing elderly behavior in social media through language use. In: Stephanidis C (ed) HCI international 2018–20th international conference, proceedings, Part II. Springer, pp 188–195
40. Monachesi P, Turco M (2017) New urban players: stratagematic use of media by Banksy and the Hong Kong Umbrella Movement. Int J Commun 11:1–18
41. Monachesi P (2020) Shaping an alternative smart city discourse through Twitter: Amsterdam and the role of creative migrants. Cities, 100
42. Monachesi P, Witteborn S (2021) Building the sustainable city through Twitter: creative skilled migrants and innovative technology use. Telemat Inf, 58
43. Morozov E (2011) The net delusion: how not to liberate the world. Perseus Books, Cambridge, MA
44. Mouffe C (2005) On the political. Routledge, London, UK
45. Mouffe C (2013) Agonistics: thinking the world politically. Verso, London, UK
46. Muramatsu N, Akiyama H (2011) Japan: super-aging society preparing for the future. Gerontologist 51:425–432
47. Nieuwenburg C, Staal EV, Winterwerp R, Zuuren TV (2011) Denkbeelden over de natuur. Utrecht University, Utrecht
48. Pandya A, Oussalah M, Monachesi P, Kostakos P (2019) On the use of distributed semantics of tweet metadata for user age prediction. Future Gener Comput Syst 102:437–452
49. Polson E (2015) A gateway to the global city: mobile place-making practices by expats. New Media Soc 17(4):629–645
50. Pratt ML (1991) Arts of the contact zone. MLA Profession, 33–40
51. Sandercock L (2003) Out of the closet: the importance of stories and storytelling in planning practice. Plan Theory Pract 4(1):11–28
52. Söderström O, Paasche T, Klauser F (2014) Smart cities as corporate storytelling. City 18(3):307–320
53. Soja D (2010) Seeking spatial justice. University of Minnesota Press, Minneapolis, MN
54. Trencher G (2019) Towards the smart city 2.0: empirical evidence of using smartness as a tool for tackling social challenges. Technol Forecast Soc Change 142(C):117–128
55. Vanolo A (2014) Smartmentality: the smart city as a disciplinary strategy. Urban Stud 5(51):881–896
56. Viitanen J, Kingston R (2014) Smart cities and green growth: outsourcing democratic and environmental resilience to the global technology sector. Environ Plan A 46(4):803–819
57. Zandbergen D, Uitermark J (2019) In search of the smart citizen: republican and cybernetic citizenship in the smart city. Urban Stud, 1–16

Mapping Contemporary Urban Challenges

An Empirical Approach to Estimating Secondary School Student's Sustainable Mobility Score

Isabel Cruz and Luís M. Afonso

Abstract Sustainable mobility is at the core of zero-carbon cities. Young people's mobility may be considered a precursor of sustainable mobility. Indeed, empirical studies show that if young people are not used to traveling by public transport by the age of eighteen, there is a lower probability that they will use them as adults. Theories of practice are the theoretical framework for understanding how young people move from home to school. We take daily practice as a part of a complex network of social relationships embedded in symbolic meanings. The aim is to propose an empirical index to estimate home-school commute sustainability scores based on a survey focused on students' mobility in a Secondary School in Matosinhos Municipality. We devise an algorithm to estimate students' sustainable mobility scores on a scale from 0 to a maximum of 10 by considering the diversity of students' mobility modalities, the GPS distance from home to school, and the "reasonability" of doing the commute by walking. Results of a random representative sample (n = 280) corresponding to 32.2% of this school's population show that 43.4% of students present a sustainable mobility pattern (score > 6), and 94.2% of these sustainable students got a high score (8). The use of public transport highly influenced this score. For this school, the rescaled estimated global sustainable index is 56.1%. This value corresponds to the proposed global sustainable mobility empirical index estimate, defined by the mean of all student scores divided by the maximum sustainability score (10).

Keywords Students' mobility · Daily practice · Sustainable mobility score · Sustainable global index

---

I. Cruz (✉) · L. M. Afonso
Matosinhos Municipality Living Lab, Porto, Portugal
e-mail: imsilva@gmail.com

L. M. Afonso
e-mail: lma@isep.ipp.pt

L. M. Afonso
School of Engineering, Polytechnic of Porto, Porto, Portugal

ISRC-Interdisciplinary Studies Research Center, Porto, Portugal

Z. Allam (ed.), *Sustainable Urban Transitions*, Urban Sustainability,
https://doi.org/10.1007/978-981-99-2695-4_4

1 Introduction

There is a large consensus regarding the link between environmental quality and mobility practices. Mobility has historically been a great contributor to CO_2 emissions and is at the centre of sustainability discussion. In order to improve people's lives quality urban policies are being implemented to reduce CO_2 emissions focusing on socio-technical innovations and behaviour change. Knowing that sustainability is far beyond the CO_2 emissions, there is an increasing interest in measuring mobility sustainability in alternative and expedited ways. Bearing this in mind, this research intends to tackle two fundamental questions: How to measure sustainability? How to follow its dynamics in time, space, and between generations?

The system of empirical heuristical indicators was constructed to estimate students' mobility sustainability. The development of these indicators is empirically based on a survey of 280 Matosinhos secondary school students to estimate home-school-home commute sustainability heuristic scores to answer the research questions. A global index is then calculated to estimate the overall school sustainability by the mean value of the sustainable score of all surveyed students allowing for comparison among schools or the same school at different times. The evolution of the global sustainability index at different times can capture transitions relevant to sustainable consumption sociology.

The indicators constitute a flexible estimation tool that can be applied to other samples in other times and spaces, making the construction of time series of indexes possible. These time series are valuable tools to unveil sustainability's flow of daily mobility practices allowing the analysis of sustainable transitions between collectives and generations.

This research is organized into six sections and intends to approach the answer to these questions. A theoretical framework based on practice theories contributions is developed in Sect. 2 to support sustainable indicators construction and interpretation. Based on the concepts presented, a methodological approach is established in Sect. 3 for developing two sustainability mobility estimators. The method of construction of the proposed estimator is carried out in Sect. 4. The results of applying the proposed indicators to a case study are presented and analyzed in Sect. 5. The research closes with Sect. 6, highlighting the main conclusions and contributions.

2 Theoretical Framework for Sustainable Mobility Analysis

Zero carbon cities are becoming ubiquitous objectives for municipality governance. Innovations technologies vis a vis social practices are core issues for several programs [5–7, 13, 14] that are being developed in the context of smart cities and Living Labs to help achieve sustainable mobility goals. These programs have raised several questions. Spaargaren focuses on citizen-consumer [12] and companies to analyse behaviour changes and technological innovations that improve sustainable consumption practices. Shove [6, 7] argues that the sustainable consumption focus is on enabling sustainable practices analysed through dynamics of everyday practices, ways of life, and the relation between material elements of practices and sociotechnical systems [5, 8, 9]. In this chapter, we aim to address two questions: How to measure sustainability? How to follow its dynamics in time, space, and between generations? The focus on these particular questions leads us to propose an empirical index as a tool to estimate and compare sustainability scores. The development of this flexible tool articulating theoretical framework and empirical observation regarding sustainable mobility is the main contribution of this research. Moreover, this tool can be used in different times and spaces to unveil sustainable mobility dynamics, contributing to a better understanding of these social practices and city governance.

The estimation of sustainable mobility is a rather complex and multidimensional issue with no simple answers. For instance, sustainability relates to physical, social, and mental health that is probably impossible to measure with arbitrary precision. Bearing this in mind, we believe in a robust heuristic approach to an estimation of sustainability alternative to the measurement of the CO_2 emissions of each agent. This estimate is based instead on the concept of sustainability in its broadest form, in which the weight for each mobility modality is generally defined, considering that agents can use multiple mobility modes. In this line of thought, walking is generally the most sustainable modality, travelling by car is the less sustainable modality, and walking and using the car is more sustainable than just using the car. So, the estimated measure will use different weights for different mobility modalities defined in a heuristic way.

The development of this tool is based on a case study of Secondary School students' home-school commute. The choice of this empirical field is due mainly for three reasons: firstly, home-school mobility has a relevant impact on CO_2 emissions and represents a significant share of the city's total mobility; secondly, by analysing how students move in car-oriented cities, we can perceive parents' contribution to more sustainable and equitable mobility practices; thirdly, although not in a consensual way, young people's mobility is considered a predictor for adults future mobility, [1, 2]. The information provided by this tool, in a given moment, is interpreted according to practice theories.

The theoretical framework is centred on practice theories' contribution to sustainable consumption analyses. Theories of practice offer an opposed analysis to the model of the sovereign consumer by focusing on sharing daily rituals that develop and transform in time and place and on practical consciousness [15]. As so, instead of centring on the individual consumer, the sociology of consumption emphasises the complexity of social practice [10], focusing on practice organisation and the 'moments of consumption enjoined' ([16], p. 146). Notably, we intend to address practices theories' contributions regarding the social practices concept and the emphasis on social practice and daily routine over consumer choices.

Although there is no single way to define social practice, finding a minimum denominator definition is possible. According to Southerton and Yates [11], practices are 'entities', as rituals imbued with meaning that is unconsciously assimilated and developed in specific contexts. Practices have a narrative and a pathway embodied through the performances [4, 8, 16]. Warde [16] defines social practice as a shared set of activities that includes a complex arrangement between 'material, embodied, ideational and affective elements. For this author, practices comprise understandings, procedures, and engagements that shape ways of 'doings and sayings' comprising 'practical activity and its representations' ([16], p. 134). Other authors emphasise meaning, competence, and material as practice elements [8]. The material and functional properties of goods and services are also considered relevant in consumption analysis [8]. Other authors like [3, 4] consider 'practical understanding', 'rules', 'teleoaffective structures' and 'general understandings' as practice elements. 'General understandings' influence how practices are done, are revealed by the performance and arrange the teleoaffective structures of practices. Welch and Warde [17] consider 'general understandings' as a broader element common to several practices as the sustainability concept and emphasise its use in practice analyses.

Anchor on practice theories, we analyse Secondary School students' home-school commute as 'entities', considering that moving from home to school and back is more 'a moment' in daily practice than a consumption behaviour, as [16] points out. Taking children to school is a daily routine that involves a complex arrangement between material (e.g. car, bicycle, urban transport, fuel, technologies), understandings (e.g. skills, practical knowledge and ideas to move and how to move in the city, to drive and how to drive, meanings of parenting/childhood, children's mobilities skills), procedures (e.g. road code, children rights, routines organisation and articulation) and engagements (e.g. the affective value of children, pro-environmental commitments). The analyses must take into account parent–child mobility practices, the compatibility between parents' labour schedules and school schedules, transport availability and social equity, the mobility modalities that can be used according to the family budget, and environmental concerns, namely how the sustainability concept influences mobility practices, among other 'general understandings.

Another relevant practice theory contribution to sustainable mobility analysis is the agency concept central to understanding the relations between consumption practices and citizen-consumer roles. Although the structure and agency dualism is not yet solved in the scope of practices theories, we consider [3, 4] perspective. The author argues that 'social order' and 'individuality' are practices outcome and that agency is a consequence of practices and dispositions [3]. This perspective has been reinforced by other authors [8] that prioritise social practice and daily routine over consumer choices. It is also a strong argument pointed out for understanding why relying on consumers' choices is one of the leading causes of the failure of political intervention regarding behaviour change.

Developing an empirical index, a subject of this research, constitutes a valuable tool that, when used to compare sustainability scores, helps the sustainable mobility dynamics analysis, reinforcing [3] idea that agency is a consequence of practices and dispositions.

3 Methodological Approach

The methodological approach is based on a case study in a Matosinhos secondary school focused on students' daily home-school commuting mobility practice. A survey was applied in the school year 2018/19 to 10th, 11th and 12th-grade students from 29 May to 17 July 2019. The school's population in grade 10 is 334; in grade 11, it is 323; in grade 12, it is 213, for a total of 870 students enrolled (N = 870). After class-by-class dissemination of the research project and the survey to parents and students, 400 students obtain parental consent to participate in the project (45.9%). However, only 300 students (34.5%) responded to the questionnaire (n = 300), which corresponds to a non-probabilistic sample, of which 280 responses are validated (32.2%).

The data relevant to the study was obtained from a more extensive survey. The proposed sustainability index was developed from students' responses to questions focused on collecting data regarding the home-school-home journeys (transport modalities used) and the distance from home to school. Data analysis was conducted using IBM SPSS 26. Different mobility patterns were determined, leading to the development of empirical-based heuristic indicators: MOB_SUST_1, MOB_SUST_2 and MOB_SUST. These indicators are an alternative to sheer CO_2 emission measuring. Instead, they are embedded in general understandings of sustainability, sustaining different weighting for different mobility modes.

The development of these indicators is presented in Sect. 4.

4 Developing a Heuristic Approach for Estimating student's Sustainable Mobility

This section proposes two classification variables, or indicators, for each student's pair of journeys—home-to-school and school-to-home noting that the students can combine several mobility modes. The first one (Q1R) is a binary variable taking zero value if the students' travel is not sustainable or taking a value one if the trip is sustainable. The second is an individual sustainable score ranging from 0 to 10, being zero the least sustainable score and ten the maximum score in terms of sustainability. The student's scores are obtained from a table where the scores are heuristic based on city environment knowledge and general sustainability understanding. The development of the proposed indicators is based on an empirical approach centered on a secondary school mobility case study.

In this analysis, two variables that estimate sustainable mobility are proposed. The binary variable Q1R refers to the home-school-home journey taking the value of zero when the commuting is not sustainable or the value of one when commuting is sustainable. The binary variable Q1R_1 refers to the home-school journey, and the binary variable Q1R_2 refers to the school-home journey.

Table 1 and the expressions for Q1R_1, Q1R_2 and Q1R establish the mobility score according to student's mobility pattern.

In Table 1 m_i's and n_i's are binary variables assuming value one if the i-th mode is used by the student or zero otherwise. The expressions for home-school and school-home journeys sustainability scores are given by

$$Q1R\_1 = s_1 m_1 \times s_2 m_2 \times s_3 m_3 \times s_4 m_4 \times s_5 m_5 \times s_6 m_6,$$

and

$$Q1R\_2 = s_1 n_1 \times s_2 n_2 \times s_3 n_3 \times s_4 n_4 \times s_5 n_5 \times s_6 n_6.$$

The variable Q1R takes the value zero (not sustainable) if any routes are not undertaken sustainably. Otherwise, it takes one since all the ways are carried out sustainably. Thus, Q1R is equal to the product of Q1R_1 and Q1R_2.

However, the binary approach provided by Q1R, Q1R_1 and Q1R_2 variables is insufficient to estimate the degree of sustainability. To address this issue, we propose an enhanced estimating variable denoted "sustainable mobility estimates" (MOB_SUST, MOB_SUST_1 and MOB_SUST_2). These new variables aggregate more information than Q1R variables, assuming values from one (minimum sustainability) to ten (maximum sustainability). The variable MOB_SUST (home-school-home) is a sustainable mobility indicator of higher resolution, resulting from the sum of MOB_SUST_1 (home-school) and MOB_SUST_2 (school-home) presenting an ascending score on sustainable mobility.

Considering that, on average, 5 km is travelled by walking in approximately 60 min and attending to the geographical features of the surrounding region, we have established the distance of 1.5 km, corresponding to 18 min, as a reasonable limit for a student to walk from home to school. Students are then heuristically classified into two categories. The first category includes those students who, given

Table 1 Classification table for Q1R_1 and Q1R_2 variables

Mode	Walk	Bicycle	Skate/Trot	Moto/Scooter	Public Transp.	Car/Taxi
Sustainability score (s_i)	1	1	1	0	1	0
Student's mobility home-school (m_i)	m_1	m_2	m_3	m_4	m_5	m_6
Student's mobility school-home (m_i)	n_1	n_2	n_3	n_4	n_5	n_6

the distance from home school, are expected to make the journey by walking. The second category comprises students who are not expected to make the journey home to school by walking, given the home-school length.

Each student's mobility pattern is identified and represented by a number. Considering that each student can use up to six mobility modalities in their home-school-home journey, there are 128 theoretical possibilities (possible patterns), including the distance factor.

For each mobility pattern, a heuristic score is proposed based on the mobility pattern of each student. Table 2 is based on heuristic rules corresponding to the six main mobility modes and their combinations, considering the observed mobility modes, distance, and other factors. Specifically:

Walking, cycling, or skateboarding/trotting corresponds to 4 points if the distance is less than 1.5 km.

Walking, cycling, skateboarding/trotting corresponds to 5 points for distances greater than 1.5 km.

Riding by car/taxi or motorbike at less than 1.5 km corresponds to 0 sustainability points.

Riding by car/taxi or motorbike, at distances greater than 1.5 km, corresponds to 1 point.

Riding public transportation (PT) at distances less than 1.5 km corresponds to 3 points.

Riding PT for a distance greater than 1.5 km corresponds to 4 points.

Table 2 Typical students' mobility patterns and the corresponding heuristic sustainability scores

Distance greater than 1.5 km		Distance less or equal to 1.5 km	
Mobility pattern	Score	Mobility pattern	Score
Just walking	5	Just walking	4
Just public transportation	4	Just car/taxi	0
Just car	1	Walking and bicycle	4
Just moto/scooter	1	Walking and skate/trot	4
Walking and public transportation	4.5	Walking and public transportation	3.5
Walking, bicycle, and public transportation	4.5	Walking and car	2
Walking, public transp. and car	3.3	Walking, bicycle, and car	2.5
Walking and car	3		
Moto/scooter and public transportation	2.5		
Moto/scooter, public transp. and car	2.5		
Car and public transportation	2.5		
Car and moto scooter	1		

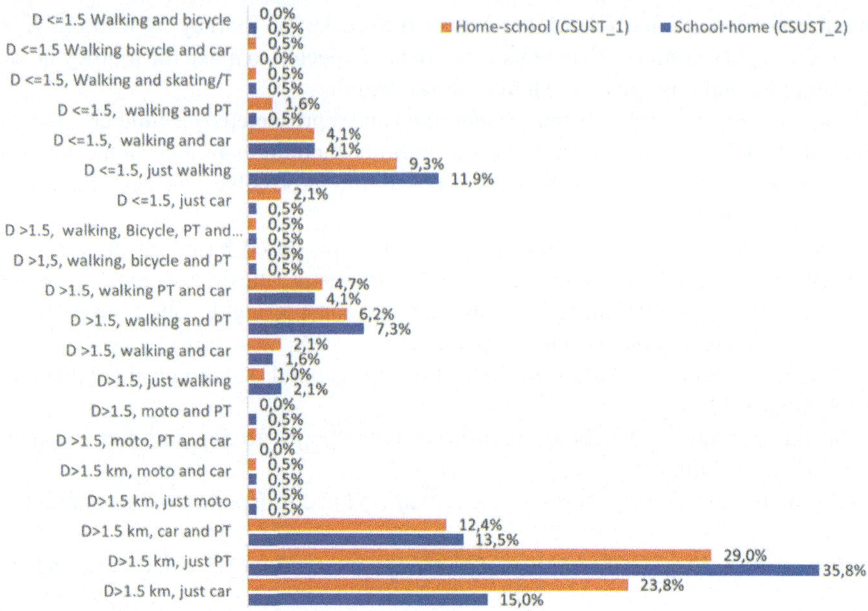

Fig. 1 Student's empirical mobility pattern distribution

These rules can and should be adapted depending on the city's particular features.

The home-school and school-home mobility patterns are extracted from the data. The sustainable mobility patterns are represented by numbers and coded in the new variables home-to-school (CSUST_1) and school-to-home (CSUST_2). The empirical distribution of mobility patterns captured in the case study, are presented in Fig. 1

.

The mobility pattern identification is crucial for constructing enhanced mobility sustainability indicators. The results provided by these indicators will be presented and discussed in Sect. 5.

5 Results and Discussion

Results from applying the proposed estimated tools to the school's survey are presented and discussed in Sects. 5.1 and 5.2.

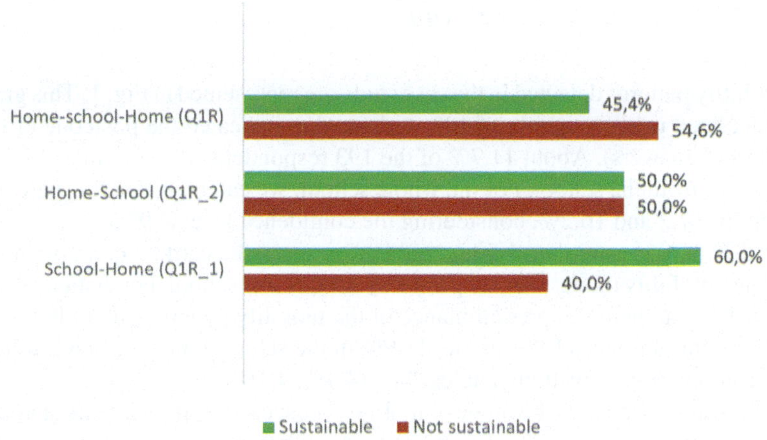

Fig. 2 Students' mobility Q1R_1, Q1R_2 and Q1R indicators

5.1 *Q1R Indicator Results*

The results provided by the classifier variables Q1R, Q1R_1 and Q1R regarding the case study are depicted in Fig. 2.

Regarding home-school journey is estimated by variable Q1R_1 that 50% of the students commute in a sustainable way. In the school-home trip the corresponding percentage, given by Q1R_2, rises to 60%.

It is estimated that the school-home journey is more sustainable than the home-school journey. It is also estimated that 45.4% of the students sustainably commute from home-school-home.

Differences in sustainable mobility in the two journeys result from several constraints that enfold daily practices and their organisation that must be analysed according to practice theories' contributions. Namely, the compatibility between labour and school schedules can promote an 'economies of scale'[1] regarding car use on the journey from home to school and increase public transport use for students on the journey back home. This possibility also highlights young people's autonomy in commuting from home to school. More than a question of safety and parenthood, parents seem concerned if their children get on time to school and attend class. The 'general understandings' of workers' duties (e.g., showing up on time, meeting work expectations) seems to unfold these practices. On the journey back home, incompatibility between schedules is associated with a decrease in these concerns.

[1] "Economies of scale" is a term connotated with cost diminution due to a high capacity use.

5.2 Mob_Sust Indicator Results

The mobility patterns detected in the case study are represented in Fig. 1. This graphic was calculated using a sample of 193 students that indicated the postcode (193 out of 296 valid answers). About 11.7% of the 193 respondents live within a radius less than 1.5 km from the school. For the whole school, we can infer that this percentage lies between 7.2 and 16.2%, considering the confidence level at 95%.

For distances greater than 1.5 km, public transport, the car/taxi or a combination of both are mobility modes that are primarily used in this school. For distances less or equal to 1.5 km, there is a predominance of the mobility pattern walk only (9.3% in the home-school route, 11.9% in the, 11.9% in the school-home journey), followed by the combination of walking and car/taxi (4.1%, 4.1%).

Public transportation is the mode of mobility most used by the students, and its use is more significant in the school-home trajectory than in the inverse, home-school. It is followed by the car/taxi.

The use of the car/taxi is more significant in the home-school journey than in the converse journey. The combined use of cars/taxis and public transport ranks third, with greater use on the school-home journey than on the home-school trip. These results reinforce those previously presented and corroborate the idea that the students' mobility is more sustainable in the school-home journey than in the school-home journey.

As stressed, this significant vector of analysis emphasises that organisation and the flow of daily practices analyses are conditioned by factors such as the 'economies of scale', the compatibility between the parents' working and the children's school schedules, and the daily uses of the car and its symbolic meaning, among others. It also reinforces social equity and justice issues regarding sustainable mobility.

It should also be noted that for distances greater than 1.5 km, the combination of the mobility modalities walking and public transport is the fourth most used, with less significant values than the three previous modalities, and different in the two journeys. This combination is more used in the school-home journey. In the fifth position comes the combination of walking, public transport and car/taxi, mainly used in the home-school journey.

Sustainability mobility scores estimates MOB_SUST_1 and MOB_SUST_2 are represented in Fig. 3.

The home-school sustainability indicator (MOB_SUST_1) has an average (index) of 2.68 and a standard deviation of 1.38. The school-home sustainability (MOB_SUST_2) indicator has an average (index) of 2.68 and a standard deviation of 1.32.

The index of 2.93 for school-home is greater than the index for home-school (2.68), indicating that the school-home journey is more sustainable than the home-school journey. The standard deviations do not differ significantly, indicating similar dispersion.

Once again, the heuristic tool emphasises "economies of scale" and the compatibility of schedules (work/school) effect on sustainable mobility patterns.

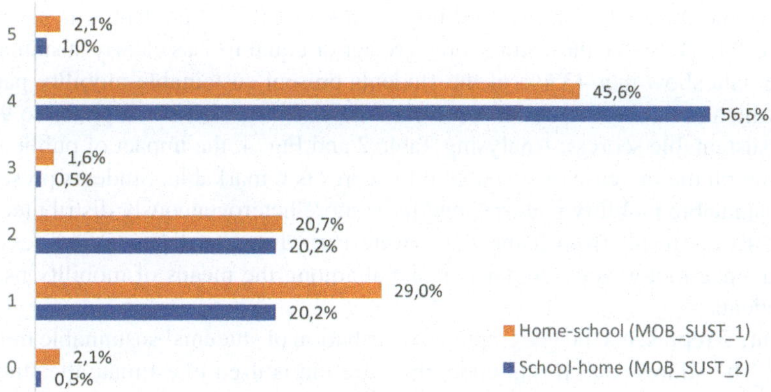

Fig. 3 Home-school and school-home estimated sustainability scores

Figure 4 shows the mobility scores regarding sustainability in the home-school-home journey. The scores are based on heuristics on an increasing numerical scale of sustainability of the combination of mobility modalities practised by students. This heuristic is also a function of the distance the student resides from the school, according to Table 2.

The school's sustainable mobility index is given by the mean value of all mob_sust scores corresponding to 5.61 points in a maximum of 10. The standard deviation corresponds to 2.4 and the variation coefficient to 0.43.

The rescaled index corresponds to 56.1%.

About 45% of the student's journey home-school-home is estimated as sustainable by the Q1R indicator. So, the percentile 45 of MOB_SUST values can be used as

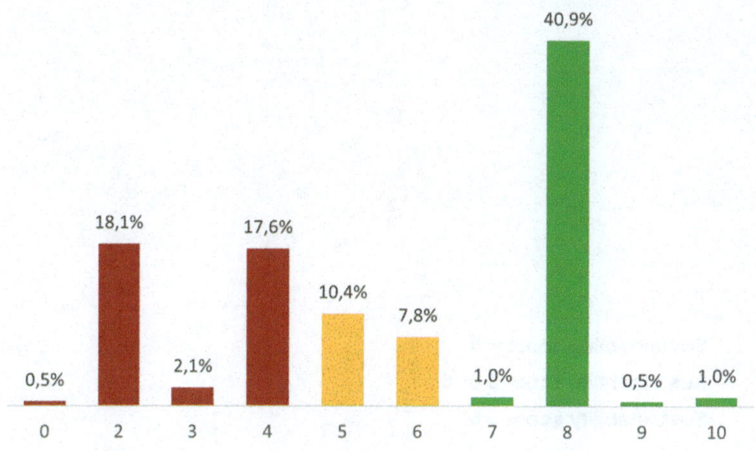

Fig. 4 Sustainable mobility score estimate for home-school-home journey

a reasonable threshold for sustainability. Given that the 45 percentile is seven, we interpret MOB_SUST indicator scores greater or equal to 7 as clearly sustainable.

The data show that 43.4% of the students present sustainable mobility patterns (>6), which are concentrated in score class 8 (40.9%, which corresponds to 94.2% of the sustainable scores). Analysing Table 2 and Fig. 4, the impact of public transportation on the increase of sustainability scores is remarkable. Students presenting non-sustainable mobility patterns tend to be more heterogeneously distributed. The lowest scores result from using the private car when the distances are less than 1.5 km since motorcycle/scooter is residual among the means of mobility used by the students.

Figure 5 represents the geographical distribution of students' sustainable mobility scores. The address and postal code gps' location is used to estimate the students' locations. A score greater than six is interpreted as "clearly sustainable"; a score lesser than five is interpreted as "clearly not sustainable". Scores between 5 and 6 are interpreted as possibly not sustainable. This interpretation is used to improve Fig. 5 readability.

The geographical distribution of sustainability scores represents the urban mosaic of sustainability areas. The mob_sust estimation tool combined with cartographic representation has the potential to identify areas to be addressed by local governance, such as lack of public transportation infrastructure, traffic security issues, active mobility campaigns, among many others.

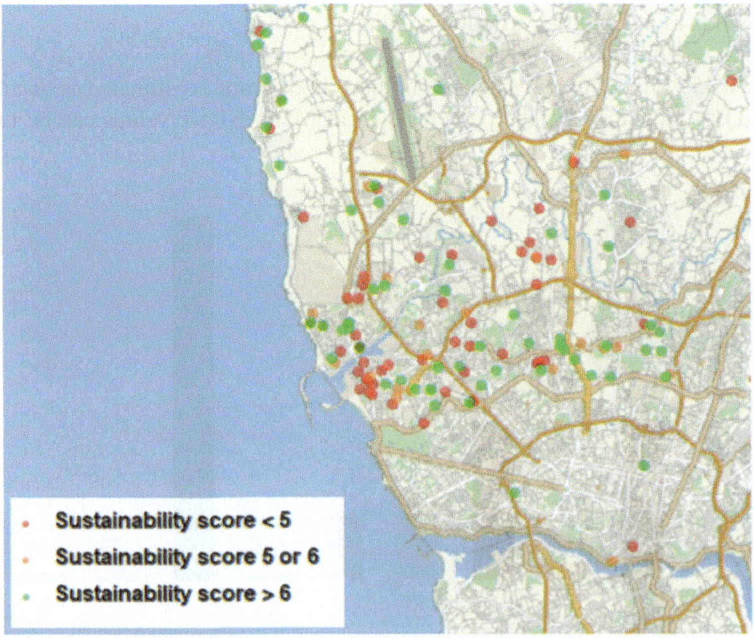

Fig. 5 Geographical distribution of students' sustainable mobility scores (MOB_SUST)

Mob_sust estimation can be combined with other estimates such as social class, level of education and geographical distribution to identify useful patterns. In this case study, the geographical distribution captured an interesting concentration of low sustainability scores in the most prosperous Matosinhos urban area, reinforcing equity and justice issues.

6 Conclusions

This pioneering research developed a heuristic-based tool for estimating students' sustainable mobility based on a secondary school case study yielding an innovative contribution to the sustainable mobility field analysis. Mobility patterns were captured, and a system of indicators was devised.

The conclusion presents the proposed answers to two main questions that motivated this research: How to measure sustainability? How to follow its dynamics in time, space, and between generations?

How to measure sustainability is addressed not by an exact measure somewhat impossible to devise but instead by a flexible approach consubstantiated in the development of an enhanced heuristic estimator of sustainability. A method of estimator construction based on student's mobility pattern identification and heuristic classification was also devised. This generalized approach can readily be applied to other subjects and samples.

How to follow sustainability dynamics in time, space, and between generations is addressed by a successive application of the estimation tool at different times and spaces and samples. A time series of index values can be generated to help unveil the organization and flow of these sustainable daily mobility practices. This important feature allows the application of the estimation tool to the analysis of sustainable transitions between collective and generations.

The construction of the classifier variable MOB_SUST was based in a case study that involved the definition of a heuristic method to determine and classify the mobility patterns for each student and the corresponding global sustainable mobility index for the school. The overall rescaled estimated global sustainable index of the school is 56.1% and is highly influenced by the use of public transport.

This index allows comparisons with other schools. The use of public transport is the factor that most contributes to increasing the value of the variable MOB_SUST (score) and, in opposition, the use of a private car, especially for short distances, is the most penalising factor in the score given by the variable MOB_SUST. Students are more sustainable in the school-home journey, where they are more autonomous and use more public transport, than in the reverse journey. Public transport is also more used for distances greater than 1.5 km. The economy of scale (home-school) and the incompatibility between school and work schedules (school-home), parenthood, symbolic value of the car, among others possible explanations for the greater use of the car for the home-school journey and public transport for the school-home journey.

In sum, the analysis of the dynamics of the transitions within and towards sustainable mobility will be facilitated by calculating the indicators at various time instants and comparing the results. Data will then be generated to analyse practices' organization and flow over several generations. On the other hand, it will also be possible to cross the data provided by the indicators with other variables of interest, such as social class and education level, to detect regularities fostering the formulation of research hypotheses. Finally, monitoring and assessing public policies regarding mobility sustainability can be aided by these indicators and indexes calculated in timely moments.

References

1. Brown AE, Blumenberg E, Taylor BD (2016) A taste for transit? Analyzing public transit use trends among youth. J Public Transp 19(1)
2. Fujii S, Gärling T (2005) Temporary structural change: a strategy to break car-use habit and promote public transport. Paper presented at the international conference of traffic and transport psychology
3. Schatzki T (2002) The site of the social. Pennsylvania State University Press, University Park, PA
4. Schatzki T (1996) Social practices: a wittgensteinian approach to human activity and the social. Cambridge University Press, New York
5. Shove E (2010) Beyond ABC: climate change police and theories of social change. Environ Plan A 42:1273–1285
6. Shove E (2004) Efficiency and consumption: technology and practices. Energy Environ 15(6):1053–1065
7. Shove E (2003) Comfort, cleanliness and convenience: the social organization of normality. Berg, Oxford
8. Shove E, Pantzar M, Watson M (2012) The dynamics of social practice: everyday life and how it changes. Sage, Thousand Oaks, CA
9. Shove E, Walker G (2010) Governing transitions in the sustainability of everyday life. Res Policy 39:471–476
10. Shove E, Warde A (2002) Inconspicuous consumption In: Dunlap RE, Buttel FH, Dickens P, Gijswijt A (eds), Sociological theory and the environment. Rowman & Littlefield, London, pp 230–241
11. Southerton D, Yates L (2014) Exploring food waste through the lens of social practice theories: some reflections on eating as a compound practice. In: Ekstrom K (ed) Waste management and sustainable consumption: reflections on consumer waste. Routledge, London, pp 133–149
12. Spaargaren G, Oosterveer P (2010) Citizen-consumers as agents of change in globalizing modernity: the case of sustainable consumption. Sustainability 2:1887–1908
13. Spaargaren, G. (2006), 'The ecological modernization of social practices at the consumption junction', Discussion paper for the ISA-RC-24 conference Sustainable Consumption and Society, June, Madison, WI
14. Spaargaren G (2003) Sustainable consumption: a theoretical and environmental policyperspective. Soc Nat Res Int J 16(8):687–701
15. Warde A (2014) After taste: culture, consumption and theories of practice. J Consum Cult 14(3):279–303

16. Warde A (2005) Consumption and theories of practice. J Consum Cult 5(2):131–153
17. Welch D, Warde A (2017) How should we understand 'general understandings'? In: Hui A, Schatzki TR, Shove E (eds) The nexus of practice: connections, constellations and practitioners. Routledge, London

The Challenges to Sustainable Urban Future in a Rapidly Changing World: A Case Study of Izmir, Turkey

Ebru Ersoy Tonyaloğlu and Engin Nurlu

Abstract The notion of sustainability is founded on the environmental, economic, and social issues under the conditions created by human–environment interactions in the twentieth century. From then on, it has been a focus of contemporary research as a common goal in urban ecology, urban planning and design, and environmental management because day by day most of the world is becoming more and more urban. The urbanization process leads unprecedented demand for natural resources with restricted financial and human resources. So, securing a sustainable urban future has become an important topic on a global scale to cope with urbanization pressures on nature, biodiversity, climate, society, economy, transportation, ecosystem services etc. Today, whilst the most cities are striving toward becoming more sustainable, there is often limited success in implementing the core principles of sustainability. For urban environments to have a sustainable future, they require integrated planning approaches which include all aspects of urban environments and set the linkages between social, environmental, political, technical, and economic aspects. Landscapes in Turkey have a long history of being utilized by different civilizations. This has led to fundamental human induced changes and environmental issues throughout the country. However, facilitating a sustainable urban future in Turkey is constrained by insufficient knowledge of the state of socio-ecological systems as in many parts of the world. Therefore, this chapter reviews the key environmental, economic, social, political, and aesthetic challenges in urban sustainability in Izmir, Turkey using the sustainable landscapes framework proposed by Selman (Sustain: Sci, Pract Policy 4(2):23–28, 2008). We share our perspectives on the need to focus on different dimensions of sustainability in Turkey's landscapes and some key points in landscape planning and management strategies.

Keywords Izmir/Turkey · Socio-ecological systems · Land use change · Landscape planning

E. Ersoy Tonyaloğlu (✉)
Department of Landscape Architecture, Aydın Adnan Menderes University, Aydın, Turkey
e-mail: ebru.ersoy@adu.edu.tr

E. Nurlu
Department of Landscape Architecture, Ege University, İzmir, Turkey
e-mail: engin.nurlu@ege.edu.tr

Z. Allam (ed.), *Sustainable Urban Transitions*, Urban Sustainability,
https://doi.org/10.1007/978-981-99-2695-4_5

1 Urban Sustainability and the Landscape

The phenomenon of city refers to "a permanent and dense settlement under continuous development which meets the needs of the society, such as settlement, accommodation, commuting, working, resting and entertainment, and where people mostly engaged with non-agricultural activities" [6, 26]. In terms of human history, cities have been crucial to the development of societies and have contributed to the formation of today's civilizations. Cities have played a key role in the emergence of social development and political revolutions and have always been the central living space of humanity and various civilizations. Even though cities have such a prominent place in the history of humanity, there is another aspect of them in terms of nature and environment. One of the main factors in the deterioration and (or) depletion of the environment is the production and consumption opportunities offered by the cities and the fact that the capacity of nature to meet these demands and to eliminate the produced wastes has been exceeded [4, 27, 37]. Today, damage to the natural environment (sometimes irreversible) has become a threat to humanity to meet the needs of the increasing population in urban areas for many countries around the world. This is the fact that the opportunities and services offered by cities to humanity come at a price that is paid by nature. Cities are the centers of economic and social developments and the places where most of the world's population live and therefore the main environmental problems originate. Whilst the increase in the opportunities of cities with the industrial revolution has accelerated the population they attract as well as the urbanization process, towards the second half of the twentieth century, the spread, concentration, and uncontrolled energy consumption of cities have led to an increase in environmental problems. Afterwards, Agenda 21 decisions were taken regarding the necessity of planning and designing cities in a more sustainable way because of United Nations (UN) Conference on the Environment (Stockholm-1972) and the UN Environment and Development Conference (Rio Conference-1992) held in 1972 and 1992, respectively. Next, the 2030 Agenda for Sustainable Development which is a plan of action for people, planet and prosperity was taken regarding the global challenge and an indispensable requirement for sustainable development. The (17) Sustainable Development Goals (SDGs) and 169 targets which were came into effect on 1 January 2016, are integrated and indivisible and balance the three dimensions of sustainable development: the economic, social, and environmental. Within those targets, SDG 11 aims to make cities and human settlements inclusive, safe, resilient, and sustainable [40]. Since then, the phenomenon of urban sustainability has become one of the most challenging and urgent issues facing humanity. In broad terms, urban sustainability defined as ensuring the effective use of natural resources, economic growth and welfare in a way that minimizes waste generation without stealing the rights of future generations [35]. Looking at the relationship between the environment and cities, it is obvious that sustainable cities can only be established through well-planning and designing based on the environmental quality in natural and artificial spaces [41].

Selman mentions that sustainable landscapes can be defined by 5 basic principles. These consist of environmental, economic, social, political/managerial, and aesthetic sustainability. Environmental sustainability in landscape is explained with the concepts of connectivity of habitats, healthy functioning of other environmental system components, protection of biodiversity, ecosystem services such as microclimate and carbon storage, and ecosystem health. Of course, while referring to all these concepts, Selman also emphasizes that environmental sustainability can only be possible with the provision of multifunctionality, landscape resilience and healthy functioning ecosystem service mechanisms that can guarantee the welfare of people [36]. When evaluated in terms of urban sustainability, it is obvious that the development and implementation of urban blue-green infrastructure planning approach is significant in ensuring environmental sustainability. Economic sustainability, often associated with visual quality and attractiveness and tourism, is related to the intrinsic properties of landscapes and the multiple benefits and functions that are spontaneously provided. With these features, economic sustainability is linked to market and non-market mechanisms and does not exhibit the same quality in all landscapes. For example, while some landscapes are the centers of visual attraction, they also stand out with agricultural production and rich biodiversity, the absence of these qualities, especially in some destructed landscapes, would also pose an obstacle to economic sustainability. Social sustainability is discussed in terms of participation and inclusion in accessing numerous services, with a decision-making system that considers the needs, preferences, and landscape perceptions of people from various cultural and social groups as well the other living organisms. Of course, at this point, the necessity of urban facilities and services to protect the physical/mental health and well-being of all people emerges. Political/administrative sustainability is possible through the establishment of an effective governance structure in public and private spaces. The preparation and implementation of spatial plans through the construction of this governance structure in cities would support an inclusive and participatory decision-making mechanism and the provision of multifunctionality in urban areas. Finally, aesthetic sustainability is not only related to the existence of visually unique areas, but also to the healthy functioning of all environmental components in the entire landscape, cultural landscapes that are ordinary damaged but give spiritual and visual pleasure to people, as well as a sense of belonging. To put it briefly, all these sustainability principles are the basic characteristics that work as an inseparable whole in ensuring sustainability in urban landscapes. The interruption/lack of one of these principles will appear as an important problem in the construction of a sustainable urban future.

2 Natural Resources in Izmir

İzmir is a historical port city located in the Aegean coastal region in the west of Turkey, between 37°45′ and 39°15′ northern latitudes, 26°15′ and 28°20′ eastern latitudes with an approximate area of 12,000 km^2 (Fig. 1). One third of the city consists of forest areas, followed by agricultural areas and grassland-meadow. Around 6% of the city consists of urban areas, especially high-density built-up areas [23]. İzmir is the third largest cities in Turkey with a vibrant topographic structure and rich biodiversity. The most important reasons for this are the presence of mountains stretching perpendicular to the sea, the basins and corridors formed by the rivers between these mountains (Bakırçay Basin, Gediz Basin, Küçük Menderes Basin, Peninsula and Körfez Basins), the high mountains surrounding the city (Bozdağ, Aydın Mountains, Nif Mountains, Madra Mountain, Kızıldağ, Spil Mountain, Yamanlar Mountain, Dumanlıdağ), ecosystem diversity (forest, maquis, phrgiana, steppes, alpine belt, wetlands, deltas) and long coastline (629 km) formed by different vegetation and wetlands formed by plains. The city of Izmir also allows many plant species (e.g., *Pinus brutia, Pinus nigra, Quercus coccifera, Olea europea, Cistus salvifolius, Sarcopoterium spinosum*, and *Phragmites australis*) to grow in its rich ecosystems that host many of the endangered, critical, vulnerable animal species (e.g. *Monachus monachus* (EN), *Caretta caretta* (EN), *Knipowitschia ephesi* (CR), *Streptopelia turtur* (VU), and *Clanga clanga* (VU)) according to the IUCN. Bakırçay, Gediz and Küçük Menderes rivers reaching the sea from the boundaries of İzmir contribute to the biodiversity of the city with the delta and river ecosystems they form [19].

In urban ecosystems, which are located at the intersection of diverse natural and semi-natural ecosystems in the immediate vicinity, the diversity of native and foreign

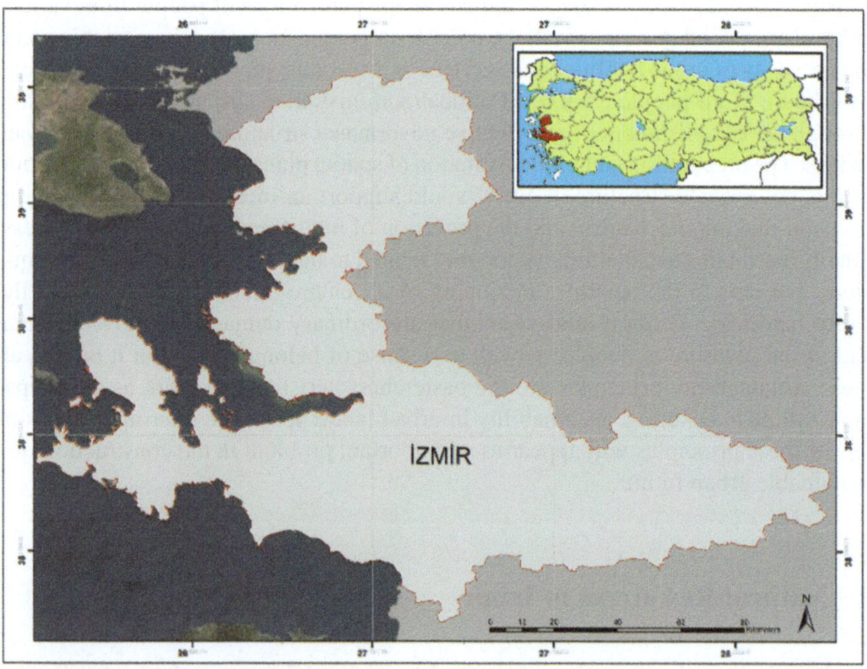

Fig. 1 Location of the study area

species is high. There are four basic layers that make up the blue and green infrastructure in the city of Izmir [20]. The first of these consists of the sea-land intersection formed by the coastal areas, beaches, lagoon, delta, and wetlands along with the marine ecosystems in the Izmir Bay. Today, however, these areas are under pressure due to intense construction and various problems caused by the increasing population. The second layer of the blue-green infrastructure in the city of İzmir includes the coastline, large-scale urban green areas (large urban parks), stream corridors and the metropolitan center, around Izmir Bay and its immediate surroundings. The most important problems encountered in this region are the insufficiency of urban green areas, their fragmented structure, and the lack of urban green areas with different typology and characteristics [16]. This situation causes an inequality in the society in terms of urban green space access and use. Another layer of the blue-green infrastructure consists of large regional parks, afforestation areas and agricultural areas, which are located on the outer periphery of the dense construction and where the city is fringed from place to place. These areas, on the other hand, are under the threat of structuring and fragmentation with the effect of intense urbanization and the recent pandemic. Finally, the fourth layer of the blue-green infrastructure includes areas where large-scale agricultural activities are conducted, forests, freshwater ecosystems, mountain steppes, and areas formed by maquis and phyrigana communities. The most critical issues experienced in these areas consist of problems such as excessive use of water, fertilizers and pesticides in agriculture, forest fires, pollution, hunting, fragmentation, and degradation of important habitats.

3 Cultural Resources in Izmir

In the city of İzmir, which has a long history of 8000–9000 years and has been exposed to human-nature interaction throughout this period. The city of Izmir got its name as a residential area from the word of ancient Smyrna [1]. It is thought that the first known settlement area in the city of Izmir was established in Yeşilova Mound (Yeşilova Höyüğü) on fertile alluvial soils within the borders of today's Bornova district [7]. This was followed by the Tepekule settlement, which was established in the present-day Bayraklı district five thousand years ago. The traces of these dates can still be observed in the city of Izmir, which later served as a settlement area for many civilizations. Agora (Konak), Pergamon (Bergama), Ephesus (Selçuk), Smyrna (Bayraklı), Metropolis (Torbalı) are the leading ones among these. Traces of religious places, aqueducts and bridges, inns, baths, and fountains belonging to different civilizations can be seen in the city. Pergamon Multi-layered Cultural Landscape (2014) and Ephesus (2015) located in the city of Izmir are now on the UNESCO World Heritage List. In addition, Historic Town of Birgi (2012), and Historical Port City of İzmir (2020) are also on the UNESCO World Heritage Tentative List [42, 43].

Today the city of Izmir is the third most populous cities in Turkey with almost 4.5 million inhabitants in 2021 and has always been an attraction center at Aegean region with its location and weather [39]. With its population, the city of İzmir constitutes

5.23% of the total population of Turkey. In the city of Izmir, which started to develop as a port city in the eighteenth century and had a low population density with a rural character before the 1960s, dense and high-rise construction have started to spread from the coastal to the interior parts of the city in the twentieth century [5, 15]. Today, composed of thirty districts the metropolitan city of Izmir is still growing and becoming heavily built and densely populated [16].

4 Current Challenges to Urban Sustainability in Izmir

The central urban area of the city of Izmir, located around the bay, constitute the densest construction and populated region (Balçova, Bayraklı, Bornova, Buca, Çiğli, Gaziemir, Karabağlar, Karşıyaka, Konak and Narlıdere). According to 2018 CORINE land cover data, this region with one third of artificial surfaces, is facing the pressure and problems created by the increasing construction and transportation networks, as it tries to meet the diversifying needs of sectors such as settlement, recreation, trade, industry, tourism, and mining [12, 28, 29, 33]. The rapid urban expansion and population growth in the city of Izmir have reduced the existing natural vegetation and caused ecological fragmentation and deterioration of natural ecosystems. Despite the high population density, in 2019, the amount of green space available per capita in İzmir is 8.6 m^2. The fragmented and scattered structure of urban green areas create important constraint in terms of both environmental and social sustainability [16]. This problem brings with leads other environmental issues such as the lack of high quality urban green areas and connections between them, uniform green area typology, competition in land use prioritization, and regulation of land use density. Of course, increasing population has led to important social issues like the lack of access to services, areas of economic deprivation as well as effecting public health through the pollutant contamination to the drinking water, air, and food [46].

Public spaces (e.g., urban parks, squares, roads, and streets) are crucial for urban environments as they integrate people and places and constitute the backbone of public life in cities. In all these, roads and streets are the basic components for the public to perform all kinds of activities, movements, and social interactions. However, a recent study on the space quality of streets found that the city of Izmir could not meet the required specifications for the public in terms of the continuity of pedestrian, access, and function; visual and functional diversity; attractiveness and appearance; comfortable experience, and human scale [8]. It is well-known that the greater the connectivity of public spaces stronger the social interaction among people [2, 34]. Urban green areas improve the physical, social, and mental health of individuals and support the biodiversity in cities [9, 14]. However, if we want to achieve all these benefits, they should provide need-based and spatial equity to vulnerable and minority groups in the society [30]. According to the results from previous assessment of the spatial equity of neighborhood parks in Izmir indicated that the high-income groups have the highest park size capacity with highest accessibility whereas lower income and vulnerable groups have the opposite [30]. At this point, it is crucial for the city of İzmir to ensure social sustainability with appropriate landscape planning and design strategies in the areas where these low-income groups, who are currently disconnected from nature but most in need of the benefits of the natural environment.

Meanwhile, there are large natural and semi-natural areas in the immediate vicinity of the central city area of İzmir. Whilst transportation infrastructure leads increasing isolation and fragmentation on those areas, the presence of road networks also encourages the development of new houses and facilities around them [12, 32]. In Turkey, hot temperatures with low humidity in the Aegean Region in summer periods pose a great danger for forest areas Kavzaoglu [24, 25]. Besides, the fast-blowing wind and negligence cause the fires to reach much more dangerous dimensions. In the city of Izmir, where 2775 ha of forests were burned in only 2020, an annual average of 1496 ha of forest area was lost in 285 forest fires in the last 10 years [31]. As another example, we have lost around 5000 ha forest area in Gaziemir, Buca and Karabağlar district of İzmir in an only one forest fire occurred on the 18th of August 2019. Accordingly, forest fires in the city of Izmir are the most important natural disasters causing environmental, social, economic and problems and putting an extra pressure on those valuable natural and semi-natural areas [3]. Together with all these issues, because the connection of these areas with urban green areas is weak or disconnected, it is hard to obtain the multi-functions and benefits of a green infrastructure [16, 22].

Unfortunately, climate change is another issue in the city of Izmir putting pressure on forests, groundwater resources and thermal comfort. For instance, water resources and water quality are under threat due to the lack of use of native species in urban green areas, excessive use of water in agriculture, lack of infrastructure in terms of wastewater treatment, reuse of gray water and rainwater harvesting, and current leaks in the urban water network [45]. Additionally, as a result of the flood that was effective in the northern part of Izmir Bay in 2019 and 2020, the ground floors of residential and commercial buildings, and streets and roads were flooded and severely damaged [10]. This situation caused the region not only to be damaged economically, but also to be injured culturally, socially, economically, and aesthetically. Moreover, with the increasing temperatures and the urban heat island effect, the areas where people can live comfortably in the central urban part of Izmir are getting narrower day by day [38]. A recent study in the city of Izmir found that the average land surface temperature (LST) of Izmir in July (for the period 2014–2017) was between 23.23 °C and 39.73 °C. Whilst the lowest LST values found in the central city where dense impervious surfaces cover the land (e.g., highways, roads, and industrial, commercial and residential buildings), the lowest values were the land is covered by vegetation and water (e.g., green areas, agriculture zones and water bodies) [11]. Altogether, these issues highlight the importance of a healthy functioning governance mechanisms for integrating ecological functioning and ecosystem integrity, their diverse benefits, and functions for all living things into the perception, valuation and interpretation of people and other living things of urban landscapes.

5 Paths to the Sustainable Urban Future in the City of Izmir

A more sustainable urban future that maintains multiple benefits to citizens and other living organisms requires an urgent understanding how ecological, social, economic, cultural, and infrastructural components interact to each other, as well the character and quality of a city. Such an understanding would be possible by engaging communities, researchers, and all the other shareholders to build both the scientific knowledge and social movement necessary to accomplish a sustainable urban transition vision for the future cities. Along with the negative effects of climate change all over the world, the pandemic conditions experienced in the last 3 years have made living conditions more difficult and have caused people and local governments to reconsider their relations with nature. As in all countries and cities, İzmir emerges as a city struggling with all these problems. In addition to these, forest fires in recent years have become notable events those need to be taken into consideration for sustainable urban development and transformation for the city of İzmir. The city of Izmir aims to be a green city in the future, maintaining its city-wide urban green infrastructure, offering close-to-home green areas for all residents, protecting threatened wildlife and climate adaptation capacity with nature-based solutions.

6 Multifunctionality

The long-term sustainability vision of the city of Izmir is to be a climate-friendly and safe city in the ecological and social inclusive sense. In this context, Izmir Metropolitan Municipality (IMM) develops action plans and strategies in line with the Izmir Metropolitan Municipality Strategic Plan (2020–2024) for the sustainability of Izmir (high quality of life for citizens, provide economic vitality and protect the environment) and has carried out and implemented projects by participating in international research networks (e.g., European Bank-EB, European Union-EU, etc.). By 2030, İzmir aimed to become a city that is more resistant to natural disasters, with an elevated level of welfare and biodiversity, and where fair sharing can be achieved by developing life strategies in harmony with nature. In this context, the Izmir Green Action Plan and the Sustainable Energy and Climate Action Plan were established in 2021. "Izmir Sustainable Energy and Climate Action Plan-SECAP" for Izmir's climate change adaptation and greenhouse gas reduction strategy by IMM, and "Izmir Green City Action Plan" for actions (including actions) for Izmir Green Infrastructure Strategies- GCAP has been prepared. SECAP aims to develop a Green City Action Plan (GCAP), which aims to identify and prioritize priority environmental problems, and to provide a more nature-friendly future for İzmir in line with the IMM Strategic Plan Vision [17, 18]. With these complementary plans, IMM has determined the road map of the city of İzmir, which includes climate and other environmental issues until 2030.

Since the existing spaces in urban spaces are limited and different land uses are in competition with each other, efficient use of a space should be taken as a basis in urban planning. Multifunctionality in urban areas requires the provision of multiple benefits, in other words, ecosystem services in distinct parts of an urban setting. Despite the high potential of the city of Izmir, it also contains important limitations in providing multi-functionality. İzmir is a city with an approximate population of 4.5 million people, developed service, industry and coordination sectors, and a demand for intense consumption. Also, as the third largest metropolis, extreme poverty is concentrated in urban areas with dense housing. In the heavy urbanized parts of Izmir city, there is a serious shortage of space for new land uses and associated functions. The requirement for promoting divergent functions in city of Izmir are generally mentioned in every policy document and strategy, but unfortunately their integration and implementation into dense urban contexts are limited by the constraints of local political environments, a lack of knowledge and complexity in different managerial bodies. So, we really think about our priorities and opportunities to realize them. A first step to achieve this can be the creation of a dynamic, up-to-date, and open-access urban information system. In this way, we can balance competing priorities, solve issues, and maximize the contrasting functions of a space with an informed perspective. In addition to this, the existing SECAP and GCAP strategies need to be strengthened to cover the whole city, and practices that can help reduce carbon emissions should be extended throughout the city.

7 Urban Greening and Nature

The main goal of İzmir's Strategy for Living in Harmony with Nature 2021–2030, based on GCAP and SECAP, is to grow the city's economy by preserving and maintaining its nature. In this context, IMM have conducted EU projects to share experience and knowledge on an international scale to adapt climate change and test the usefulness of nature-based solutions on the city scale. For instance, with the 'New Strategy for Re-Naturing Cities through Nature-Based Solutions-URBAN GreenUP' project (EU-HORIZON2020 funded research project 2017–2022), IMM aimed to develop, implement and replicate its city plan, mitigate the adverse effects of climate change, improve air quality and water management strategies, and increase the sustainability of the city of Izmir through effective and innovative nature-based solutions together with a group of European and non-European partner cities [13]. IMM has implemented 19 innovative Nature Based Solutions (NBSs) through non-technical interventions in the selected pilot sites throughout the city of Izmir by (1) Urbanization Re-Naturing (producing solutions that cover large urban areas and reduce the effects of climate change), (2) Single green infrastructures (combating environmental problems in certain urban areas) (3) Water interventions. These interventions include pedestrian routes, bicycle routes, pollination aid areas and applications related to urban watershed forestry [44].

In terms of urban greening and natural resources we need to: implement the Izmir's strategy of living in harmony with nature in an effective and equal way throughout the city; conduct studies to reduce the effects of invasive species (which are used in open and green areas throughout the city) on land and marine ecosystems; integrate of ecosystem and biodiversity values into national and local planning and development processes. In addition, we need to determine and monitor the effects of human activities on different ecosystems in the city (especially forests and wetlands), protect these ecosystems against manmade hazards and ensure their sustainable use. In dense urban structure, it is hard to increase the amount of green and open spaces. For this reason, raising public awareness of the importance of ecosystem and biodiversity values in İzmir in terms of local development, poverty reduction and ensuring equality in access to resources; and integrating these values into national/local planning and development strategies and reports will be a crucial step in terms of the protection and sustainability of the urban green areas.

8 Climate Change Mitigation

With the climate crisis, the importance of improving ecosystem services such as ecosystem health and integrity in cities, urban temperature regulation, reducing flow water and flood control for sustainable cities has begun to be emphasized. Such an overarching objective can be achieved through the implementation of a blue-green infrastructure approach. In this framework, IMM has conducted another EU funded research project between 2017 and 2019 (Green Re-vision: A Framework for the Resilient Cities). The project has focused on creating climate models for the period of 2050 and 2100 in and around İzmir and a create a guide to provide suggestions within the scope of adaptation to climate change. The overall aim of this international project is 'to create a resilient urban area in a context of climate change by using/enhancing/supporting the potential of green infrastructure. The overarching goal will be planned through three specific objectives and consequent results; (a) developing a framework that makes a city more resilient city to the impacts of climate chance, (b) increasing the technical capacity to enhance capacity to enhance capacity building of target groups and stakeholders on climate change action and (c) increasing public awareness on climate change by disseminating the knowledge and experience' [21].

The largest share in urban greenhouse gas emissions in Izmir comes from the industrial sector, fuel, and electricity consumption. For this reason, it is necessary to increase the more efficient use of resources and to support clean and environmentally sustainable industrialization and to support small-scale producers operating in the local agriculture (with the presence of various agricultural products) and livestock sectors. On the other hand, Izmir, which has a remarkably high potential for clean energy and renewable energy, is in a prominent position for clean energy in Turkey. Izmir province is one of the important geothermal centers in the Aegean region in terms of hot water resources and usage. In this context, it is necessary to increase the number of renewable energy facilities by using the renewable energy potential of Izmir (such as Biogas and Electricity production facilities, electricity generation from landfill gas, energy generation from waste and solar power plants). In addition, to reduce carbon emissions in the city, the number of electric buses should be increased, and the tram, metro, bicycle, and pedestrian networks should continue to be developed. Besides, trainings that raise awareness about climate change and its effects, increase information capacity, and mechanisms that will increase effective planning and management capacity regarding climate change should be supported by increasing the number of participants. At the same time, because İzmir is a poor city in terms of water (with less than 1.000 m^3 of usable water per capita per year); protecting natural areas and water basins, increasing green areas; reducing the production of solid waste through prevention, reduction, recycling, and reuse will also help us in the fight against climate change.

9 Inclusivity, Equity and Community Engagement

Inclusivity, equity, and community engagement are important criteria for the sustainable development pathways. Firstly, in the city of Izmir, strengthening the cultural, social, and economic ties between people living in urban and rural areas and ensuring social cohesion will support the creation of sensitive, inclusive, participatory, and representative decision-making mechanisms at all levels. For example, half of the population of İzmir is made up of women. However, despite significant efforts, women's participation in decision-making mechanisms is not at the desired/sufficient level. Therefore, women's full and effective participation in the decision-making processes of political, economic, and social life and equal opportunities for women to be leaders at all levels in decision-making mechanisms should be ensured. In addition, necessary policies and legally applicable legislation should be adopted and implemented to ensure gender equality. On the other hand, only a small portion of people have a good access to adequate, safe, and affordable housing and basic services. This issue requires efforts to provide safe and accessible housing, transform slums into healthy housing areas, increase efficiency in public transportation, increase public green spaces, improve urban planning and management in a way that is both participatory and inclusive. Finally, municipalities and governorships in Izmir provide aid to the poor and vulnerable with different practices and try to meet the needs of the poorer segments of the society. However, considering the population struggling with poverty throughout the city; it is clearly seen that social aid systems should be developed in terms of education, health and social security, and the total public expenditures spent on social aid services should be increased.

10 Concluding Remarks

For a city to be sustainable or to evolve into a sustainable future, many questions need to be answered about the multidimensional dimensions of cities (environment, economics, society, governance, aesthetics) and how the spaces of these dimensions will be designed, planned, and managed [36]. Here, one of the most crucial questions of Izmir would be how technical solutions could be tailored to enhance relational context in the city of Izmir among Science (both academic and public), community (their requirements, perceptions, and place attachment), non-governmental organizations, and governance (local and regional governance,) over various scales and settings. The city of Izmir is still under planning with both strategy and action level goals and implementing examples of NBSs, aiming for building a more sustainable urban system, in which wellbeing of both residents, environmental issues, cultural-ecological-economic relationships and a vision of democratic and participatory practice are taken into consideration. It is obvious that all these efforts will have positive results for the city of Izmir. However, monitoring the results of the practices conducted with these projects is also a necessity in order to understand how satisfactory the results achieved with the objectives to be achieved are. Strengthening the economic competitiveness between different land uses and sectors is another aspect for a sustainable urban future in the city of Izmir. On the other hand, there is a long way to go to achieve all of these.

References

1. Akurgal E (2016) Anadolu uygarlıkları, Net Turistik Yayınları, İstanbul
2. Amin A (2008) Collective culture and urban public space. City 12(1):5–24
3. Atak BK, Tonyaloğlu EE (2020) Evaluating spectral indices for estimating burned areas in the case of Izmir/Turkey. Eurasian J Forest Sci 8(1):49–59
4. Benton-Short L, Short JR (2013) Cities and nature. Routledge
5. Can I (2010) Urban design and the planning system in Izmir. J Landsc Stud 3(2010):181–189
6. Caves RW (ed) (2004) Encyclopedia of the city. Routledge
7. Derin Z (2020) İzmir-Yeşilova Höyüğü. İzmir Araştırmaları Dergisi 12:1–10
8. Doğan U (2021) A comparison of space quality in streets in the context of public open space design: the example of Izmir, Barcelona, and Liverpool. J Urban Aff. https://doi.org/10.1080/07352166.2021.1919018
9. Dwyer JF, McPherson EG, Schroeder HW, Rowntree RA (1992) Assessing the benefits and costs of the urban forest. J Arboric 18(5):227–234
10. Ercanlı Ç, Savaşır G (2022) Kentsel kıyı alanlarında taşkın riskine yönelik uyum stratejilerini İzmir Karşıyaka kıyı bandı örneği ile değerlendirmek. Megaron 17(2):274–291
11. Erdem U, Çubukçu KM, Sharifi A (2021) An analysis of urban form factors driving urban heat Island: the case of Izmir. Environ Dev Sustain 23(5):7835–7859
12. Ersoy E (2019) Assessment of road-induced landscape fragmentation and implications for landscape planning: the case of İzmir province. Gümüşhane Üniversitesi Fen Bilimleri Dergisi 9(4):699–709
13. European Commission (2022) Cordis EU research results. https://cordis.europa.eu/project/id/730426. Accessed 30 July 2022

14. Gómez-Baggethun E, Gren Å, Barton DN, Langemeyer J, McPhearson T, O'Farrell P, Andersson E, Hamstead Z, Kremer P (2013) Urban ecosystem services. In: Elmqvist E, Fragkias M, Goodness J, Güneralp B, Marcotullio PJ, McDonald RI, Parnell S, Schewenius M, Sendstad M, Seto KC, Wilkinson C (eds) Urbanization, biodiversity and ecosystem services: challenges and opportunities. Springer, Dordrecht, pp 175–251
15. Güngördü A, Güner HE (2019) 1922 Büyük İzmir yangını sonrası İzmir Cumhuriyet Meydanının oluşumu ve mekânsal gelişiminin İncelenmesi. İstanbul Aydın Üniversitesi Dergisi 11(2):111–130
16. Hepcan Ş (2013) Analyzing the pattern and connectivity of urban green spaces: a case study of Izmir, Turkey. Urban Ecosyst 16(2):279–293
17. IBB & EBRD (2020) Izmir sustainable energy and climate action plan. https://www.skb.gov.tr/wp-content/uploads/2021/10/Izmir-Surdurulebilir-Enerji-ve-Iklim-Eylem-Plani_ENG.pdf. Accessed 30 July 2022
18. IBB & EBRD (2020) Izmir green city action plan. https://ebrdgreencities.com/assets/Uploads/PDF/b5cbbe2fd1/Izmir-GCAP-report_FINAL-ISSUED-ENG-002.pdf. Accessed 30 July 2022
19. Izmir Governorship Provincial Directorate of Environment and Urbanization (2020) Izmir province 2019 environmental status report. https://webdosya.csb.gov.tr/db/ced/icerikler/-zmr_2019_cevre_durum_raporu-20200908103247.pdf. Accessed 30 July 2022
20. IMM (2021) Izmir'in Doğa ile Uyumlu Yaşam Stratejisi 2021–2030. http://skpo.izmir.bel.tr/Upload_Files/FckFiles/file/2020/izmir_doga_stratejisi.pdf Accessed 30 July 2022
21. IMM (2022) Green re-vision: a framework for the resilient cities. IMM Publications. https://direnclikent2019.izmir.bel.tr/en/Documents/18. Accessed 30 July 2022
22. Kaplan A (2010) Green infrastructure as a means to deliver a multi-scale approach for urban sustainability. In: Proceedings of the Fabros conference on landscape and greenway planning 3(1), Article 46. https://scholarworks.umass.edu/fabos/vol3/iss1/46. Accessed 30 July 2022
23. Kara F (2017) Spatio-temporal analysis of land cover changes of Izmir province of Turkey using landsat TM and OLI imagery. J Geogr, Environ Earth Sci Int 10(4):1–14
24. Karaca AC, Güllü MK (2019) Detection of forest fire in Menderes district using a superpixel segmentation based search method. J Faculty Eng Archit Gazi Univ 34(2):1061–1076
25. Kavzoglu T, Erdemir MY, Tonbul H (2016, July) Evaluating performances of spectral indices for burned area mapping using object-based image analysis. In: 12th International symposium on spatial accuracy assessment in natural resources and environmental sciences, vol 5, No 8, 5–8 July, Montpelier, France
26. Keleş R (1998) Kent bilim terimleri sözlüğü. İmge Kitabevi Yayınları, Ankara
27. McNeill JR (2006) Population and the natural environment: trends and challenges. Popul Dev Rev 32(S1):183–201
28. Nurlu E, Kesgin Atak B, Barut I (2015a) Analyzing the degree of landscape fragmentation in Izmir, Turkey from 1984 to 2009. In: Efe R, Bizzarri C, Cürebal İ, Nyusupova GN (eds) Environment and ecology at the beginning of 21st century. St. Kliment Ohridski University Press, Chapter 39, pp 545–555
29. Nurlu E, Doygun H, Oguz H, Atak BK (2015b) Simulating the impacts of future policy scenarios on urban land use in Izmir Metropolitan Area using the SLEUTH urban growth model. In: Efe R, Bizzarri C, Cürebal İ, Nyusupova GN (eds) Environment and ecology at the beginning of 21st century, St. Kliment Ohridski University Press, Chapter 11, pp 166–189
30. Özkan SP (2019) GIS besed spatial equity mapping and park provision at neighborhood scale: İzmir case, Unpublished doctoral dissertation, Izmir Institute of Technology, Izmir, Turkey
31. Pakdemirli B, Küçük Ö, Bayraktar Z, Takmaz S (eds) (2021) Ekoloji ve ekonomi ekseninde Türkiye'de orman ve ormancılık. Sonçağ Akademi, Uzun Dijital Matbaa, Ankara
32. Ribeiro MC, de Lara Muylaert R, Dodonov P, Ciocheti G, Magioli M, Martello F, Rocha A, Borges BD, Carvalho C, Kanda CZ, Rodriguez-Castro KG (2016) 4. Dealing with fragmentation and road effects in highly degraded and heterogeneous landscapes. In: Gheler-Costa C, Lyra-Jorge MC, Verdade LM (eds) Biodiversity in agricultural landscapes of southeastern Brazil, De Gruyter Open, pp 43–64

33. Salata S, Özkavaf-Şenalp S, Velibeyoğlu K (2022) Integrating ecosystem vulnerability in the environmental regulation plan of Izmir (Turkey)—What are the limits and potentialities? Urban Science 6(1):19
34. Sampson RJ, Raudenbush SW (1999) Systematic social observation of public spaces: a new look at disorder in urban neighborhoods. Am J Sociol 105(3):603–651
35. Saunders WSA, Becker JS (2015) A discussion of resilience and sustainability: land use planning recovery from the Canterbury earthquake sequence, New Zealand. Int J Disaster Risk Reduct 14(Part 1):73–81
36. Selman P (2008) What do we mean by sustainable landscape? Sustain: Sci, Pract Policy 4(2):23–28
37. Stone B, Vargo J, Habeeb D (2012) Managing climate change in cities: will climate action plans work? Landsc Urban Plan 107(3):263–271
38. Tonyaloğlu EE (2020) Spatiotemporal dynamics of urban ecosystem services in Turkey: the case of Bornova, Izmir. Urban Forest Urban Green 49:126631
39. TUIK (2022) The results of address based population registration system, 2021. https://data.tuik.gov.tr/Kategori/GetKategori?p=Nufus-ve-Demografi-109. Accessed 30 July 2022
40. United Nations (2015). Transforming our world: the 2030 agenda for sustainable development. https://sustainabledevelopment.un.org/post2015/transformingourworld/publication. Accessed 30 July 2022
41. Whitehead M (2003) (Re) analysing the sustainable city: nature, urbanisation and the regulation of socio-environmental relations in the UK. Urban Stud 40(7):1183–1206
42. UNESCO (2022a) World heritage list. https://whc.unesco.org/en/list/. Accessed 30 July 2022
43. UNESCO (2022b) Tentative lists. https://whc.unesco.org/en/tentativelists/?action=listtentative&state=tr&order=states Accessed 30 July 2022
44. UrbanGreenUP (2022) New strategy for renaturing cities through nature based solutions. Izmir'deki uygulamalar. URBAN GreenUP Project. https://www.urbangreenup.eu/kdocs/2049804/ugu-izmir-infopack_digital.pdf. Accessed 30 July 2022
45. Velibeyoğlu K, Yazdani H, Baba A (2018) Groundwater in local development strategies: case of Izmir. Water Sci Technol: Water Supply 18(4):1339–1349
46. Yılmaz ES, Yılmaz S (2019) A review on urbanization, pollution and biodiversity in İzmir. Int J Environ Trends (IJENT) 3(1):31–38

Nature-Based Solutions for Climate Adaptation in School Environments: An Interdisciplinary Assessment Framework

Isabel Ruiz-Mallén, Francesc Baró, Mar Satorras, Funda Atun, Nathalie Blanc, Sarah Bortolamiol, Lidia Casas, Céline Clauzel, Elsa Gallez, Eddy Grand-Meyer, Àgueda Gras-Velázquez, Ivelina Ivanova, Tim Nawrot, Paula Presser, Diana Reckien, and Filka Sekulova

Abstract Nature-based solutions (NbS), defined as actions supported by nature that simultaneously provide environmental and socio-economic benefits in sustainable and resilient ways, can play an important role in systemic transitions leading to climate resilient cities. Indeed, when incorporated in school environments (both within school settings and their surroundings), NbS interventions seem a promising path for boosting climate change adaptation while providing multiple co-benefits,

I. Ruiz-Mallén (✉) · F. Sekulova
Internet Interdisciplinary Institute (IN3), Universitat Oberta de Catalunya (UOC), Barcelona, Spain
e-mail: ruiz_mallen@uoc.edu

F. Baró · E. Gallez
Department of Geography, Vrije Universiteit Brussel (VUB), Brussels, Belgium

Department of Sociology, Vrije Universiteit Brussel (VUB), Brussels, Belgium

M. Satorras
Institut d'Estudis Regionals i Metropolitans de Barcelona (IERMB), Barcelona, Spain

F. Atun · P. Presser · D. Reckien
Faculty Geo Information Science and Earth Observation, Department of Urban and Regional Planning and Geo-Information Management, University of Twente, Enschede, The Netherlands

N. Blanc · C. Clauzel
UMR LADYSS CNRS, Université Paris Cité, Paris, France

S. Bortolamiol
UMR LADYSS CNRS, Université Panthéon-Sorbonne, Paris, France

L. Casas
Social Epidemiology and Health Policy (SEHPO) Department of Family Medicine and Population Health (FAMPOP), University of Antwerp, Antwerp, Belgium

E. Grand-Meyer · À. Gras-Velázquez · I. Ivanova
Science Education Department, European Schoolnet (EUN), Brussels, Belgium

T. Nawrot
Centre for Environmental Sciences, Hasselt University, Hasselt, Belgium

Z. Allam (ed.), *Sustainable Urban Transitions*, Urban Sustainability,
https://doi.org/10.1007/978-981-99-2695-4_6

such as enhanced wellbeing, health and learning opportunities for children as well as increased carbon storage (climate change mitigation). However, the potential of these interventions has not yet been consistently assessed from an interdisciplinary perspective that can give a complete picture of how school-based climate NbS can lead to transformative, radical urban change and the main barriers to its implementation. This chapter responds to this knowledge and methodological gap by developing a framework for a holistic assessment of the health, socio-cultural, environmental and educational co benefits of implementing climate NbS in school environments, with a focus on children as primary beneficiaries. To do so, this chapter proposes and develops the concept of Nature based Climate School Shelters as an innovative strategy for urban sustainability transitions.

Keywords Climate change adaptation · Climate shelter · Nature-based solutions · School greening · Urban resilience · Transformative urban change

1 Introduction

To create more livable cities and mitigate climate risk, the promotion, design and implementation of Nature-based Solutions (NbS) are gaining momentum both in scientific and policy agendas [25, 39, 45]. Indeed, global policy agendas, including the New Urban Agenda [70] and the United Nations Sustainable Development Agenda 2030 [69], are promoting the operationalization of NbS as adaptation and mitigation actions to foster rapid, systemic transitions leading to carbon neutral and climate resilient cities [19, 57]. NbS are defined as actions supported by nature that simultaneously provide environmental, social and economic benefits in sustainable and resilient ways [26]. Besides advancing climate resilience, urban NbS interventions can also provide relevant co-benefits such as changes in behaviors and lifestyles, enhanced urban biodiversity and public health, social cohesion or empowerment of local communities [23, 51].

Following the call from UNICEF [71] and the UN Sustainable Development Goal (SDG) #11 on Sustainable Cities and Communities, cities are requested to ensure equitable access to safe and inclusive green and public spaces, especially for vulnerable groups such as children. However, access to urban nature is not equitable as it has been documented that children, and particularly those in color, low-income communities, could obtain more wellbeing and health benefits if they had more opportunities to engage with natural environments [63]. Furthermore, children are identified as highly vulnerable to climate change effects, such as extreme heat, because their adaptation capacity is less effective, and they experience increased climate-sensitive exposures than adults [60]. High exposure to extreme temperatures negatively impacts children's cognitive performance, as has been shown by an international study using

Centre for Environment and Health-Department of Public Health and Primary Care, Leuven, KU, Belgium

PISA test results [50]. These authors suggest that increased exposure to heat may affect children's capacity to learn but also teachers' capacity to teach. According to the European Climate and Health Observatory [24] 43% of schools in European cities are located in areas at least 2 °C hotter than the regional average, thus increasing heat-related risks to school community members.

Urban NbS, such as tree planting in schoolyards, can reduce children's exposure to these severe climatic conditions [2, 3]. Scientific evidence shows how incorporating NbS in school environments seems a promising path for boosting climate change adaptation and for providing multiple co-benefits ranging from new learning opportunities for children to expanded social interactions and physical activities, reduced stress, improved mood, and other health benefits [9, 11, 16, 72, 73, 79] while overcoming disparities in residential access to urban nature [5]. All this evidence highlights the need to design, implement and maintain NbS to reduce children's vulnerability to rising temperatures and enhance their overall wellbeing, both within school settings (e.g. in schoolyards) and also in school surroundings (e.g., streets and other public spaces) [63].

Moreover, the New Urban Agenda commits to child-responsive urban planning, understanding children as stakeholders to be involved in urban climate planning and development [70]. Despite children's early engagement in urban planning is still a challenge for city governments, even within school environments [32], several pioneer and promising experiences from cities worldwide are striving for the implementation of climate NbS in school environments through the engagement of children as critical voices in the co-design process (e.g., *Opération Ré-création* in Brussels[1]).

However, the potential of co-designed school-based climate NbS to act as transformative urban change models towards city-wide resilience, focusing on the documented co-benefits (i.e., enhanced health and safety, equal accessibility, biodiversity conservation, inclusive governance and quality education), remains largely underexplored. It is crucial to understand on a comparative and interdisciplinary basis how NbS in school environments can deliver urban changes that are transformative, i.e., disrupting current pathways by radically altering existing urban structures, cultures, and practices [29, 78], while unravelling the challenges and barriers confronting these interventions and the underlying reasons behind. To fill this gap, this chapter develops a holistic assessment framework of the multiple co-benefits related to implementing climate NbS in school environments, focusing on children as primary beneficiaries. To do so, we propose and develop the concept of Nature-based Climate School Shelters as an innovative strategy for urban sustainable transitions.

[1] https://environnement.brussels/thematiques/espaces-verts-et-biodiversite/lecole/pour-une-cour-de-recreationvegetalisee#:~:text=Pour%20y%20rem%C3%A9dier%2C%20Bruxelles%20Environnement,nature%20rafraichissante%20dans%20le%20quartier.

2 Nature-Based Climate School Shelters: An Interdisciplinary Assessment Framework

We propose and advance the concept of "Nature-based Climate School Shelters" (hereafter NbCSS) as innovative strategies relying on NbS that are implemented within school settings and their surroundings for responding to climate change. Our approach to NbCSS goes beyond the idea of climate shelters as locally confined safe havens in terms of heat stress or flood risk, as for instance it is the case of cooling centers, primarily associated with air conditioned public buildings [76]. In contrast to such short-term maladaptive solutions, we argue that NbCSS can offer a more sustainable and resilient strategy in the long term as we conceive them as *co-created nature-based, enabling environments for responding to climate change that ensure wellbeing, social equity, learning and inclusivity regards schoolchildren, other school community members and the wider neighborhood community of actors around school centers.*

Based on previous evidence on the multiple co-benefits of transforming schoolyards and wider school environments into greener and climate resilient spaces [5, 6, 9, 16, 72, 73], NbCSS are expected to provide co-benefits regarding children's health, safety, environmental justice, biodiversity conservation, participatory decision-making and quality education.

To holistically analyze the potentials and challenges of NbCSS multiple co-benefits regarding urban sustainability and climate resilience, we propose an assessment framework based on a social-ecological approach that combines the lenses of bio-medical, natural, social and education sciences to explore six dimensions of analysis: (1) children's cognitive and behavioral development; (2) perceptions of safety and other qualities; (3) accessibility and equity; (4) biodiversity practices, (5) inclusive governance, and (6) education for climate solutions (Fig. 1). In what follows, we conceptualize and describe each dimension of analysis and provide methodological details on how to address them.

2.1 *Children's Cognitive Development and Wellbeing*

Childhood is a critical period in life when organs are still developing. Thus, environmental exposures during this early life period can impact health, not only during childhood but also in adulthood (Baker and Osmond 2021). Epidemiological studies show that contact with nature and urban green spaces is associated with better health and wellbeing over the lifespan [56]. In particular, the brain is especially vulnerable to the effects of the environment [35]. Although the brain starts developing in-utero, some cognitive functions closely related to learning and school achievement (e.g., working memory and attention) develop across childhood and adolescence [49, 67]. Natural environments provide children with unique opportunities such as inciting engagement, risk-taking, discovery, creativity, mastery and control, strengthening sense

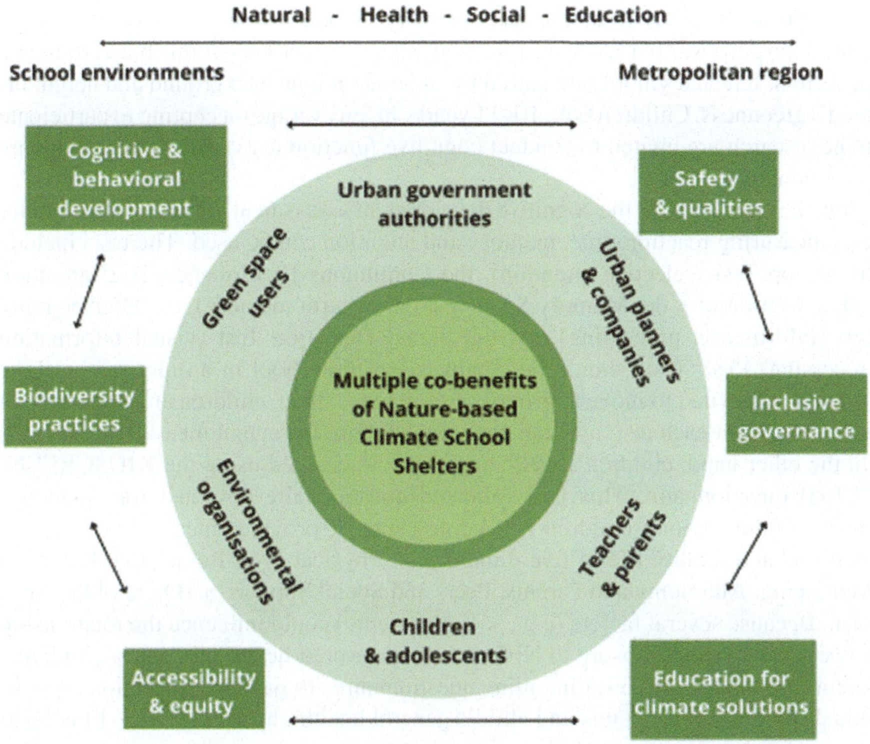

Fig. 1 An analytical framework for a holistic assessment of the health, socio-cultural, environmental and educational co-benefits of implementing Nature-based Climate School Shelters (own elaboration based on the COOLSCHOOLS research project, www.coolschools.eu)

of self, inspiring basic emotional states including sense of wonder, and enhancing psychological restoration, which can positively influence cognitive development and academic performance [12]. Indeed, results from previous epidemiological studies suggest that children living close to green spaces have better cognitive development [9, 22]. In addition, the results of a study conducted in primary schools in Barcelona show that the proportion of green spaces surrounding schools could have a beneficial impact on pupils' cognitive development [18]. School interventions to transform schoolyards into NbCSS add green areas to the urban environment. Therefore, based on this previous evidence, such interventions may positively contribute to children's cognitive development and health in general. However, scientific evidence on the effects of NbCSS on children's health is still minimal [79].

To explore this unaddressed issue, we propose investigating the associations between exposure to NbS in school environments (both in and surrounding schools, see Sect. 2.3 for the methodological approach) and the cognitive development and wellbeing of the children attending these schools.

For the health study, a sample of ten primary schools in one city makes it possible to fulfil the purpose. To reduce the potential impact of socio-economic background on the results, selected schools are paired by socio-economic background and neighborhood's greenness. Children (i.e., 10–11 years old, on average) accepting to participate in the research are invited to conduct cognitive function tests and answer wellbeing questionnaires.

On the one hand, for the cognitive development assessment, a battery of computer tests measuring reaction time, memory and attention can be used. The tests include the Stroop Test (selective attention), the Continuous Performance Test (attention and concentration), the Memory Span Test (short-term memory), the Digit Symbol Test (information processing) and the Signal Detection Test (visual information processing) [58]. Tests should be conducted at the school in a quiet room where children are invited to move to groups of four. Once there, children are instructed on how to perform each test, and can try it once before the actual measurements start. On the other hand, children's wellbeing can be measured using the KIDSCREEN-27 [65] questionnaire. This is a validated questionnaire designed for epidemiological studies among children and adolescents. It provides measures of generic health related quality life in five dimensions: Physical Well-Being, Psychological Well-Being, Autonomy and Parents, Peers and Social Support, and School Environment. Because several factors (e.g., socioeconomic) could influence the relationship between children's exposure to NbS and the measured health outcomes, children's parents are also asked to fill in online questionnaires to gather information on individual socioeconomic status and child's general health physical activity. Finally, to account for the exposure to NbS outside school, parents are asked to provide residential addresses, information on commuting modes and times, and outdoor activities outside school hours.

2.2 *Perceptions of Safety and Other Qualities*

The type, design, and conditions of greening urban spaces, including school environments, very majorly influence how potential advantages are realized. For example, the distribution of green areas, the relative number of paved areas, inadequate maintenance, and insufficient play space can reduce the positive impact that NbCSS features can bring to pupils and other users [38, 48, 72]. Studies referring to the perception of schoolyards and their greenery show that certain qualities and characteristics, such as wilder forms and the use of colored flowers [14], are preferred. In addition, users' contribution to the design of these spaces (see also Sect. 2.5 on inclusive governance for an in-depth view of this analysis) increases the perception of safety [41] and raises awareness of the qualities that NbS can bring to school environments [1].

However, more evidence is needed concerning the perceived impact of NbCSS [79], e.g., related to thermal comfort and climate change-related aspects [11], as well as preferences linked to users' gender and age [14]. Studies conducted in Dutch schools, for instance, indicate that the change in quantity and quality of greening after NbS interventions in school environments are modestly reducing the potential of NbS to act as climate shelters and provide co-benefits to users [72, 74]. It is also shown that assessing the awareness and understanding of the complex contribution of NbCSS needs a system's thinking approach and an understanding of the perception of different actors [33].

Our framework here addresses this caveat by paying particular attention to the role of quality and safety of different types, designs, and features of NbCSS across different actors and the potential of these interventions to initiate socio-cultural transformations in school environments and wider neighborhoods. One way to do such analysis is by implementing the interactive approach of Fuzzy Cognitive Mapping (FCM) with users. To identify variables to be tested in the FCM, a previous literature search on the perceived quality and safety characteristics of greenery as NbCSS is needed. It can be a computer-assisted literature search that uses supervised machine learning techniques to efficiently conduct large-scale analyses of primary documents [62].

The result of these variables is thus implemented for the use of FCM interviews [53], potentially asking parents, teachers, and pupils of schools that have undergone the implementation of a NbCSS recently about their perception of the quality and safety of school NbS and school environments, what would make them feel safer, and what else would increase the use of NbS in and around school environments. In addition, children can also be interviewed by using FCM to find out their perceptions of the quality and quantity of greenery as NbCSS. These data contribute to finding out which qualities of NbCSS and other characteristics of the environment or neighborhood could be decisive in leading to a change in the perception of NbCSS and its safety, induce a higher use of the NbCSS within and around schools, including their pedagogical use (see more about this topic in Sect. 2.6), and hence have the potential for a socio-cultural transformation of the neighborhood and school environment.

2.3 Accessibility and Equity

While ambitious urban greening strategies based on NbS are increasingly being adopted by municipal policy-makers and planners, scholars raise awareness of the potential (re-)production of social and environmental inequalities produced by the implementation of NbS interventions [36]. The consideration of social vulnerability and environmental justice aspects in urban NbS assessments and greening planning instruments is still scarce, even though it is globally encouraged by the UN SDG #11.7, which envisions "*by 2030, to provide universal access to safe, inclusive and accessible green and public spaces, in particular for women and children, older persons and persons with disabilities*" [69, p. 26]. Research on (unequal) access or exposure to urban green and blue spaces and other NbS by vulnerable social groups-notably children—still remains focused on residential metrics [47, 75]. However, other foundational amenities represent an essential place in their daily urban experience. For instance, children spend considerable time in schools and school environments such as nearby parks or playgrounds [31]. But the distributional inequalities in access to NbS from a school-based perspective are still rarely explored and mainly dominated by American case studies that mostly focus on academic performance [13, 52, 81]. A few studies analyzing access to green spaces around schools explore European case studies, notably Barcelona [5] and London [61], while cross-cities studies in Europe are still lacking [5]. Moreover, the potential of green schoolyards and green school surroundings to reduce social disparities in access to green spaces for children and the local community is rarely considered [82].

In response to the knowledge gaps mentioned above, this analytical dimension focuses on the exposure and access to NbCSS by examining the spatial, temporal and use patterns of NbS and other nature-based initiatives within and around school environments using a socio environmental equity perspective. The underlying research hypothesis is that the distributional patterns of NbCSS are associated with the socio-demographic characteristics of schoolchildren and school types, involving potential environmental and climate justice implications. To test this hypothesis, a mix-method approach can be used, based on spatial quantitative and qualitative research methods implemented in one or more cities.

First, geospatial analysis is conducted to map the spatial patterns of NbCSS in the studied city/ies. This analysis can be done through very high-resolution (VHR) remote sensing imagery (5 and 25 cm) to identify, map and characterize NbCSS, both within and around sampled school compounds. Indicators on NbCSS typology (e.g., educational gardens, street greenery, urban parks) and NbS structure (tree, shrub, herbaceous cover, water bodies) are generated through remote sensing techniques. Potentially derived climate benefits (e.g., air purification, water retention, carbon storage, noise reduction) are estimated by applying available state-of the-art coefficients to the vegetation structure indicators (e.g., based on Derkzen et al. [20] and other potential sources). It is recommended to conduct a cross-city comparative analysis to focus on the distributional patterns and temporal dynamics of green and blue land covers and tree canopy cover around schools.

The next step consists of analyzing the associations between the NbCSS indicators quantified in the previous step and established schools' socio-demographic vulnerability variables (e.g., percent of low-income pupils' households, percent of pupils whose parents have low educational attainment, percent of pupils who live in a neighborhood with educational delay or disadvantage, etc.) and environmental vulnerability variables (e.g. building density, noise level, urban heat island risk, air pollution and flood risk). Controlling for the school size (number of enrolled students), associations are evaluated using different statistical techniques, including bivariate, multivariate and cluster analysis. Potential differences across cities and school types (public, charter, private) can be considered in the analyses to unravel in(equity) patterns.

When possible, map-based surveys can be conducted to assess the use patterns and perceptions related to NbCSS by different user groups (e.g., school staff -mainly teachers-, schoolchildren parents and external green schoolyard users) during and after school hours, also considering potential enablers and barriers to the implementation and upscaling of these interventions. The surveys use a dedicated participatory GIS (PPGIS) tool and can be circulated online and/or face-to-face. Qualitative information gathered through these map based surveys contributes to the understanding of the ways and reasons why children and the local community access (or not) to green schoolyards and green school surroundings.

2.4 Biodiversity Management Practices

From the perspective of urban politics of adaptation to climate change, biodiversity is an asset to evaluate the effects of implementing NbS in school environments, particularly schoolyards. Biodiversity refers to all the interactions between the different living forms and between them and their environment. Analyzing biodiversity is, therefore, not only a matter of counting the number of species, but also of assessing the diversity of functional relationships within a complex system. Depending on the composition and organization of vegetation, NbCSS can contribute to urban connectivity by providing functional intermediate habitats for different fauna species to connect larger and/or higher quality habitats [42]. However, this is more a theoretical assumption than a fact [10]. Indeed, biodiversity levels could vary according to schoolyards' configuration and surrounding environments' management practices [44]. Also, the degree of stakeholders' involvement in the different stages of NbCSS implementation (co-design, project support, awareness-raising, etc., see also next Sect. 2.5 on inclusive governance for more detail) could impact their level of biodiversity knowledge, which can in turn impact biodiversity levels.

To check these unresolved hypotheses, the analytical framework presented here focuses on examining biodiversity practices in NbCSS through a methodology that confronts the ecological dimensions of schoolyards' biodiversity with the practices and representations of a diverse set of stakeholders who access and use these NbS (i.e., children, parents, teachers, city biodiversity managers, etc.). Our methodological approach links the quantity and diversity of different biodiversity components (i.e., plants, arthropods), their representations by these different actors, and the variety of knowledges and practices generated by their interaction through the greenery of school environments. Biodiversity issues are conceived as a communication hub between different dimensions of the urban space, from the social to the morphological and from the economic to the ecological, as well as between the different actors (private, public, associative sectors), and at different levels (school, neighborhood, city, region) [34]. The analysis of urban biodiversity is undertaken with the idea of evaluating the place of biodiversity in urban schoolyards, but it is also a means for analyzing ordinary practices that contribute to rethinking the links between the different forms of living beings in cities.

To monitor management practices in relation to species diversity and quantity through time, as well as to the spatial and morphological configurations of the schools and surrounding neighborhoods, a series of biodiversity inventories and interviews with diverse stakeholders are conducted in each school in each studied city (an appropriate sample includes five to ten schools per city). In parallel, spatial analysis at different scales (school, neighborhood, city) is conducted to assess NbCSS contribution to urban functional connectivity by combining landcover/land-use maps and species movement abilities [17]. From these datasets, ecological and social networks of species and stakeholders are modelled and potential interlinkages between biodiversity and governance can be identified. To do that, connectivity metrics are computed and vegetation management practices in the schoolyards are censused and statistically related to each other to test the influence of connectivity and practices on biodiversity levels. Overall, results are mapped to spatially represent biodiversity territories within school environments, which provides an initial diagnosis of the biodiversity presence and management practices in the NbCSS. Further, to actively engage children in biodiversity monitoring and build capacities among the school community, it is recommended to launch a citizen science program through workshops with school students and their parents, as well as teachers.

2.5 Inclusive, Reflexive and Multi-level Governance

Environmental governance is a complex process, which can be approached or analyzed as a combination of: (i) existing rules/norms, (ii) diverse actors/institutional structures, and (iii) multiple formal and informal processes, including decision-making, policy creation, conflict-resolution, and the negotiation of values [8]. To better understand the governance of climate-adapted and green schoolyards is helpful to add two more ingredients to the tripartite structure above-mentioned: power dynamics and dominant discourses [4]. Power is understood as a multi-dimensional concept here, which beyond influence in a broad sense, also entails the capacity to mobilize and direct resources/capital, while discourses underpin the views and narratives (the epistemology) around how problems are defined and approached, framing for each particular context where governance is being enacted. Analyzing the governance of the NbCSS from a mix-methods perspective can help us understand how naturalized and climate resilient playgrounds are jointly produced, by whom, for whom and with what implications. This implies focusing both on the results (such as multi functional, inclusive, educational and climate-resilient schoolyards) and the processes (such as transparent communication, participation, inclusive and democratic decision-making and potentially, conflict-resolution or value-negotiation).

On the one hand, when studying the governance of these transformed schoolyards, particular attention needs to be paid to the diversity of actors initiating, preparing, performing, evaluating and maintaining the initiative and the ways they have been engaged in this process at various points of time. This can be done through a Social Network Analysis supported by in-depth interviews with identified stakeholders. A number of studies indicate that while multi stakeholder involvement and participation (of students, teachers, school communities, neighborhood organizations, public institutions, funders and architects) is standard in the preparation phase of green schoolyards [28], engagement in the management and maintenance part starts to dwindle [32]. Moreover, multi stakeholder involvement does not necessarily imply that all actors have equal capacity to influence the process. Attention thus needs to be paid to how the results of the participative processes are interpreted, and then translated at the level of school ground implementation, and how the voices of children inform the development works [16]. A common assumption underpinning much of the schoolyard transformations by using NbS in Europe and North America is that technical and implementation details should be left to professionals (e.g., urban planners or landscape architects), limiting children's engagement only in the conceptual design [21, 55]. The implementation phase, however, has important pedagogic implications which can be employed as educational opportunities (see also Sect. 2.6 for the educational potential of NbCSS). In this regard, governance also implies understanding how the newly transformed space is being used, and integrated into the day to-day school life beyond recess times. Many studies suggest that for outdoor learning, and green schoolyards to be successful along the lines of the multiple benefits mentioned above, schools need to value these as a means of achieving curricular goals, rather than as an add on initiative, or a purely aesthetic exercise [80]. Understanding the arrangements conducive to the multi-dimensional success of NbCSS is another key question that needs to be addressed in the context of governance analysis.

On the other hand, the study of governance interacts with the other dimensions of our assessment framework, exploring how well they have been considered, produced and replicated, spotting points of friction, potential barriers, lock-ins, or amplifiers. One peculiarity of the NbCSS is their immanent, living and evolving nature, which requires non-conventional maintenance—where a profound understanding of ecology and play are fundamental for the manifold benefits of the intervention to be sustained and retained [46]. Such organization however implies a unique governance configuration in each school context [74].

Exploring governance, finally, requires studying the institutional norms and discourses that frame existing school environments, such as existing city policy frameworks, or legislation initiatives with a look at their eventual evolution and upscaling potential. A fundamental element here is to explore issues of (environmental) justice, meaning the distribution and application of schoolyard greening across multi-ethnic working-class neighborhoods, whose access to ample parks and vegetated areas has been precluded [5, 7].

In sum, the analysis of the governance of NbCSS can shed light on how their implementation can become even more effective, more equitable, more adaptive, more robust, more democratic and more sustainable, and eventually reduce some important and unexpected set backs in the future.

2.6 Education for Climate Solutions

The last dimension of our assessment framework revolves around the potential of NbCSS regarding educational transformations at schools. Such educational transformations rely on the extended use of NbCSS as learning spaces to foster quality education on climate change in thematic areas such as wellbeing, safety, biodiversity conservation, inclusive decision making and other attributes documented in previous sections. The pedagogical use of NbS in schoolyards and surrounding green areas needs to be supported by providing the school communities with tools and resources, but mainly through the training of school teachers. In that way, teachers are assisted in introducing the co-benefits of NbCSS to students, while also being engaged in their management [15, 30, 66, 68].

Teachers' capacity building in using NbCSS for educational purposes relies on a competence learning approach in sustainability education and a holistic understanding of climate education [27, 40, 54, 77]. In particular, and to address un(equal) access issues, such capacity building process embraces an intersectional approach integrating different axes of inequality, e.g., gender, class, age, ethnicity, race, etc. [43], and their complex interactions.

To leverage teachers' capacity building, we propose a multifaceted strategy starting with the design of guidelines and resources to help teachers introduce NbCSS co-benefits to their students, while engaging them in the implementation and management of NbCSS in their schools. These guidelines can be inspired by the participatory action-research activities organized to assess the other dimensions of our analytical framework (i.e., children's health, safety, equal accessibility, biodiversity management, inclusive governance). Examples of such activities include, among others, participatory mapping of NbCSS use by children and citizen science programs for biodiversity monitoring involving students and teachers (see Sects. 2.3 and 2.4, respectively). In addition, educators can be supported through the well-practiced approach of creating a tailor-made Massive Online Open Course—MOOC, in this case on using climate shelters interventions in schools for educational purposes, whereby participant teachers design action plans outlining various implementation strategies of NbCSS content.

Further, the identification of best practices and lessons learnt from the other dimensions of our assessment framework is key to promoting teachers' capacity building because these best practices can be disseminated and upscaled through a wide-reach campaign to encourage teachers worldwide in producing and sharing their ideas between each other while promoting their action plans. International organizations can play a key role in supporting this action, such as the International School Grounds Alliance (ISGA) and the European Schoolnet, the latter of which is a network of 34 European Ministries of Education. The involvement of stakeholders at policy level through the creation of a set of recommendations for the Ministries of Education aims to ensure that the resources are taken up by those developing STE(A)M education strategies and by other decision-makers on urban climate adaptation, as well as to contribute to the application of the research in sustainability teaching and learning [37]. In sum, the educational approach in mind leverages good practices and learnings on the transformation capacities of school climate shelters, promotes them to educators through capacity-building activities and targeted dissemination activities, and ensures the outreach to education policy-makers, particularly Ministries of Education.

3 Conclusion: Implications for NbCSS Research and Practice

NbCSS can be an innovative strategy for urban sustainable transitions. The NbCSS framework described here is conceived as an analytical tool for providing evidence on the assessment of the multiple co-benefits of implementing these interventions, with a focus on children as primary beneficiaries, and throughout a holistic perspective. In previous sections, it has been shown how those school environments transformed into climate shelters by using NbS are spaces where multiple rationalities, and functionalities intersect, so their results, or impacts, cannot be analyzed in mono-dimensional terms. Key aspects to be considered when assessing the potentials and challenges of NCSS revolve around: children health and wellbeing; safety and other qualities; accessibility and equity; biodiversity management; inclusive governance; and education for climate solutions (Table 1).

We argue that such a holistic, inclusive, collaborative, and reflective assessment framework that puts children at the center of the research on climate NbS in school settings is not only needed to co-produce relevant knowledge for user stakeholders, but also to respect children's rights according to international urban development agendas [64] and provide them with the agency to envision radical alternatives for urban change towards climate change adaptation.

The NbCSS assessment framework is useful for academic purposes and can also directly contribute to practice. Engagement and capacity building processes with school communities, municipal authorities, urban planners, local associations and businesses, environmental NGOs, and green space users are fostered and consolidated throughout its application. Such learning processes, in turn, provide new insights into the transformative potential derived from NbCSS. In doing so, transition pathways for sustainable urban development can be fostered while ensuring actors' inclusive engagement to contribute to democratizing urban resilience decision-making [59].

Table 1 Summary of the main research questions and methodological topics for each key dimension of the NbCSS assessment framework

Six key assessment dimensions of NbCSS	Main research questions	Methodological approaches
Cognitive and behavioural development	Do NbCSSS have a beneficial impact on pupil's cognitive function development and well-being?	Cognitive function tests and well-being questionnaires to children
Accessibility and equity	How NbCSS are spatially distributed across schools and the urban fabric? Which are the structural and functional characteristics of NbCSS (vegetation types, ecosystem services provided, etc.)? What are the equity implications of NbCSS distribution considering socio demographic vulnerabilities and school types? What are the use patterns and perceived benefits/disservices related to NbCSS?	Geospatial analysis of NbCSS and map based surveys to NbCSS users
Perception of safety and other qualities	Which qualities of NbCSS are perceived as high quality and preferred by users? Which qualities of NbCSS are associated with safety concerns? Which types of NbCSS and in which quantities show largest potential as climate shelters?	Computer-assisted literature search Fuzzy Cognitive Mapping (FCM) with NbCSS users
Biodiversity management practices	What are the abundance and the diversity of small fauna in NbCSS? Do connectivity and vegetation management practices have an influence on biodiversity? What is the degree of the involvement of actors in the different stages of NbCSS implementation?	Biodiversity inventories and interviews with diverse stakeholders in schools Multi-scale spatial analysis to assess NbCSS contribution to urban functional connectivity
Inclusive, reflexive and multi-level governance	What are the NbCSS collaborative arrangements and participative processes at the level of design and implementation?	Social Network Analysis supported by in-depth interviews with identified stakeholders
	What is the role of children/students and other stakeholders (parents, other entities) in NbCSS maintenance? What are the key factors of success regarding NbCSS governance? How are NbCSS being used and by whom?	Institutional and policy analysis

(continued)

Table 1 (continued)

Six key assessment dimensions of NbCSS	Main research questions	Methodological approaches
Education for climate solutions	What good practices and lessons learnt can be identified from the other key dimensions of NbCSS? How can teachers and other relevant stakeholders benefit from the identified good practices and lessons learnt in introducing NbCSS co-benefits? What recommendations might be beneficial to policy-makers with relation to achieving a school's transition towards climate shelters?	Guidelines and resources for helping teachers introduce NbCSS co-benefits Training courses targeting teachers (MOOC) Dissemination campaign promoting action plans and recommendations for Ministries of Education

Acknowledgements This work builds on the COOLSCHOOLS project (www.coolschools.eu) and has received funding from the European Union's Horizon 2020 research and innovation programme under grant agreement No. 101003758, the Spanish Research Agency (AEI), Innoviris (Brussels Capital Region), Dutch Research Council (NWO), The Research Foundation—Flanders (FWO), and Agence Nationale de la Recherche (ANR).

References

1. Anderson CC, Renaud FG (2021) A review of public acceptance of nature-based solutions: the 'why', 'when', and 'how' of success for disaster risk reduction measures. Ambio 50:1552–1573
2. Antoniadis D, Katsoulas N, Papanastasiou D, Christidou V, Kittas C (2015) Evaluation of thermal perception in schoolyards under Mediterranean climate conditions. Int J Biometeorol 60(3):319–334
3. Antoniadis D, Katsoulas N, Kittas C (2018) Simulation of schoolyard's microclimate and human thermal comfort under Mediterranean climate conditions: effects of trees and green structures. Int J Biometeorol 62(11):2025–2036
4. Arts B, Leroy P, van Tatenhove J (2006) Political modernisation and policy arrangements: a framework for understanding environmental policy change. Public Organiz Rev 6:93–106
5. Baró F, Camacho DA, Pérez Del Pulgar C, Triguero-Mas M, Anguelovski I (2021) School greening: right or privilege? Examining urban nature within and around primary schools through an equity lens. Landsc Urban Plann, 208
6. Baró F, Camacho DA, Pérez Del Pulgar C, Ruiz-Mallén I, García-Serrano P (2022, in press) Nature-based climate solutions in European schools: a pioneering co-designed strategy towards urban resilience. In: (pxx-x) Ruiz-Mallén I, March H, Satorras M (eds) Urban resilience to the climate emergency: unravelling the transformative potential of institutional and grassroots initiatives. Springer, Cham
7. Bates C, Bohner A, Gerstein DE (2018) Green schoolyards in low-income urban neighborhoods: natural spaces for positive youth development outcomes. Front Psychol 9(805)
8. Bennett NJ, Satterfield T (2018) Environmental governance: a practical framework to guide design, evaluation and analysis. Conserv Lett 11:e12600

9. Bijnens EM, Derom C, Thiery E, Weyers S, Nawrot TS (2020) Residential green space and child intelligence and behavior across urban, suburban, and rural areas in Belgium: a longitudinal birth cohort study of twins. PLoS Med 17(8):e1003213
10. Blanc N, Glatron N, Lamarche T, Rankovic A, Sourdril A (2017) Governance of urban nature. J Urban Res. https://journals.openedition.org/articulo/3212
11. Bohnert AM, Nicholson LM, Mertz L, Bates CR, Gerstein DE (2021) Green schoolyard renovations in low-income urban neighborhoods: benefits to students, schools, and the surrounding community. Am J Commun Psychol
12. Boeve-de Pauw J, Halbac-Zamfir R (2020) Environmental citizenship in the context of primary non-formal education. Springer, Cham, pp 179–91
13. Browning MHEM, Rigolon A (2019) School green space and its impact on academic performance: a systematic literature review. Int J Environ Res Pub Health 16(3). https://doi.org/10.3390/ijerph16030429
14. Campagnaro T, Vecchiato D, Arnberger A, Celegato R, Da Re R, Rizzetto R, Semenzato P, Sitzia T, Tempesta T, Cattaneo D (2020) General, stress relief and perceived safety preferences for green spaces in the historic city of Padua (Italy). Urban For Urban Greening 52:126695
15. Chang CH (2014) Preparing teachers for climate change education. Climate change education: knowing, doing and being. Routledge, Oxon, pp 103–123
16. Chawla L, Keena K, Pevec I, Stanley E (2014) Green schoolyards as havens from stress and resources for resilience in childhood and adolescence. Health Place 28:1–13
17. Clauzel C, Godet C (2020) Combining spatial modeling tools and biological data for improved multispecies assessment in restoration areas. Biol Cons 250:108713
18. Dadvand P, Nieuwenhuijsen MJ, Esnaola M, Forns J, Basagaña X, Alvarez-Pedrerol M et al (2015) Green spaces and cognitive development in primary schoolchildren. Proc Natl Acad Sci 112(26):201503402
19. de Coninck H, Revi A, Babiker M et al (2018) Strengthening and implementing the global response. In: Global Warming of 1.5 °C. An IPCC special report on the impacts of global warming of 1.5 °C above pre-industrial levels (…) to the threat of climate change, sustainable development, and efforts to eradicate poverty. Cambridge University Press
20. Derkzen ML, van Teeffelen AJA, Verburg PH (2015) REVIEW: quantifying urban ecosystem services based on high-resolution data of urban green space: an assessment for Rotterdam, the Netherlands. J Appl Ecol 52(4):1020–1032
21. Derr V (2017) Urban greenspace as learning laboratory and participatory space. Urban 658 Des Plan
22. Dockx Y, Bijnens EM, Luyten L, Peusens M, Provost E, Rasking L et al (2022) Early life exposure to residential green space impacts cognitive functioning in children aged 4 to 6 years. Environ Int 161
23. Dumitru A, Wendling L (2021) Evaluating the impact of nature-based solutions: a handbook for practitioners. Eur Commhttps://doi.org/10.2777/2498
24. European Climate and Health Observatory (2022) Exposure of vulnerable groups and facilities to flooding and heat. https://climate-adapt.eea.europa.eu/observatory/evidence/projectionsand-tools/exposure-of-vulnerable-groups-to-climate-risks
25. European Commission (2015) Towards an EU Research and Innovation policy agenda for nature-based solutions & re-naturing cities. https://doi.org/10.2777/765301
26. European Commission (2015) Nature-based solutions & re-naturing cities. Final report of the horizon 2020 expert group on 'Nature-Based Solutions and Re-Naturing Cities'. Publications Office of the EU, Luxembourg
27. FFF (2022) Fridays for future climate education: our demands. https://fffclimateedu.com/our-demands/. Accessed 14 Jul 2022
28. Flax L, Altes RK, Kupers R, Mons B (2020) Greening schoolyards-an urban resilience perspective. Cities 106https://doi.org/10.1016/j.cities.2020.102890
29. Frantzeskaki N, Hölscher K, Bach M, Avelino F (eds) (2018) Co-creating sustainable urban futures. A primer on applying transition management in cities. Future City 11. Springer, Cham

30. Fronza V, Gras-Velazquez A (Dec 2020) Sustainability in formal Education: ways to integrate it now. Scientix Observatory, p 160
31. García Serrano P, Urda Peña L, Leal P, de Blas M (2017) Estudio Micos: Entornos Escolares Saludables. Bases para una estrategia de intervención de patios, accesos y entornos escolares. Conference: Jornadas de Innovación Educativa "I&EDU". https://www.researchgate.net/publication/317836292
32. Giezen M, Pellerey V (2021) Renaturing the city: factors contributing to upscaling green schoolyards in Amsterdam and The Hague. Urban For Urban Greening 63https://doi.org/10.1016/j.ufug.2021.127190
33. Gómez Martín E, Giordano R, Pagano A, van der Keur P, Máñez Costa M (2020) Using a system thinking approach to assess the contribution of nature based solutions to sustainable development goals. Sci Total Environ 738:139693
34. Gough A, Chi-Kin Lee J, Po Keung Tsang E (eds) (2020) Green schools globally stories of impact on education for sustainable development. Springer
35. Grandjean P, Landrigan PJ (2014) Neurobehavioural effects of developmental toxicity. Lancet Neurol 13:330–338
36. Haase D, Kabisch S, Haase A, Andersson E, Banzhaf E, Baró F, Brenck M, Fischer LK, Frantzeskaki N, Kabisch N, Krellenberg K, Kremer P, Kronenberg J, Larondelle N, Mathey J, Pauleit S, Ring I, Rink D, Schwarz N, Wolff M (2017) Greening cities–to be socially inclusive? About the alleged paradox of society and ecology in cities. Habitat Int 64:41–48https://doi.org/10.1016/j.habitatint.2017.04.005
37. Holdsworth S, Thomas I, Hegarty K (2013) Sustainability education: theory and practice. In: Stevenson RB, Brody M, Dillon J, Walls AEJ (eds) International handbook of research on environmental education. Routledge, New York, pp 349–358
38. Jansson M, Abdulah M, Eriksson A (2018) Secondary school students' perspectives and use of three school grounds of varying size, content and design. Urban For Urban Greening 30:115–123
39. Kabisch N, Frantzeskaki N, Pauleit S, Naumann S, Davis M, Artmann M, Haase D, Knapp S, Korn H, Stadler J, Zaunberger K, Bonn A (2016) Nature-based solutions to climate change mitigation and adaptation in urban areas: perspectives on indicators, knowledge gaps, barriers, and opportunities for action. Ecol Soc 21(2). https://doi.org/10.5751/ES-08373-210239
40. Kidman G, Chang C-H, Wi A (2020) Defining education for sustainability (EfS): a theoretical framework. In: Kidman G, Chang C-H, Wi A (eds) Issues in teaching and learning of education for sustainability: theory into practice. Routledge, Oxon, pp 1–15
41. Kondo MC, South EC, Branas CC (2015) Nature-based strategies for improving urban health and safety. J Urban Health 92(5):800–814
42. LaPoint S, Balkenhol N, Hale J, Sadler J, van der Ree R (2015) Ecological connectivity research in urban areas. Funct Ecol 29(7):868–878
43. Lee L, Luykx A (2007) Science education and student diversity: race/ethnicity, language, culture and socioeconomic status. In: Abell SK, Lederman NG (eds) Handbook of research on science education. Routledge, USA, pp 171–198
44. Lepczyk CA, Aronson MFJ, Evans KL, Goddard MA, Lerman SB, MacIvor JS (2017) Biodiversity in the city: fundamental questions for understanding the ecology of urban green spaces for biodiversity conservation. Bioscience 67(9):799–807
45. Liu HY, Jay M, Chen X (2021) The role of nature-based solutions for improving environmental quality, health and well-being. Sustainability (Switzerland) 13(19):1–56
46. Maas J, Muller R, Hovinga D (2014) Groene schoolpleinen: Succes-en faalfactoren bij een duurzaam ontwerp van groene schoolpleinen [Green schoolyards: success and failure factors in sustainable design of green schoolyards]. Working paper
47. Markevych I, Feng X, Astell-Burt T, Standl M, Sugiri D, Schikowski T, Koletzko S, Herberth G, Bauer CP, von Berg A, Berdel D, Heinrich J (2019) Residential and school greenspace and academic performance: evidence from the GINIplus and LISA longitudinal studies of German adolescents. Environ Pollut 245:71–76

48. Massey WV, Perez D, Neilson L, Thalken J, Szarabajko A (2020) Observations from the playground: common problems and potential solutions for school-based recess. Health Educ J 80(3):313–326
49. Østby Y, Tamnes CK, Fjell AM, Walhovd KB (2011) Morphometry and connectivity of the fronto-parietal verbal working memory network in development. Neuropsychologia 49(14):3854–3862
50. Park RJ, Behrer AP, Goodman J (2021) Learning is inhibited by heat exposure, both internationally and within the United States. Nat Hum Behav 5:19–27
51. Raymond C, Raymond CM, Frantzeskaki N, Kabisch N, Berry P, Breil M, Nita MR et al (2017) A framework for assessing and implementing the co-benefits of nature based solutions in urban areas. Environ Sci Policy 77:15–24
52. Requia WJ, Li L, Amini H, Roig HL, James P, Koutrakis P (2022) Nationwide assessment of green spaces around 186,080 schools in Brazil. Cities, 121
53. Reckien D (2014) Weather extremes and street life in India—Implications of Fuzzy Cognitive Mapping as a new tool for semi-quantitative impact assessment and ranking of adaptation measures. Glob Environ Chang 26:1–13
54. Rieckmann M (2018) Key themes in education for sustainable development. In: Leicht A, Heiss J, Byun WJ (eds) Issues and trends in education for sustainable development. UNESCO Publishing, France, pp 39–60
55. Rigolon A, Derr V, Chawla L (2015) Green grounds for play and learning: an intergenerational model for joint design and use of school and park systems. Handbook of green infrastructure. ElgarOnline
56. Rojas-Rueda D, Nieuwenhuijsen MJ, Gascon M, Perez-Leon D, Mudu P (2019) Green spaces and mortality: a systematic review and meta-analysis of cohort studies. Lancet Planet Heal 3(11):e469–77. http://www.ncbi.nlm.nih.gov/pubmed/31777338
57. Romero-Lankao P, Bulkeley H, Pelling M et al (2018) Urban transformative potential in a changing climate. Nat Clim Chang 8:754–756
58. Saenen ND, Provost EB, Viaene MK, Vanpoucke C, Lefebvre W, Vrijens K et al (2016) Recent versus chronic exposure to particulate matter air pollution in association with neurobehavioral performance in a panel study of primary schoolchildren. Environ Int 95:112–9. https://pubmed.ncbi.nlm.nih.gov/27575366/
59. Sharifi A (2016) A critical review of selected tools for assessing community resilience. Ecol Ind 69:629–647
60. Sheffield PE, Landrigan PJ (2011) Global climate change and children's health: threats and strategies for prevention. Environ Health Perspect 119(3):291–298
61. Shoari N, Ezzati M, Doyle YG, Wolfe I, Brauer M, Bennett J, Fecht D (2021) Nowhere to play: available open and green space in greater London schools. J Urban Health 98(3):375–384. https://doi.org/10.1007/s11524-021-00527-0
62. Sietsma AJ, Ford JD, Callaghan MW, Minx JC (2021) Progress in climate change adaptation research. Environ Res Let
63. Stevenson KT, Moore R, Cosco N, Floyd F, Sullivan W, Brink L, Gerstein D, Jordan C, Zapalatosch J (2020) A national research agenda supporting green schoolyard development and equitable access to nature. Elem Sci Anthr 8:10
64. Sugar S (2021) The necessity of urban green space for children's optimal development. Discussion paper. UNICEF. https://www.unicef.org/documents/necessity-urbangreen-space-childrens-optimal-development
65. The KIDSCREEN-27-kidscreen.org [Internet]. [cited 2022 Jun 29]. https://www.kidscreen.org/english/questionnaires/kidscreen-27-short-version/
66. Timm JM, Barth M (2020) Making education for sustainable development happen in elementary schools: the role of teachers. Environ Educ Res, 1–17.
67. Ullman H, Almeida R, Klingberg T (2014) Structural maturation and brain activity predict future working memory capacity during childhood development. J Neurosci 34(5):1592–1598
68. UNESCO (2021) Teachers have their say: motivation, skills and opportunity to teach education for sustainable development and global citizenship. UNESCO, France

69. United Nations (2015) Transforming our world: the 2030 agenda for sustainable development. https://sustainabledevelopment.un.org/post2015/transformingourworld/publication
70. UN-Habitat (2020) The new urban agenda. https://unhabitat.org/about-us/new-urban-agenda
71. UNICEF (2021) The climate crisis is a child rights crisis: introducing the children's climate risk Index. https://data.unicef.org/resources/childrens-climate-risk-index-report/
72. van Dijk-Wesselius JE, Maas J, Hovinga D, van Vugt M, van den Berg AE (2018) The impact of greening schoolyards on the appreciation, and physical, cognitive and social-emotional well being of schoolchildren: a prospective intervention study. Landsc Urban Plan 180:15–26
73. van Dijk-Wesselius JE, van den Berg AE, Maas J, Hovinga D (2020) Green schoolyards as outdoor learning environments: barriers and solutions as experienced by primary school teachers. Front Psychol 10:2919
74. van Dijk-Wesselius JE, Maas J, van Vugt M, van den Berg AE (2022) A comparison of children's play and non-play behavior before and after schoolyard greening monitored by video observations. J Environ Psychol 80. https://doi.org/10.1016/j.jenvp.2022.101760
75. Vanaken GJ, Danckaerts M (2018) Impact of green space exposure on children's and adolescents' mental health: a systematic review. Int J Environ Res Pub Health 15(12). https://doi.org/10.3390/ijerph15122668
76. Widernyski S et al (2017) The use of cooling centers to prevent heat-related illness: summary of evidence and strategies for implementation. Climate and health technical report series, centers for disease control and prevention
77. Wiek A, Redman A (2022) What do key competences in sustainability offer and how to use them. In: Vare P, Lausselet N, Rieckmann M (eds) Competences in education for sustainable development: critical perspectives. Springer, Switzerland, pp 27–34
78. Wolfram M (2016) Conceptualizing urban transformative capacity: a framework for research and policy. Cities 51:121–213
79. Zhang Y, Mavoa S, Zhao J, Raphael D, Smith M (2020) The association between green space and adolescents' mental well-being: a systematic review. Int J Environ Res Public Health 17(18):6640
80. Zhang ZKT, Stevenson KLM (2021) Exploring geographical, curricular, and demographic factors of nature use by children in urban schoolyards in Raleigh, NC, USA. Urban For Urban Greening 65. https://doi.org/10.1016/j.ufug.2021.127323
81. Zhang Z, Martin KL, Stevenson KT, Yao Y (2022) Equally green? Understanding the distribution of urban green infrastructure across student demographics in four public school districts in North Carolina, USA. Urban For Urban Greening 6:127434. https://doi.org/10.1016/j.ufug.2021.127434
82. Zhao S, Zhou S, Noonan DS (2019) Environmental justice and green schools—assessing students and communities' access to green schools. Soc Sci Q 100(6):2223–2239. https://doi.org/10.1111/ssqu.12715

Urban Transition in Monsoon Asian Megacity: Will Paddy Fields in the Rural–Urban Fringe Persist? Insight from Jakarta-Bandung Mega Urban Region, Indonesia

Didit Okta Pribadi, Khalid Saifullah, La Ode Syamsul Iman, Muhammad Nurdin, and Andi Syah Putra

Abstract Megacity emerges as urbanization occurs beyond the urban core towards adjacent peri-urban and rural regions. Unlike megacities in other parts of the world, Monsoon Asia megacities are expanding to highly populated rural areas where villagers mostly work on rice farming. Rice has become a major staple food as it is produced well under suitable climate and soil conditions in the Monsoon region. As it is labor intensive, incrementally, a strong sense of community and densely rural areas are grown. In such a situation, the expansion of Monsoon Asian megacities has led to the urban–rural collision, forming a specific feature called desakota where urban and rural land uses are intermingled in parallel with soaring population density. Paddy fields become preferable for urban development as the land is usually flat, inexpensive, and close to nearby cities, thus having better access to infrastructures and services. Some scholars argued that desakota is a temporary phenomenon during the urban transition. It is predicted that rice farming will vanish after the completion of urbanization. Contrarily, the loss of paddy fields has raised public and government concerns about maintaining rice production as the central pillar of food security. Accordingly, the Indonesian government established Law No 41/2009 to protect paddy fields from further conversion via spatial planning policy at the district/municipality level. Still, incorporating paddy fields in urban land use is problematic as it is still regarded as rural land use type. Besides, the multifunctional agricultural land, including paddy fields, that potentially support urban resiliency and sustainability is less understood. Paddy fields continue to decline via Land

D. O. Pribadi (✉)
Research Center for Behavioral and Circular Economics, National Research and Innovation Agency (BRIN), Jakarta, Indonesia
e-mail: didi017@brin.go.id; didit.pribadi@tum.de

D. O. Pribadi · K. Saifullah · L. O. S. Iman · M. Nurdin · A. S. Putra
Center for Regional Systems Analysis, Planning and Development (CRESTPENT/P4W), IPB University, Bogor, Indonesia

Z. Allam (ed.), *Sustainable Urban Transitions*, Urban Sustainability,
https://doi.org/10.1007/978-981-99-2695-4_7

Use Cover Changes (LUCC) which are not yet known in detail; thus, it is difficult to be controlled. This chapter aims to portray the process of rice field conversion in terms of landscape dynamics by using the Landscape Dynamic Typology model (LDTtools) applied to time series of land use maps 2017–2021 generated from satellite image analysis. The study was done in the North corridor of Jakarta Bandung Mega Urban Regions (JBMUR), a growing conurbation between Jakarta and Bandung Metropolitan Areas, consisting of three districts of rice producers, namely Bekasi, Karawang, and Purwakarta. The results show different types of rice field conversion, including (1) *Number of patches (NP) decremental by loss*, (2) *Fragmentation by loss*, and (3) *Loss*. The first one was likely caused by a larger scale of development supported by governments' policies. The second one likely occurred due to a lack of monitoring and control of land use changes, especially around big-scale development. The third one was likely caused by the declining profit of rice farming. These three patterns of paddy field conversion have threatened sustainable food production land (Lahan Pangan Berkelanjutan/LP2B) that has been planned for enhancing local and regional food security. Strategies and recommendations were formulated based on site-specific issues of paddy fields conversion in LP2B region. Maintaining and managing LP2B is important to support the resiliency and sustainability of future Monsoon Asia megacities.

Keywords Monsoon Asia megacities · JBMUR · Urban transition · Paddy field · LP2B · Landscape dynamics

1 Introduction

Asia continent has experienced the most accelerated expanding Megacities (i.e., cities with more than 10 million population) compared to other parts of the world [27]. From 34 megacities globally, Asia and the Pacific have 21 megacities [30]. Most of these megacities are located in deltas and coastal areas [2, 4]. Within this context, Asian megacities in the monsoon climate are the most vulnerable to flooding due to heavy rainfall intensity [4]. Tokyo, Shanghai, Beijing, Delhi, Karachi, Dhaka, Bangkok, Manila, Jakarta, etc., are some Asian Monsoon Megacities that are increasingly threatened by the more intense rainfall and long-term drought owing to global climate changes [26, 28]. It is exacerbated by the sea-level rise that exposes cities' populations to coastal flooding [3, 17].

Prior to the rise of megacities, monsoon delta areas have become suitable land for wet rice agriculture [18]. Fertile soil coupled with abundant water support higher rice production [18, 25], thus, people's need for rice as the main staple food can be satisfied. Rice farming is organized based on the water cycle, it utilizes excessive water in the rainy season for planting and growing rice. This is the people's way of turning flood plain into food production. In parallel, labor management is needed as much work should be done periodically, especially in field preparation, transplantation, weeding, and harvest [25]. Outside of those periods, farmers move to the nearby city to work and earn additional income thus, it tightens the link between cities and adjacent rural areas and supports employment for the city development [18, 25].

As rice farming is a labor-intensive business, agricultural-rural areas in Monsoon Asia are inhabited by a dense population [21]. Hence, unlike urban growth in western countries that expands to less dense rural areas, the urban expansion in Monsoon Asia has collided with populated rural areas. This formed a distinct feature called desakota as urban and rural land uses are intermingled [14]. Desakota is a term originating from the Indonesian language describing the coexistence of desa (rural) and kota (urban) in the peri-urban. In the planning context, desakota triggers a debate about whether it will vanish as replaced by the development of new urban centers in the peri-urban [10], or it will persist as a specific regional system in Monsoon Asia Megacities [15, 20].

In the case of the Jakarta Metropolitan Area (JMA), accelerated urban expansion followed by tremendous economic growth from the 70 s to the late 90 s converted large tracks of rice fields [7]. At that time, Asia has become the factory of the world as they compete to invite global industries for accelerating economic growth, especially by offering cheap labor [5]. In Indonesia, cheap labor was created by suppressing rice prices [29] thus, rice farming has turned into a less profitable business. As a result, rice farmers prefer to sell their land rather than run their farming business [7]. On the other hand, rice fields have become a preferable land use for urban expansion as they are flat, cheap, and have a better link with the nearby city.

The situation changed when Asia was suddenly hit by the economic crisis in 1998 and global climate change issues have gradually risen; thus, farming has gotten spaces for persisting in JMA. Farming saved urbanites from hunger during the time of crisis [20], while the vulnerability of cities amidst climate changes has increased the need for ecosystem services one of which can be provided by agricultural land [16, 20]. Particularly, besides food provision, paddy fields are capable of supporting the regional capacity to contain water, thus retaining the increasing flood risk in JMA [22].

Still, the continuity of urban expansion is uncontrolled due to economic growth-oriented development. Economic growth was pushed to compensate for an economic setback at the time of crisis. Simultaneously, dams and water channels were massively built to tackle flood risk. The extended urbanization leads to conurbation between two adjacent metropolitan areas, Jakarta and Bandung Metropolitan Area, forming Jakarta-Bandung Mega-Urban Region (JBMUR) [6, 21]. The expansion was led by the development of industrial estates which were then followed by residential and commercial areas [10]. Unlike the previous industrial development which was not supported by services and infrastructures, industrial estates got full support from the government. Toll road development, called Cipularang, is one of the infrastructures aimed to facilitate the extension of JMA urbanization, which then enhances the connection with Bandung Metropolitan Area [21].

The crucial issue arises as the urban corridor of JBMUR cuts across the primary land for rice production. Increasingly rice farming conversion has threatened food security not only in JBMUR but also in West Java Province [21]. Regarding the issue, the National Government has enacted Law Number 41/2009 to sustain food production land. It mandates district/municipality governments to draw sustainable food production land (Lahan Pangan Berkelanjutan/LP2B) in their spatial planning policy, especially for protecting paddy fields from further loss. However, this regulation is hard to implement due to accelerated urban growth and a lack of policy instruments to protect paddy fields. Farming is still regarded as a rural economic activity that gets less priority in district/municipality that experiences rapid urban transition.

In this chapter, we aim to portray the process of paddy field conversion and assess the future existence of paddy fields in the JBMUR corridor. It is critical as the corridor crosses one of the most important rice granaries areas in Indonesia. The urban transition should be distinctively managed by accommodating paddy fields as part of the urban land use. This plays a key as the growth of JBMUR is threatened by resiliency and sustainability issues. A comprehended knowledge of the dynamics of paddy fields in relation to urban growth and development is required. Thus, the government could design policy instruments to protect paddy fields as an integrated part of urban agendas to comprehensively address social-economy-ecological issues.

JBMUR has two corridors, the North and South corridors, where both cross the rice fields regions. However, the North corridor is more critical as the area has large productive land for rice farming. Besides, this corridor has a more accelerated urban expansion boosted by toll road development [21]. Therefore, we focus our study on the paddy fields dynamic in the North corridor of the JBMUR landscape.

2 Methods

2.1 Study Area

This study focuses on the North corridor of JBMUR, which is passed by the toll road Jakarta-Bandung. The area includes three districts, namely Bekasi, Karawang, and Purwakarta. It is located on the Northern coast of Java, known as one of the important rice granaries in Indonesia. Lowland and plain areas are dominant, except for hilly topography in Southeastern Purwakarta. The map of the study area is shown in Fig. 1.

2.2 Data

We used Sentinel-2 satellite imagery to produce land cover maps for 2017 and 2021. The period was chosen considering the availability and clarity of the images (i.e. free from cloud cover). Sentinel-2 imagery has 10 m spatial resolution for Blue (B2), Green (B3), Red (B4), and NIR (B8) Band, which can produce a semi-detailed map at a scale of 1:25.000. Sentinel-2 has 10 days of the temporal resolution taken by

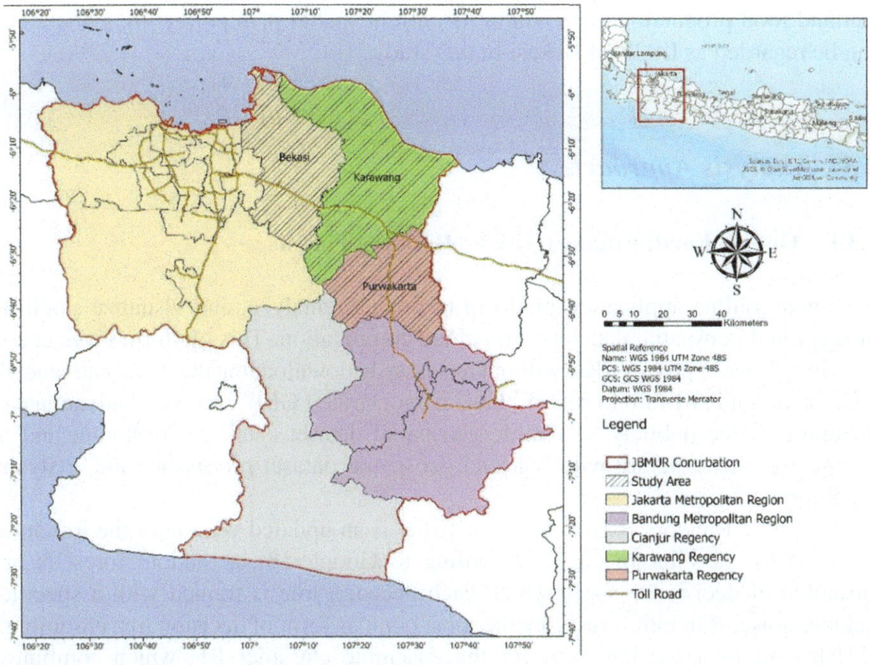

Fig. 1 The study area

one satellite or 5 days taken by two satellites, and it is useful for making a mosaic map to address cloud cover problems. The Sentinel-2 MSI (Level-1C) images were collected and classified by using Google Earth Engine (GEE).

Furthermore, we collected 6 GIS (Geographic Information System) datasets from various institutions as follows: (1) Administration base map 2021 at a scale of 1:25,000 from Geospatial Information Agency/BIG; (2) Toll road map 2022 from Ministry of Public Works And Housing/KemenPUPR; (3) National road map from KemenPUPR; (4) Spatial plan maps (Rencana Tata Ruang Wilayah/RTRW) of Bekasi, Karawang, and Purwakarta District from Ministry of Agrarian Affairs and Spatial Planning/National Land Agency (KemenATR/BPN); (5) Paddy field map dataset of 2013 from KemenATR/BPN; and (6) Sustainable food agriculture land (LP2B) map from each District Government.

Particularly for LP2B, the area contains paddy fields that should be protected from land conversion to enhance food security. It has been enacted by each district government via spatial plan policy (RTRW). The map of LP2B in Karawang was extracted from the local government web-GIS under the regulation of Karawang District Number 1/2018. While LP2B maps in Purwakarta and Bekasi District were extracted from their spatial plan/RTRW under the regulation of Purwakarta District Number 11/2012 and the regulation of Bekasi District Number 12/2011. Particularly for Bekasi District, actually, LP2B has not been enacted yet. However, there is a wetland food production zone within its current spatial plan policy; thus, the zone can be regarded as LP2B of Bekasi in this study.

2.3 Analysis Approach

2.3.1 Google Earth Engine (GEE)—Random Forest

GEE is an online application platform to process, analyze, and visualize satellite imagery and geospatial data based on cloud computation. This platform gives users the capabilities to process data without necessarily downloading the data or advanced computational hardware to run. [9] have demonstrated GEE's powerful advantages, including a free publicly accessible geospatial dataset catalog and storage,and a computing infrastructure with big data geospatial dataset processing and analysis capabilities.

The Smile Random Forest (SRF) classifier is an updated version of the Random Forest (RF) classifier in GEE. According to Google [8], a random forest is an ensemble of decision trees in which each decision tree is trained with a specific random noise. Random forests are the most popular form of decision tree ensemble. [24] has summarized the basis for the ensemble classifier RF, which combines output from multiple decision trees to decide the label for new input data based on maximum votes. There are a few advantages of RF, including its robustness to noise and outliers,better performance than other classifiers which use ensemble methods such as bagging and boosting; and good classification results when used in various applications such as urban landscape classification and land cover classification on multi-temporal and multi-frequency SAR data [24].

In this study, a supervised classification method with a smile random forest classifier was used to produce 5 land cover classes: (1) Water body, (2) Built-up area, (3) Open space, (4) Paddy field, and (5) Tree vegetation. Shetty [24] and Utomo et al. [31] have analyzed the good accuracy performance of random forest classifiers in GEE for Land Use Land Cover (LULC) classification. It was also reflected in our series of classification results that range from 94 to 99% calculated from the GEE confusion matrix function. The resulting land cover maps were then combined and post-processed with a baseline map of paddy fields 2013 obtained from KemenATR/BPN. The map was used to correct and validate paddy fields in our land use maps, especially to distinguish paddy fields from fishpond aquaculture that in the image shows similar appearances, located in the coastal area of Bekasi and Karawang District.

2.3.2 LDTtool

LDTtool is a toolbox in Arcgis developed by Machado et al. [13]. It is applied to analyze the changes in the landscape over time by considering landscape composition and configuration. Landscape composition exhibits the kind of land use/land cover type that exists in the landscape and its amount. While landscape configuration exhibits the shape and location of land use/land cover type. This analysis was based on the change of composition and configuration of single land use/land cover within a certain period of time. There are two variables that are calculated, including the change in areas (A) and the number of patches (NP). The results of the analysis are classified into 9 categories, as can be seen in Table 1.

In running LDTtool analysis, we should determine the size of the regular grid to minimize patch intersection, but the resolution should be adequate for studying paddy field dynamics in the JBMUR corridor. Minimizing patch intersections aims to avoid the increasing number of patches because of cutting by grids. We set the minimum patch size of 5000 m^2 as the average area of rice farming in JMA (part of the JBMUR region) is equal to or less than 0.5 ha [20]. In addition, we used the grid or square size of 500 × 500 m or 25 ha (ha) to minimize patch intersection.

As a result, besides the dynamic landscape composition and configuration of paddy fields, we can identify the number of squares and areas for every type of gain and loss resulting from the LDTtool (see Table 1). As noted, a square is a rectangle area of 25 ha, while areas are wide of paddy fields patches within squares. Then an overlay would be done between the LDTtool results and the map of LP2B. It is important to detect what kind of conversion risk is faced by LP2B across the landscape. This can be used as a basis for formulating appropriate policies to sustain LP2B.

Table 1 LDTtool results

Type	Area (A)	Number of patches (NP)	Interpretation
A	$\Delta A = 0$	$\Delta NP = 0$	No change
B	$\Delta A = 0$	$\Delta NP > 0$	Fragmentation per se
C	$\Delta A = 0$	$\Delta NP < 0$	Aggregation per se
D	$\Delta A > 0$	$\Delta NP = 0$	Gain
E	$\Delta A < 0$	$\Delta NP = 0$	Loss
F	$\Delta A > 0$	$\Delta NP > 0$	NP increment by gain
G	$\Delta A > 0$	$\Delta NP < 0$	Aggregation by gain (NP decrement by gain)
H	$\Delta A < 0$	$\Delta NP < 0$	NP decrement by loss
I	$\Delta A < 0$	$\Delta NP > 0$	Fragmentation by loss (NP increment by loss)

3 Results

3.1 Land Use Cover Changes (LUCC) in the North Corridor of JBMUR

From 2017 to 2021, regarding 5 land use classes, Bekasi District experienced gains in fishpond aquaculture and built-up area and losses of paddy fields and open spaces. While in Karawang District, fishpond aquaculture and open spaces have undergone gains and losses, but for open spaces, losses were more dominant. Built-up areas became the only land use that was extended in Karawang, but tree vegetation, paddy fields, and water body have declined. In Purwakarta District, tree vegetation and open spaces have experienced gains and losses. In addition, paddy fields have been converted, and contrarily built-up areas only experienced gains. In detail, the results of LUCC analysis can be seen in Fig. 2.

A more detailed analysis was conducted to assess the change in paddy fields in three districts (see Fig. 3). In Bekasi, paddy fields were mainly converted into built-up areas, then fishpond aquaculture. In Karawang, paddy fields were mainly converted into built-up areas, followed by open space and fishpond aquaculture. While in Purwakarta, the majority of paddy fields were converted into built-up areas. It can be highlighted that paddy fields have experienced largely losses due to built-up area expansion. As noted, paddy fields conversion to fishpond aquaculture likely occurred as an impact of the sea-level rise and shoreline retreat on the Northern coast of Java [12, 26].

Regarding the spatial distribution of paddy field conversion, Fig. 4a shows that the losses of paddy fields in Bekasi occurred in the western part that shares boundaries with Jakarta as the city core of JMA. Furthermore, paddy field conversion was also found along the side of the toll road, which continues to Karawang (see Fig. 4b).

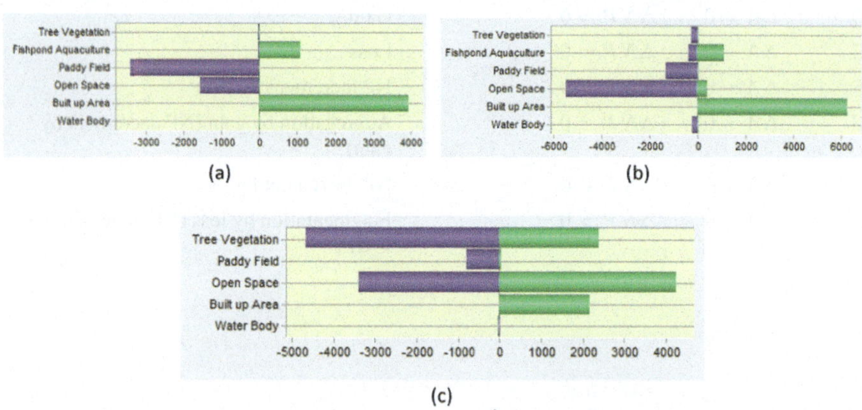

Fig. 2 Gains and *losses* of 5 land use classes (hectares) between 2017 and 2021 in Bekasi (**a**), Karawang (**b**), and Purwakarta district (**c**)

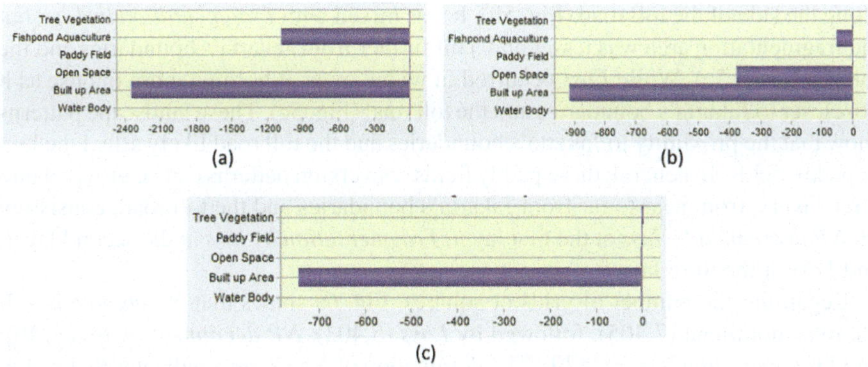

Fig. 3 Contribution to net change in paddy field (hectares) in Bekasi (**a**), Karawang (**b**), dan Purwakarta district (**c**)

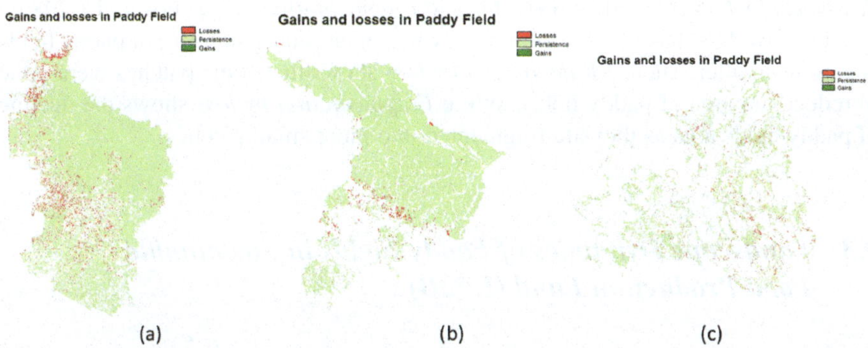

Fig. 4 Maps of paddy fields conversion 2017–2021 in Bekasi (**a**), Karawang (**b**), and Purwakarta district (**c**)

While in Purwakarta, paddy field conversion was more randomly distributed. Still, the losses along the side of the toll road were dominant (Fig. 4c). This indicated that the extended urbanization in JMA mainly occurred in relation with road network development forming a linear or ribbon pattern.

3.2 *Landscape Dynamics of Paddy Fields*

The results of LDTtools show that paddy fields in the North corridor of JBMUR have experienced loss with different landscape compositions and configurations. According to Fig. 5a, there are three types of paddy field conversion, including (1) *NP decrement by loss*, (2) *Fragmentation by loss*, and (3) *Loss. NP decrement by loss* was mainly found in the western part of Bekasi, close to Jakarta's boundaries, and along the side of the toll road (Fig. 5b). It was mixed with *Fragmentation by loss*, but the fragmentation area was lesser and a bit further from Jakarta's boundaries and the toll road (Fig. 5c). While *Loss* occurred in wider areas, it became more intense as it got closer to Jakarta's boundaries and the toll road (Fig. 5c). These landscape patterns show that the proximity to Jakarta's boundaries and the toll road likely affect the loss of paddy fields. In general, these paddy fields conversion patterns can be grouped into three layers, from near to far from Jakarta's boundaries and the toll road, consisting of: *NP decrement by loss* at the first layer, *Fragmentation by loss* at the second layer, and *Loss* at the third layer.

Regarding the number of grids or squares, Fig. 6a shows that *No change* is still the most dominant (7,405), followed by *Loss* (5,403), *NP decrement by loss* (740), and *Fragmentation by loss* (429). The domination of *No Change* indicates that paddy fields in 7,405 squares have a lower risk of being converted. Whereas paddy fields in 5,402, 740, and 429 squares were threatened by *Loss*, *NP decrement by loss*, and *Fragmentation by loss*, respectively. Regarding the area, *Loss* covered 2,984.64 ha, *NP decrement by loss* covered 1,565.67 ha, and *Fragmentation by loss* covered 1,298.65 (see Fig. 6b). *Loss* has become dominant, which means the patch size of paddy fields is getting smaller. Then, *NP decrement by loss* shows that some patches were gone to reduce the area of paddy fields. While *Fragmentation by loss* shows the decline of paddy fields area as they are fragmented into more small patches.

3.3 Landscape Dynamics of Paddy Fields in Sustainable Food Production Land (LP2B)

As mentioned in the method, an overlay between LDTtool results and LP2B (i.e. protected zone of paddy fields) is important to assess the risk of LP2B loss in the JBMUR corridor. Figure 7 shows that LP2B in Bekasi District is allocated further from Jakarta and the toll road. Therefore, *Loss*, as the third layer of paddy field conversion pattern, becomes the higher risk faced by LP2B. This can be seen by squares (Fig. 7a), area (Fig. 7b), and the map (Fig. 7c) where the *Loss* has become dominant. However, some LP2B areas are located close to the National road near the toll road. Therefore, *Fragmentation by loss* was still found as a risk that should be of concern in this LP2B area, although it is much smaller than *Loss* conversion type.

Compared to other districts in the North corridor of JBMUR, Karawang is the most important rice granary on the Northern coast of West Java Province. Most of LP2B was allocated in the North of the National road and toll road (see Fig. 8c). Small parts of LP2B were located in the middle of the district, between the National road and toll road, while others were located further in the South. The higher risk faced by LP2B in Karawang is *Loss* which is randomly distributed. However, as some LP2B were located close to the toll road, other risks should be of concern especially *NP decremental by loss* (first layer risk) and *Fragmentation by loss* (second layer risk) (see Fig. 8a, b).

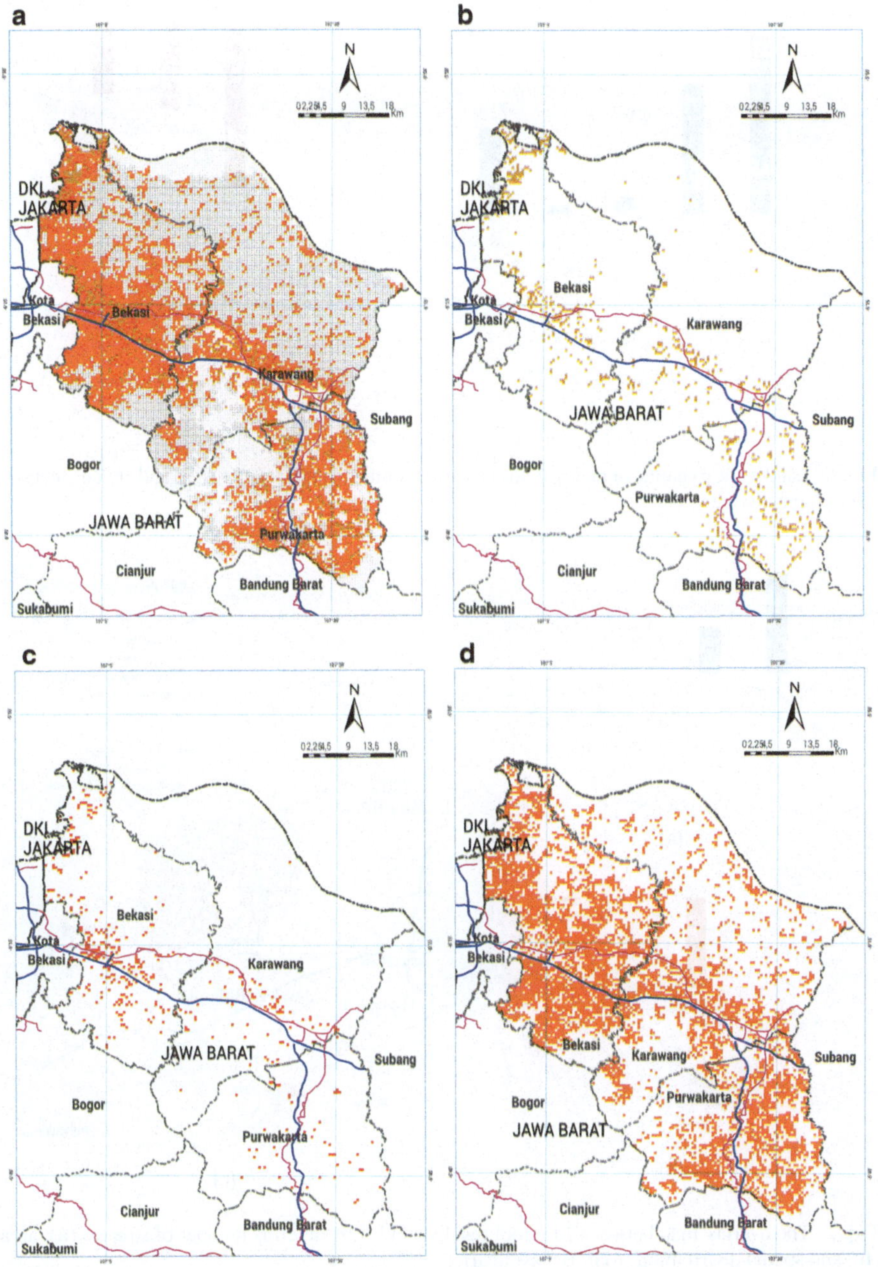

Fig. 5 The map of LDTtool results for all districts (**a**), separated into the map of *NP decrement by loss* (**b**), *fragmentation by loss* (**c**), and *loss* (**d**)

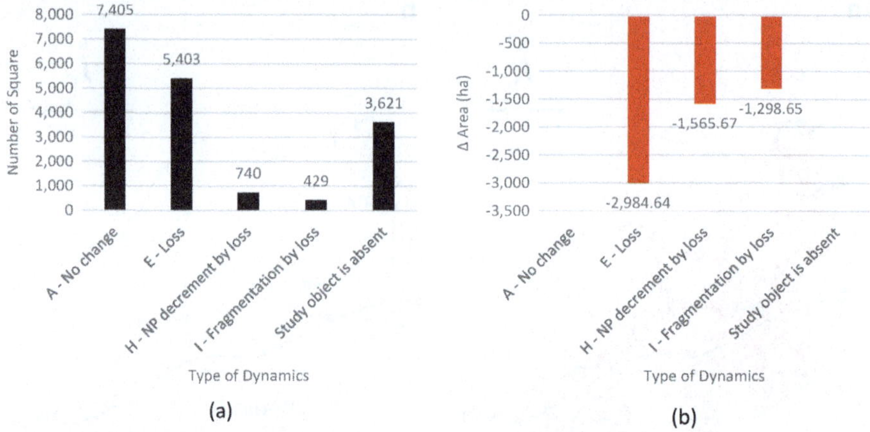

Fig. 6 Landscape dynamic of paddy fields based on numbers of grid/squares (**a**) and area in hectares (**b**) in three districts

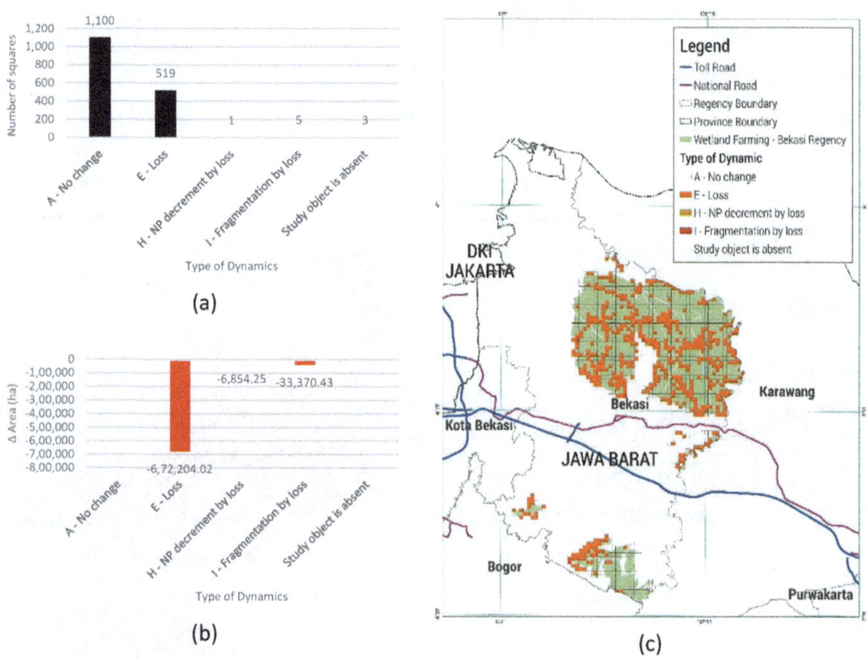

Fig. 7 The overlay map between LDTtool results and LP2B in terms number of squares (**a**), area (**b**), and spatial distribution (**c**) in Bekasi district

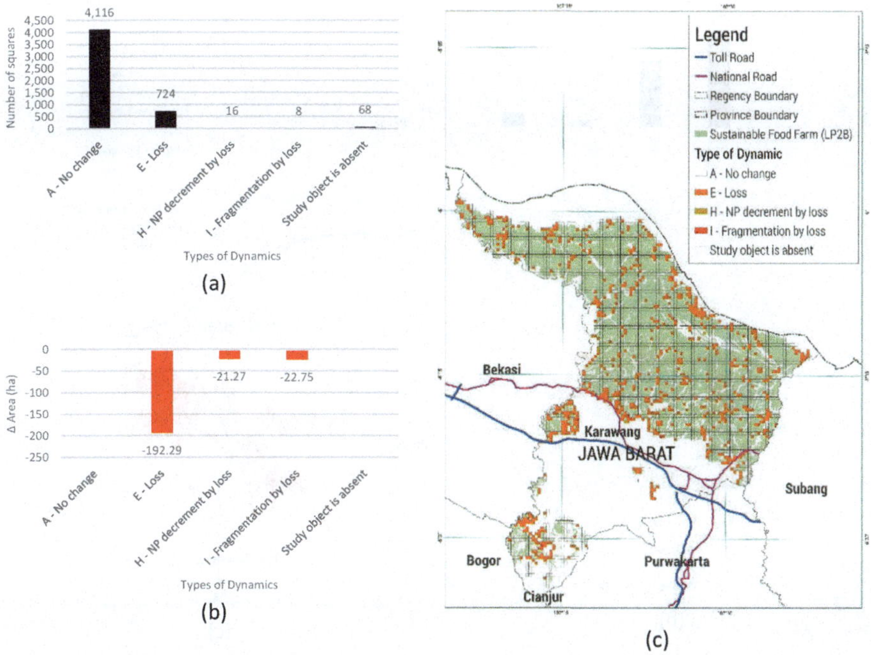

Fig. 8 the overlay map between LDTtool results and LP2B in terms of number of squares (**a**), area (**b**), and spatial distribution (**c**) in Karawang district

Purwakarta District has more intense risks in the allocated area for LP2B compared to Bekasi and Karawang (see Fig. 9c). There are no LP2B areas that were free from the risk of paddy field conversion. Purwakarta's LP2B has experienced *Loss* followed by *NP decrement by loss* and smaller *Fragmentation by loss* in terms of squares (see Fig. 9a). Whereas regarding the area, *NP decrement by loss* has become more dominant, followed by *Loss* and small *Fragmentation by loss* (Fig. 9b). Particularly for the *Loss*, the number of squares is higher than the area, showing that the *Loss* was spread to more number of squares although the converted area was smaller. Contrarily, *NP decrement by loss* shows a high value in terms of squares and area. It can be highlighted that *Loss* and *NP decremental by loss* should be of concern to maintain LP2B in Purwakarta.

4 Discussion

LUCC analysis in the North corridor of JBMUR shows that paddy fields were mainly converted into built-up areas; thus, it indicates the expansion of urban areas in this corridor. On the other hand, paddy fields on the Northern coast of Java, especially in the Northern Bekasi and Karawang, were converted into fishpond aquaculture.

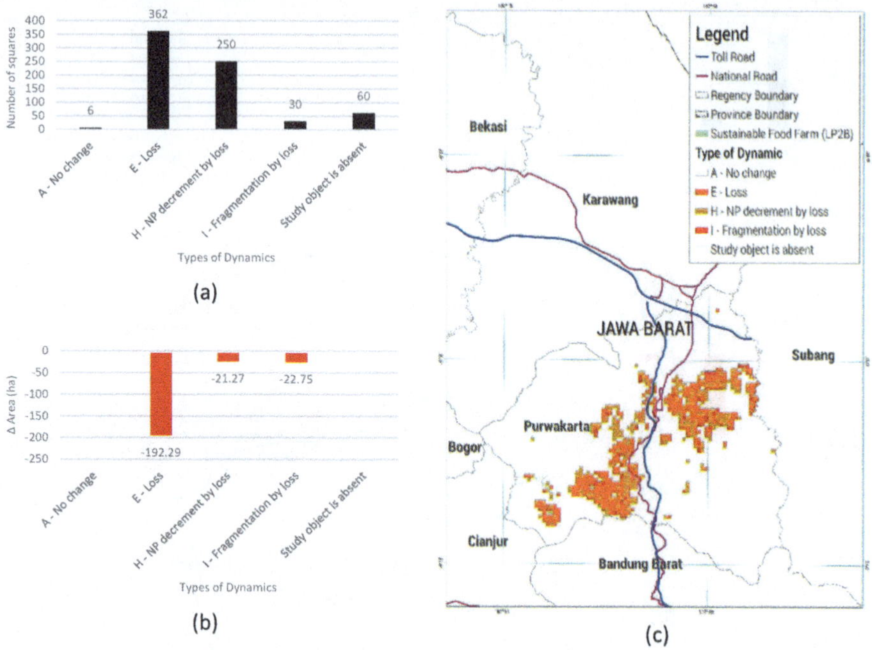

Fig. 9 The overlay map between LDTtool results and LP2B in terms of number of squares (**a**), area (**b**), and spatial distribution (**c**) in Purwakarta district

It indicates the effect of climate changes where the rise of sea level has expanded the suitable areas for fishpond aquaculture. Within this situation, maintaining paddy fields has got a lot of pressure either from the terrestrial or coastal.

Regarding paddy fields conversion to built-up areas, it mainly occurred in the Western Bekasi, which shares boundaries with Jakarta, the capital city and the core city of JMA, and along the side of the toll road. It shows the influence of urban sprawling from Jakarta, while the toll road development also triggered linear or ribbon urban development in the JBMUR corridor [11]. The toll road, namely Cipularang, which connected Jakarta-Bandung, first aimed to boost industrial estate development [1, 23]. It was expected the toll road would facilitate the needs of the industrial sector for transportation and limit LUCC along the roadside. However, as explained by [10], the emergence of industrial estates was followed by the development of residential and commercial areas, that in turn triggered the conversion of paddy fields.

The results of LDTtool can be used to explain the process of paddy field conversion in the study area based on the three layers of different landscape dynamic types. Near Jakarta and the toll road, paddy field conversion was dominated by *NP decremental by loss*. It means paddy fields experienced loss in areas and numbers of patches. Thus, it can be presumed that the loss was caused by big-scale development such as industrial estates. Certainly, this kind of development was supported by a government permit.

Therefore, spatial planning policy plays a key in controlling paddy field conversion in such areas.

A bit further from the first layer, there was the second layer, where paddy fields have experienced *Fragmented by loss*. Here, paddy fields lost their areas and were fragmented into more number of smaller patches. This mainly occurred around the industrial estates where residential and commercial areas were haphazardly developed. Normally, the development has a smaller scale, thus causing fragmentation in paddy fields. In this case, monitoring and controlling land use changes by the government play a key in protecting paddy fields.

Finally, there was the third layer that covered a larger area from near to further away from Jakarta and the toll road. In this case, paddy fields have experienced *Loss* meaning the area was reduced, but the number of patches still remains. Most likely, it was caused by decreasing profit of the rice farming business; thus, some lands were gradually converted for other land use. In this case, the loss was incremental and hard to be monitored. Government programs to support the economic feasibility of rice farming are certainly required.

As every type of paddy field conversion was driven by different factors, policies to protect LP2B in each district should consider this matter. In Bekasi, LP2B has been allocated in an area far from Jakarta and the toll road; therefore, the risk of conversion was lower. The most dominant risk was *Loss*; thus, government support to develop profitable rice farming is important. It can be done not only by increasing production and productivity but also by developing multifunctionality. As explained by [19], non-commodity products and services can be developed, such as agro-tourism, agro-education, etc.,furthermore, non-marketable public benefits can be designed thus, ecosystem services produced by rice farming can be monetary valued and compensated by the government.

In Karawang, LP2B was allocated in the North of the toll road; thus, most of the paddy fields have a lower risk of conversion. *Loss* has become dominant; thus, again, the government's support to enhance the rice farming businesses is crucial. Production, productivity, and multifunctionality of rice farming have to be developed. Still, *NP decremental by loss* and *Fragmentation by loss* were found as a small portion of LP2B was located close to the toll road. This implies that formulating spatial planning policy and controlling land use changes should be improved for LP2B protection.

In Purwakarta, all the area of LP2B was threatened by the conversion risk. Compared to Bekasi and Karawang, LP2B in Purwakarta is the most vulnerable to being converted. Both *Loss* and *NP decremental by loss* were dominant. Therefore government support and improved spatial planning policy are needed to protect LP2B. Besides, monitoring and controlling land use changes are still needed as a small portion of LP2B was threatened by the risk of *Fragmentation by loss*.

5 Conclusion

This study highlights the proximity to big cities (i.e. Jakarta) and the road network (i.e. toll road) should be of concern as factors that drive paddy field conversion. There were three types of paddy field conversion consisting of *NP decremental by loss*, *Fragmentation by loss*, and *Loss*. The first one was mainly caused by government policy, such as planning for building industrial estates. The second one was mainly caused by lacking monitoring and controlling land use changes. The third one was mainly caused by less profitable rice farming.

In dealing with each type of those landscape dynamics, different approaches are needed. To minimize *NP decremental by loss*, spatial planning policies play a key in preventing large-scale development on paddy fields. So, it depends on urban planners and the urban government to make the decision. To reduce *Fragmentation by loss*, monitoring and controlling land use changes should be enhanced. While to lessen *Loss*, government support should be given, in terms of technical or policy aspects, to increase the economic feasibility of the rice farming business.

The results of this study imply that protecting LP2B needs site-specific approaches that consider the risk of paddy field conversion due to urbanization. Landscape dynamics can be used as an instrument to assess the risk. It will ease the government and other related stakeholders in Monsoon Asia Megacities to maintain LP2B as well as food security and agricultural-based ecosystem services, which are urgently required for supporting urban resiliency and sustainability. Still, this study has a limitation as socio-economic variables are not integrated yet into the model. Further research can be conducted by incorporating socio-economic variables; thus, we can comprehensively define incentives and disincentives policy to protect LP2B (paddy fields) in the future landscape of Monsoon Asia Megacities.

References

1. Anas R, Tamin OZ, Tamin RZ, Wibowo SS (2017) Measuring regional economic impact of cipularang toll road investments: using an input-output model (case study: Bandung district). Int J Civ Eng Technol (IJCIET) 8(10):796–804
2. Barragán JM, de Andrés M (2015) Analysis and trends of the world's coastal cities and agglomerations. Ocean Coast Manag 114:11–20. https://doi.org/10.1016/j.ocecoaman.2015.06.0
3. Cao A, Esteban M, Valenzuela VPB, Onuki M, Takagi H, Thao ND, Tsuchiya N (2021). Future of Asian deltaic megacities under sea level rise and land subsidence: current adaptation pathways for Tokyo, Jakarta, Manila, and Ho Chi Minh City. In: Current opinion in environmental sustainability, vol 50. Elsevier B.V., pp 87–97. https://doi.org/10.1016/j.cosust.2021.02.010
4. Chan FKS, Mitchell G, Adekola O, McDonald A (2012) Flood risk in Asia's urban mega-deltas: drivers, impacts, and response. Environ Urban Asia 3:41. https://doi.org/10.1177/097542531200300103
5. Dahiya B (2012) Cities in Asia, 2012: demographics, economics, poverty, environment, and governance. Cities 29(Suppl. 2). https://doi.org/10.1016/j.cities.2012.06.013

6. Firman T (2009) The continuity and change in mega-urbanization in Indonesia: a survey of Jakarta-Bandung region (JBR) development. Habitat Int 33:327–339. https://doi.org/10.1016/j.habitatint.2008.08.005
7. Firman T (2000) Rural to urban land conversion in Indonesia during boom and bust periods. Land Use Policy 17:13–20
8. Google Developers (2022) Decision Forests. In: Google machine learning education: advanced courses. Google Developers. https://developers.google.com/machine-learning/decision-forests. Accessed 30 Jul 2022
9. Gorelick N, Hancher M, Dixon M, Ilyushchenko S, Thau D, Moore R (2017) Google earth engine: planetary-scale geospatial analysis for everyone. Remote Sens Environ 202:18–27. https://doi.org/10.1016/j.rse.2017.06.031
10. Hudalah D, Firman T (2012) Beyond property: industrial estates and post-suburban transformation in Jakarta Metropolitan Region. Cities 29(1):40–48. https://doi.org/10.1016/j.cities.2011.07.003
11. Jatayu A, Rustiadi E, Pribadi DO (2020) A quantitative approach to characterizing the changes and managing urban form for sustaining the suburb of a mega-urban region: the case of north cianjur. Sustainability (Switzerland) 12(19):1–21. https://doi.org/10.3390/su12198085
12. Kartika FDS, Helmi M, Amirudin (2019). Meta-analysis of community's adaptation pattern with tidal flood in Pekalongan City, Central Java, Indonesia. https://doi.org/10.1051/e3sconf/201
13. Sousa-Neves N (2020) LDTtool: a toolbox to assess landscape dynamics.Environ Model Softw 133. https://doi.org/10.1016/j.envsoft.2020.104847
14. McGee TG (1991) The emergence of Desakota regions in Asia: expanding a hypothesis. In: Ginsberg N, Koppel B, McGee TG (eds) The extended metropolis: settlement transition in Asia. University of Hawaii Press, Honolulu
15. McGee TG (2005) Distinctive urbanization in the peri-urban regions of East and Southeast Asia: Renewing the debate, Jurnal Perencanaan Wilayah dan Kota16(1):39–55
16. McGee TG (2010) Building liveable cities in Asia in the twenty first century research and policy challenges for the urban future of Asia. Malays J Enviro Manag 11(1):14–28
17. Nicholls RJ, Hanson S, Herweijer C, Patmore N, Hallegatte S, Corfee-Morlot J, Chateau J, Muir-Wood R (2007) Ranking of the world's cities most exposed to coastal flooding today and in the future. https://climateadapt.eea.europa.eu/metadata/publications/ranking-of-the-worlds-cities-to-coastal-flooding/11240357. Accessed 25 Jul 2022
18. Okabe A (2017) Dynamic spaces with subjective depth. The public space in monsoon Asia. Kult-Ur Revista Interdisciplinària Sobre La Cultura de La Ciutat 4(7):151–164. https://doi.org/10.6035/kult-ur.2017.4.7.6
19. Pribadi DO, Zasada I, Müller K, Pauleit S (2017) Multifunctional adaption of farmers as response to urban growth in the Jabodetabek Metropolitan Area, Indonesia. J Rural Stud 55:100–111. https://doi.org/10.1016/j.jrurstud.2017.08.001
20. Pribadi DO, Pauleit S (2015) The dynamics of peri-urban agriculture during rapid urbanization of Jabodetabek Metropolitan Area. Land Use Policy 48:13–24. https://doi.org/10.1016/j.landusepol.2015.05.009
21. Rustiadi E, Pravitasari AE, Setiawan Y, Mulya SP, Pribadi DO, Tsutsumida N (2021) Impact of continuous Jakarta megacity urban expansion on the formation of the Jakarta-Bandung conurbation over the rice farm regions. Cities111. https://doi.org/10.1016/j.cities.2020.103000
22. Rustiadi E, Pribadi DO, Pravitasari AE, Indraprahasta GS, Iman LO (2015) Jabodetabek megacity: from city development towards urban complex management system. In: Singh RB (ed) Urban development challenges, risk, and resilience in Asian Mega Cities. Springer, pp 421–445
23. Sembiring IS, Anas R (2019) The economic benefit of toll road investment on the performance of the industry sectors in West Java Province. In: IOP conference series: materials science and engineering, vol 615(1). https://doi.org/10.1088/1757-899X/615/1/012038
24. Shetty S (2019) Analysis of machine learning classifiers for LULC classification on google earth engine. Thesis, Faculty of geo-information science and earth observation, University of Twente. http://essay.utwente.nl/83543/1/shetty.pdf

25. Smith ML, Mohanty RK (2018) Monsoons, rice production, and urban growth: the microscale management of 'too much' water. Holocene 28(8):1325–1333. https://doi.org/10.1177/0959683618771497
26. Sutrisno D, Darmawan M, Rahadiati A, Helmi M, Yusmur A, Hashim M, Shih PTY, Qin R, Zhang L (2021) Spatial-planning-based ecosystem adaptation (SPBEA): a concept and modeling of prone shoreline retreat areas. ISPRS Int J Geo-Inf 10(3). https://doi.org/10.3390/ijgi10030176
27. Swerts E, Denis E (2015) Megacities: the Asian era. In: Singh RB (ed) Urban development challenges, risk, and resilience in Asian Mega Cities. Springer, pp 1–28
28. The Esri Story Map Team (2021) The age of megacities. https://storymaps.arcgis.com/stories/a900831b442e43c79cf9eeb399d5440f. Accessed 25 Jul 2022
29. Timmer CP (2004) Food security in Indonesia: Current challenges and the long-run outlook (November 12, 2004). Center for global development working paper No. 48, Available at SSRN: https://ssrn.com/abstract=1112807 or https://dx.doi.org/10.2139/ssrn.1112807
30. UN-Habitat (2020) Global state of metropolis 2020–population data booklet. https://unhabitat.org/sites/default/files/2020/09/gsm-population-data-booklet-2020_3.pdf. Accessed 25 Jul 2022
31. Utomo DP, Handayani T, Susiloningtyas D, Mansessa MDM (2021). The spatial dynamics of mangrove forest in the Alas Purwo Banyuwangi National Park marine tourism area using remote sensing images. In: IOP conference series: earth and environmental science, vol 771, 012012. https://doi.org/10.1088/1755-1315/771/1/012012
32. Yoshida K, Azechi I, Hariya R, Tanaka K, Noda K, Oki K, Hongo C, Honma K, Maki M, Shirakawa H (2013) Future water availability in the asian monsoon region: A case study in Indonesia

Controlling the Valves: Dealing with Conflict in a Transition Initiative for Public–Private Water Governance in Amsterdam

E. A. Van Der Heijden, B. C. Van Mierlo, S. J. H. Majoor, and P. J. Beers

Keywords Urban experimentation · Transition initiative · Pluvial flood · Conflict · Learning · System change · Discourse analysis

1 Introduction

Over the past 10 years, municipalities in the Netherlands, as in many other cities worldwide, have often initiated and participated in multi-actor alliances to address their city's societal and sustainability challenges. To test new solutions and distribute responsibility on themes such as heat-stress prevention, low-carbon mobility and energy-positive districts [30, 33], many of these alliances have made use of cross-disciplinary collaboration. In this chapter, we examine one such multi-actor alliance established to explore an urban sustainability transition towards a 'rainproof' city: backed by a European subsidy, RESILIO aimed to upscale solutions for smart water retention on green urban roofs, a concept referred to as 'Blue-Green' (BG) rooftops. In addition to the complex technological and financial challenges, this innovation touches upon important governance questions as it requires rethinking accountability and responsibility between public and private actors in urban water management.

E. A. Van Der Heijden (✉) · B. C. Van Mierlo
Knowledge, Technology and Innovation, Wageningen University, Wageningen, The Netherlands
e-mail: e.a.van.der.heijden@hva.nl

E. A. Van Der Heijden · S. J. H. Majoor
Amsterdam University of Applied Science, Amsterdam, The Netherlands

P. J. Beers
Dutch Research Institute for Transitions, Erasmus University Rotterdam, Rotterdam, The Netherlands

Z. Allam (ed.), *Sustainable Urban Transitions*, Urban Sustainability,
https://doi.org/10.1007/978-981-99-2695-4_8

In July 2019, the partners of RESILIO—including the municipality, the local water authority, private technology enablers and social housing associations (i.e., the roof owners)—organised a meeting they termed a 'Hard Talk'. RESILIO had been up and running for six months and some participants were troubled by the neglect of fundamental issues in their collaboration. One of these issues, constituting a central theme throughout the Hard Talk, concerned control of the BG roofs' 'smart valves', which could be opened or closed to unburden the municipal sewerage system in the event of heavy rainfall. Hoping to encourage the future involvement of both new property owners and valve (service) providers (i.e., the technology enablers), the water authority strongly insisted on the development of an open-source support system to control the valves. In response, the technology enablers expressed reluctance to endorse open-source development plans, emphasising the quality of their existing software, while most of the other partners favoured an open-source system facilitated by the water authority as a means of creating balance between the various roof-water management partners. Three years later, in the autumn of 2022, the final outcome failed to live up to the partners' initial intentions: while the code of the support system had been made freely available, the system itself was hosted by one of the private technology enablers involved in RESILIO. Throughout this chapter, we focus on the process that brought the partners to this final result, analysing participants' difficulty in engaging with potentially conflicting interests by closely observing interactions between RESILIO partners in the three years following the Hard Talk.

During this influential Hard Talk, partners showed awareness of the need to discuss both the alliance's fundamental issues and the participating actors' potentially conflicting positions. This awareness aligns with more recent scholarly work on the importance of disagreement and dissonance in transition-oriented learning processes [8, 24]. While actors' abilities to identify and engage in potential conflict has been established as a crucial success factor in sustainability transitions, participants' strategies for enacting such disagreements, especially in collaborative and/ or consensus-oriented settings, have thus far been empirically understudied. This limits our understanding of how conflict can play a constructive role in the learning process between collaborative participants and be productive in urban sustainability transitions. We therefore examine the following central research question:

How do participants handle conflict that emerges in the context of an urban sustainability transition initiative?

To answer this question, we first introduce a discourse analytical framework in Sect. 2 that allows us examine the interaction in sustainability transition initiatives from an interpretative perspective. Discourse analysis can help to reveal opposing positions between partners and disclose discursive strategies of partners to influence the course of the initiative. Next, we use Sect. 3 to elaborate on the idiosyncrasies of RESILIO as our research case and to introduce its Decision Support System, i.e., the software application developed to control the smart valves on BG roofs to mitigate urban pluvial flooding. Section 4 presents the results of our analysis of the interactions between RESILIO partners in conflict regarding control of the valves. Finally, we conclude in Sects. 5 and 6 with a discussion of the value of our approach for understanding and fostering learning in urban transition initiatives.

2 Theoretical Background

2.1 A Discourse Analytical Perspective on Conflict: Combining Two Approaches

Sustainability transition initiatives involve multi-actor collaborations in which experimentation with technical and social innovations aims to contribute to structural change towards a more sustainable system [11, 37]. The learning process of these initiatives is often understood as an ongoing interaction of meaning making between persons and their environment [24] in which participants negotiate the meaning of their knowledge, values, interests and responsibilities [4, 37]. Discourse analysis (DA) offers several approaches that enable us to relate societal-level system change to meaning-making activities at the level of social interaction within the initiatives. Furthermore, discourse analysis can aid in revealing an initiative's often-hidden and sensitive layers of conflict [18]. In this chapter, we combine two discourse analytical approaches to help us focus on conflict in the meaning-making process between participants. The first of these is a DA approach based on the work of Laclau and Mouffe which poses that contestation lies at the heart of societal change [15, 19]. The second is a discursive psychological approach that focusses both on actors' strategies for interacting in conversation and on how this influences the direction of their learning process [21, 31].

2.2 Conflict as Incompatible Positions

The first approach understands discourse as complex structures of meaning that constitute social orders. Such discourses are never neutral, as they frame reality and transform certain aspects of reality into comprehensible narratives [35]. According to [19], discourses become dominant when they (temporarily) succeed in suppressing alternative discourses, which represent reality differently, making the essence of conflict the struggle over power between dominant and alternative discourses (2001). This understanding of conflict is reflected in the concepts of sociotechnical regimes and niches as used in transition studies literature [16], which often refers to dominant and alternative societal practices and structures. However, Laclau and Mouffe take a more political understanding of conflict in societal change, emphasising ineradicable differences in the identities, values and stakes between competing discourses. Reflected by their conception of societal change as a process of contestation in which positions of discursive dominance are continuously at stake, Laclau and Mouffe [22, p. 756] assert that '*we have to accept that every consensus exists as a temporary result of a provisional hegemony, as a stabilisation of power, and that it always entails some form of exclusion.…The idea that power could be dissolved through a rational debate and that legitimacy could be based on pure rationality are illusions, which can endanger democratic institutions.*' Conflict must therefore be understood not only as a struggle over power, but also entails a painful struggle between incompatible stakes, norms, values, routines and other matters closely entwined with the identities of the participants in conflicting discourses. The pretence of 'rationality' in debate and 'consensus' in collaboration can lead, then, to the denial of incompatibilities between, and the underestimation of losses for, certain actors within an initiative [5, 17].

Drawing on this theory of Laclau and Mouffe, we regard conflict as incompatible positions between actors—in our case, between participants' ambitions for systemic change in a transition initiative. In our analysis of the data, we use incompatible positions as our first marker to identify conflict in the transition initiative.

2.3 Discursive Strategies for Position Management

Although incompatible positions function as a first important marker for conflict, such positions can be hard to recognise in the interactions between participants and fail to reveal how participants interact on these positions. We therefore use insights from discursive psychology as a second discourse analytical approach to examine the actual interactions and strategies actors use to handle conflict in conversation. Discursive psychology, in addition to the approach of Laclau and Mouffe, alerts us to markers for the emergence of conflict in interactions between, and strategies used by, participants—whether consciously or unconsciously—to manage their position and represent their reality in the discourse of a transition initiative [14, 26, 25].

Both Laclau and Mouffe and discursive psychologists refer to moments of interaction that deserve extra attention because they indicate the negation of meaning and essential changes in the development of the discourse. In discourse analysis these moments are also referred to as critical moments or episodes [36] as they are crucial in the progression of the discourse. As described below, we distinguish two additional markers—in addition to conflict as incompatible positions, to recognise the emergence of conflict as potential critical episode in the interaction between participants and to identify the strategies of participants to deal with conflict.

Grounded in these DA perspectives, our second important marker involves the emergence of alternate meanings between participants [19, 28]. Laclau and Mouffe show how the meanings of certain words, most often those referring to important concepts in the discourse, constantly change in conversation. These 'alternations' show how the exact meanings of words are constantly up for discussion, revealing the efforts of the involved parties to influence the meaning-making process. Especially in transition initiatives, which generally aim to form a whole new discursive community in part by creating a mutual understanding of its central concepts, negotiation of meaning is distinctly important. Discursive psychology displays a similar focus on how actors react to utterances of other participants in conversation by making small changes in what was said. These small alternations in responses can be regarded as strategic actions, for example, when a request to investigate a certain issue is reformulated by the receiver to suit their own agenda.

Informing our third marker, Potter [26] and Sneijder and Te Molder [28] show how the triad of identities, stakes and accountability form an important discursive engine behind the dynamics of 'position management'. Having distinguished three strategic discursive actions performed by actors to manage their representations of reality, we find these tactics equally relevant to the dynamics of conflict in transition initiatives:

- Identity management: actions intended to influence the development of a discourse by selectively creating or promoting particular identities; for example, by strengthening or discrediting others' identities
- Stake and interest management: actions intended to influence participants' positions in a discourse by selectively employing the use of stakes; for example, by accusing others of self-serving management practices and thereby weakening their position.
- Accountability management: actions intended to influence the responsibilities and roles of participants in a discourse by selectively addressing perceived obligations either to oneself or to others; for example, by accusing others not to meet up their responsibilities and drawing the attention away of ones' own responsibilities.

2.4 *Analysing Conflict in Critical Episodes of Interaction*

Combined, our three markers constitute the discourse analytical framework used in this study to identify conflict and analyse its development, a summary of which is presented in Table 1.

Table 1 Discursive markers used to recognise conflict and identify actors' strategies for dealing with conflict

Markers used to identify both conflict and actors' strategies for coping with conflict
(1) Incompatible participant positions, e.g., on an issue
(2) Alternations:
– Ongoing alternation of the meaning of central concepts in the discourse
– Small alternations of meaning of specific words or utterances between partners, used by participants to assert their own perspectives
(3) Position management focused on:
– Identity management
– Stake management
– Accountability management

3 Case Introduction and Methods

3.1 Research Case

RESILIO (Resilience Network of Smart Innovative Climate-adaptive rooftops, 2019–2022) focused on the potential of smart Blue-Green (BG) rooftops in urban stormwater management to reduce pluvial flooding, heat stress and biodiversity loss [1, 12]. To control water flow (see Fig. 1), smart BG roofs combine water buffers with automatically operated valves to, ideally, turn the rooftops into a 'squeezable sponge' informed by weather forecasts: programmed before dry periods to retain and then slowly release rainwater, and before heavy rainfall to release all water, thereby creating extra retention capacity and reducing peak release on the central sewerage system.

RESILIO was jointly led by the municipality and the local water authority. The project was subsidised under the European Union Urban Innovative Action program. Table 2 describes the main activities and partners involved in RESILIO.

The ultimate objective of RESILIO was to develop an infrastructure of BG roofs connected through a digital network of automated valves. The roofs were provided by four different social housing associations, each of which was compensated for 80% of the total construction costs. Besides technological innovation and a feasible business case, this infrastructure required a new form of rainwater governance. The task of central rainwater management, traditionally a municipal responsibility, would need to be extended to include roof owners and technology enablers through a grid of valves managed via Decision Support System (DSS) software. Right from the start of the project, this crucial component (activity 5 in Table 2) demanded the repositioning of participating actors, implicated by the need for new BG rooftops governance protocols (activity 6).

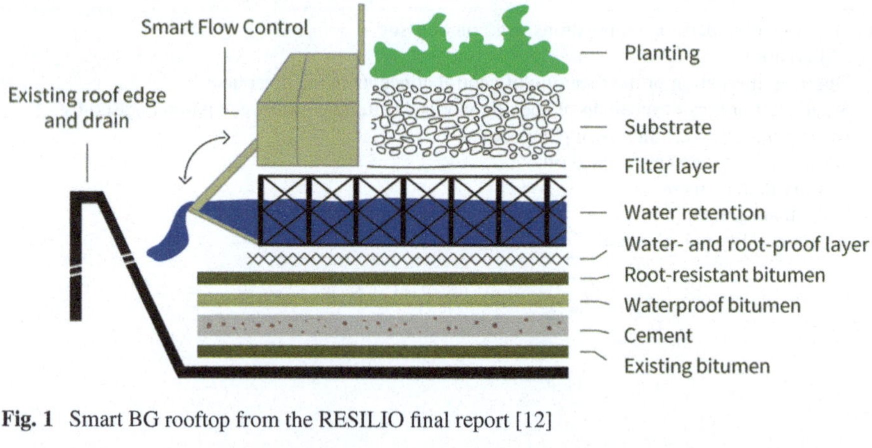

Fig. 1 Smart BG rooftop from the RESILIO final report [12]

Table 2 Main activities and partners of the RESILIO project

	Main activities	Main partners
1	Project management	Municipality: *Urban Planning and Sustainability* Water authority: *Water-System Team*
2	Creation of 10,000 m^2 of smart BG rooftops on existing social housing and private real estate; communication with research and development teams	Housing associations 1, 2, 3 and 4; Municipality: *Traffic & Public Place*; Technology enabler 1
3	Research on the climate-mitigating impact of BG rooftops, including storm water management, heat reduction, increased biodiversity and specific health benefits	Municipality: *Health Services* University: *Institute for Environmental Studies* University of applied sciences: *Urban Technology*
4	Technical innovation and product development of BG rooftops on an innovation-lab rooftop	Technology enablers 1 and 2 University: *Institute for Environmental Studies* University of applied sciences: *Urban Technology*
5	Development of an open source water-management platform (Decision Support System (DSS)) that connects to the automated valves of different suppliers	Water authority: *Water-Technology Team and IoT (Internet of Things) Team* Municipality: *Data Lab*
6	Development of BG rooftops governance protocols tailored to urban micro-level water management	University of applied sciences: *Urban Governance*
7	Social cost–benefit analysis and development of a viable business case	University: *Environmental Studies* University of applied sciences: *Economic Transformation*
8	Community involvement and communication	Technology enabler 3 Municipality University of applied sciences: *Urban Governance*

By combining micro-level rooftop data with macro information, including sewerage data and weather forecasts, the aim of the DSS was to provide individual owners and smart-valve service workers with advice on whether to open or close their valve at any specific moment for each specific rooftop water buffer, respecting roof owners' individual preferences for water use (Project Plan RESILIO). Figure 2 shows the various types of information used in the DSS.

This study focuses on participant conflict both around the design, ownership and openness of the software behind the DSS, and around the control and ownership of the BG rooftops' smart valves.

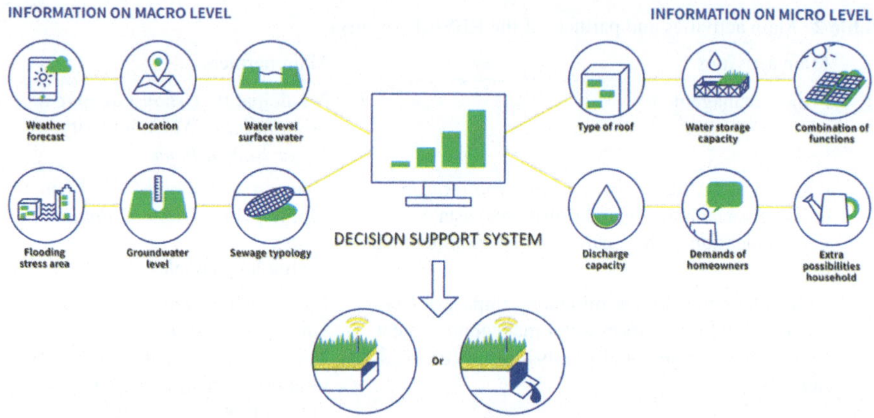

Fig. 2 Decision Support System from the RESILIO final report [12]

3.2 *Data*

Over the course of the project's three years, the team of researchers responsible for the development of the BG rooftops governance protocols collected data on RESILIO in a process of participative action research [3]. Data consisted of audio recordings and extensive notes made at regular (once or twice a month) central Project Team Meetings (PTM) (activity 1 in Table 2), and at dedicated work sessions for both the development of the DSS (activity 5) and new governance protocols (activity 6). In the end, we analysed a total of 19 meetings—16 of which had been documented with audio recordings and notes, and 3 with notes only—focusing on participants' discursive interactions. In addition, documentation in the form of the project's initial plan, midterm reviews, final report and presentations were analysed. Furthermore, we conducted 20 semi-structured interviews with RESILIO participants at the beginning of the project to learn about their motivations, roles and concerns. Both the interviews and documents were used to better understand participants' different positions and to corroborate the points of conflict identified by our analysis of participant interactions.

3.3 *Mixed Discourse Analysis: Coding and Analysing*

We used the three conflict markers (see Table 1) as a coding scheme to identify conflict and partner strategies in the data gathered on participant interaction. Based on our initial analysis of the Hard Talk in search of incompatible RESILIO positions (marker 1), we chose to focus on DSS control of the rooftop valves. Next, we examined the interviews for additional information relevant to incompatible positions on the DSS and control of the valves, and coded the agendas and notes of all PTM meetings and work sessions for the occurrence of 'moments of interaction' regarding DSS development.

To further substantiate the emergence of conflict in participant interactions and strategies, we then applied markers 2 and 3 to the audio recordings of the PTM meetings and work sessions that contained moments of interaction relating to the DSS. Specifically, we first used marker 2, the occurrence of alternations, to code the critical moments that influenced the development of the DSS, of which we identified six critical episodes that spanned multiple meetings. Within these episodes, we then analysed the most important alternations in participants' understandings of the governance protocols used to control the valves. Finally, we applied marker 3 to analyse the strategies used by participants to manage their positions, and to understand whether and how these affected the development of the discourse.

4 Results: Who Will Squeeze the Sponge?

4.1 Introduction

Urban sustainable development often entails a redefinition of the roles fulfilled by public organisations. Moreover, urban sustainability challenges tend to demand the involvement of private organisations and citizens, thus reshuffling traditional divisions of rights, duties and responsibilities [11]. In this section detailing our results, we will show how the partners of RESILIO tried to negotiate change beyond the development of technological solutions for the smart valve, necessitating a serious adjustment in urban water-management responsibilities.

In Sect. 4.2, we elaborate on participants' conflicting positions, marker 1, regarding control of the valves as these positions emerged during the Hard Talk in the early stages of the project. In the following result sections, we then focus on how conflict around control of the valve and development of the DSS continued after this influential session. Informed by marker 2, Sect. 4.3 presents an overview both of the critical episodes observed during DSS development and of the changes in discursive interactions associated with these episodes. Section 4.4 presents a small, specific selection of alternate meanings used by the water authority to deal with conflicting positions regarding the role of DSS valve control. Section 4.5 rounds off the results with our analysis of how accountability, identity and stake management, marker 3, enabled partners to manage their positions on responsibilities vis-à-vis DSS development and control of the valves.

4.2 *Emergence of Conflict: Controlling the Valves*

At the start of RESILIO, in early 2019, the first signs of conflict began to emerge in May during a project team meeting (PTM), when the DSS activity leader (see activity 5, Table 2), in contrast to the other attending partners who reported being on track, expressed concerns. Explaining the dependency of the DSS on other factors and project activities—such as the valves' physical properties and software, the research on the BG roofs' performance and the development of BG rooftops governance protocols—the activity leader suggested there were serious issues to discuss. Prior conversations with other activity leaders and partners had led the DSS leader to believe that important differences in participant stakes and ambitions were going unaddressed, potentially endangering the future of the project. To make these incompatible stakes and aspirations explicit, the DSS leader proposed a so-called Hard Talk. After several other attendees also recognised these concerns, the partners agreed to schedule the Hard Talk before the beginning of summer.

One of the most important issues discussed during the Hard Talk concerned control of the valves. Characterising the most important conflicting positions regarding this control, Table 3 presents the paraphrased remarks, further informed by interviews, of the Hard Talk participants. Following our first marker, these paraphrases are organised per incompatible position, with the left column presenting the positions most aligned with the initial project plan and the right an opposing position that could have potentially changed the RESILIO discourse.

Marker 1 thus reveals the partners' incompatible positions regarding the role of the DSS in governance protocols, a concept central to the systemic ambitions of RESILIO. In search of a new urban water-management system, the water authority felt compelled to join forces with other urban actors, such as real estate owners and technology enablers, who should share in the responsibilities associated with pluvial flooding due to climate change. In their eyes, only an open-source DSS platform could adequately facilitate the participation of other actors in a responsive, micro water-management system, i.e., the independent BG roof owners and technology enablers whose connection of their valves to the DSS would be imperative to the plan's success. In line with European funding regulations, the platform was ordered to be built on an open-source Internet of Things network to enable other parties to use, host or implement the platform in another setting after the project was finished: '*It will allow other parties in the Netherlands, and elsewhere in the world, to replicate the network-based micro water-management support system in their own public and/or private domains*' [6, p. 55].

In sum, both the Hard Talk and the interviews uncovered several points of conflicts related to the project's ambition to create a scalable BG infrastructure through which rainwater is controlled by the DSS; i.e., conflict between:

(1) The water authority—who viewed an open-source network as essential to the project's ambitions—and the technology enablers—who considered the request to share their core business unreasonable given their large investment costs.

Table 3 Paraphrased remarks by RESILIO partners indicating different positions regarding control of the valves

Participant position aligned with the initial project plan	Participant position incompatible with the initial project plan
We need to be informed on the technology behind the automated valves because we can't create a transferable model if we're missing a piece of the puzzle (water authority)	We've invested a lot of time and money in developing the valve and its software, so it's not in our interest to make that technology public. Careful consideration is imperative (technology enabler)
The valves shouldn't be controlled by technology enablers. A multi-actor decision model focused on managing the urban sewerage system as a whole should be the goal and the role of technology enablers should be to connect their valves to the DSS. (water authority)	Participating in a larger water-management system is interesting for us, but we can't just reprogram the valve according to the specificities of this particular city. We're doing our best to balance commercial and public interests. (technology enabler)
The technology enablers are currently operating in too many different positions regarding the valve: developer, innovator, producer, programmer and service provider. (water authority, municipality)	We've already developed the automated valve, giving us a front-runner position and ensuring the development of BG roofs. (technology enabler)
We're really worried that we're supposed to be developing a transferable model—using public (European) money, no less—but that, in the end, the future of the entire (Dutch) smart water grid will be dependent on a few technology enablers. (municipality)	You can't expect the technology enablers to just share their commodities. The solution doesn't have to be that complicated—the water authority should be able to work with standard protocols. (housing association)
We need to find new partners and a feasible business case to help diversify and sustain responsibilities for mitigating climate change, which includes a long-term solution for water management. A lot depends on that. (water authority, municipality)	Affordable housing is our core concern, so we can't just raise the rent because BG rooftops greatly increases the cost and complexity of maintenance. We don't want to be stuck with the valves and ongoing high maintenance costs at the end of the project. (housing associations)

(2) The water authority—who felt hampered by their dependency on the technology enablers for control of the valves—and the technology enablers—who saw the service of providing access to (the software that controls) the valves as part of their core business.
(3) The water authority—who believed the project's success depended on their ability to regulate rooftop water levels themselves—and the roof owners and technology enablers—who were legally entitled to this responsibility.
(4) The housing associations—who saw the BG rooftops' extra costs and responsibilities as impeding their primary task of providing affordable housing—the water authority—who was unable to compensate the housing associations without third-party funding—and the technology enablers—whose business model depended on income from service costs for the valve.

4.3 Critical Episodes in the Development of the DSS

While the Hard Talk made the partners' awareness of potential conflict explicit, awareness alone was not enough to facilitate changing the existing system. Many transition initiatives encounter such 'system lock-in' as participants work towards structural change. In this section, we describe how, in the three and a half years after the Hard Talk, the RESILIO partners tried both to negotiate the terms of the DSS and to cope with conflict regarding governance of the valves.

Informed by marker 2 of our discourse analytical framework, we focused on patterns of alternations in participants' discursive interactions around development of the DSS in relation to governance of the valves. Data analysis revealed different types of alternations in participant interactions, the most important of which were all related to control of the valves and concerned: (1) new terms, (2) alternate meanings for existing terms, (3) the absence or presence of discussion, and (4) changes in participant positions (marker 1). In search of these influential interactions, we identified six critical episodes in which essential alternations were used when compared to previous episodes, using the Hard Talk as our baseline of comparison. Rather than providing extensive descriptions of each alternation, the following overview of critical episodes is meant to highlight the major alternations in participant interactions over time and identifying changes in the positions of partners by the end of the project (see Table 4).

Distinguishing these alternations in critical episodes is essential to understanding both the development of conflict and the final outcome regarding the implementation the DSS. Our analysis draws attention to the water authority as having made the most significant changes in its position towards control of the valves and in their relation to other RESILIO partners. As the project progressed, their need to initiate the development of a new system in which water-management responsibilities would be shared repeatedly forced them to reconsider their own role in a responsive micro water-management system. Their biggest switch came at the end of the project, when they redefined their role as developer and host of the DSS to instead be that of an authority responsible for issuing future standards for technology enablers. Although this redefinition and hosting of the DSS by the technology enabler seems the unexpected result of the freeze of ICT projects, it also reflects the difficulties that the water authority encountered over the project to enforce the right conditions within its own organisation and position the DSS as part of their future responsibility. Furthermore, it is important to note that these six critical episodes of discursive interaction do not encompass all changes in partners' positionings. Our prolonged engagement in the project and the examination of the final report teaches us that the water authority had—partly based on their experience in RESILIO—made additional arrangements for distributing the responsibilities associated with pluvial flooding; e.g., new rainwater regulations that force property developers to create water-retention devices on all new estates.

Table 4 Overview of critical episodes and major alternations in discursive interactions between RESILIO partners

Baseline established in episode 1	Emergence of conflict
Critical episode 1	Incompatibilities in partner positions were actively addressed for the first time during the *Hard Talk*: although the smart-valve technology was still in private hands, both the water authority and the municipality expected the DSS to become a governance instrument in the form of a transferable, open-source platform tailored to the needs of each RESILIO partner. Moreover, the water authority's insistence on being granted control over the valves clashed with the technology enablers' unwillingness to municipalise their software, and with the roof owners' and technology enablers' legal right to control the valves. Development of the DSS was moving slowly, and the DSS team and other participants attributed this to its dependency on other activities and on a lack of DSS team personnel The housing associations argued several times for a readjustment of costs and responsibilities regarding the valve, as the current distribution was competing with their primal goal of providing affordable housing
Main alternation in episode 2	Introduction of a new term—Roof Water-Level Agreement—to address new aspects of the valves' control
Critical episode 2	New terms arose to address governance issues and control of the valves. One such term was the Roof Water-Level Agreement, intended to inform the DSS on each individual rooftop's use of water. With this new term, focus on the DSS as a technical instrument for valve control shifted into a discussion about the partners' different roles. Trying to keep the focus on the DSS as the main instrument used to distribute control, the DSS team emphasised the platform's need for smart KPIs (key performance indicators) that would correspond with the goals of RESILIO—KPIs they promised to develop. However, this triggered discussion and confusion around which governance responsibilities should be distributed and to whom. The focus then shifted to the governance team, who was responsible for defining the partners' roles and governance arrangements
Main alternation in Episode 3	Introduction of a new concept—the Micro Water-Management Agreement—to disconnect 'who controls the valve' from development of the DSS

(continued)

4.4 *Alternate Understandings of the DSS*

The six critical episodes reveal how the water authority had found itself in the most difficult of the partners' positions: tasked with creating a future-proof, multi-actor model for urban rainwater management, they were dependent on both property owners, such as housing associations, and technology enablers, who supplied and operated the valves. While these partners were chiefly and jointly responsible for forging structural change by constructing an agreement regarding divisions of costs, benefits, roles and responsibilities, negotiations around these divisions were difficult due to an imbalance in dependency between the partners. Moreover, alongside interactions regarding what a Roof Water-Level Agreement exactly entailed and what

Table 4 (continued)

Baseline established in episode 1	Emergence of conflict
Critical episode 3	One main goal of RESILIO was to establish a funding programme for the development of the rooftops owned by housing associations—but also for those of private citizens and business owners. To distribute the associated rights and obligations among the involved parties, the water authority and a researcher from the governance protocol team developed an instrument called the Micro Water-Management Agreement. This agreement would require roof owners who had received a BG subsidy via the water authority to request connection of their valves to the DSS. Because only the technology enablers could facilitate this connection, the water authority became both acutely aware of this dependency, and motivated to discuss the ins-and-outs of connecting the valves and its software to the DSS Meanwhile, DSS development was proceeding swiftly, but seemed to be disconnected from discussions about responsibilities, interests and roles. Over the course of the next seven months, there were no partner interactions concerning the relationship between who should be responsible for controlling the valves and development of the DSS. The governance team was increasingly solicited as the partner responsible for creating suitable governance protocols, but the partners themselves continued to have different expectations of the project and the governance team had difficulty solving this conundrum
Main alternation in episode 4	Absence of discussion and use of alternate meanings—terms of 'problems to solve' were replaced by 'productive collaboration' in discussions concerning the relation between the water authority and the technology enablers
Critical episode 4	Choosing to focus on 'progress above problems', the conflicting interests of the technology enablers and the water authority had become a 'non issue' in their interactions. During this phase of development, both parties recognised their mutual dependency and engaged in active collaboration, mentioning how the work of one reinforced the progress of the other. Finally allowing the technology enablers to connect their valves to the DSS, a new application interface was finished and the first version of the user dashboard was launched by the DSS team Nonetheless, in a joint session on the DSS, it became clear there were still no normative KPIs for determining and informing the use of rainwater and the DSS team and other participants from the water authority had difficulty embedding the DSS in their regular ICT structure
Main alternation in episode 5	Resurfacing of discussion—old positions vis-à-vis control of the valves were reconfirmed when the BG roofs were finished and the valves were ready to be connected to the DSS

(continued)

Table 4 (continued)

Baseline established in episode 1	Emergence of conflict
Critical episode 5	Just as the first roofs were about to be delivered, discussions during a PTM made clear that the BG rooftops had not yet been connected to the DSS, triggering the re-emergence of conflict. This was attributed in part to the roofs not yet being fully ready, but also to obscurities regarding the BG rooftops governance protocols. The housing associations complained that both service contracts and instructions on how to control the valves had still not been provided by the technology enablers. Hesitant about the trustworthiness of the DSS ground rules, the technology enablers expressed feeling unable to formalise the connection between their valves and the DSS in a contract. The water authority claimed to also feel hesitant about its own role towards individual roof owners, whose valves they no longer aspired to service in future developments of the DSS In response, the main RESILIO project leader announced a call for a special session—to be held immediately following summer holidays—to fix the major obstacles impeding valve connection. During the summer, however, the water authority was temporarily forced to stop all experimental ICT projects due to a negative cyber safety evaluation
Main alternation in episode 6	Change in participant positions—unable to continue with the project, the water authority changed its tactics, seeking a way to assign further DSS development to the technology enablers
Critical episode 6	Because of the water authority's freeze on experimental ICT projects, the partners of RESILIO were forced to find a quick solution for the DSS that would be viable for at least the next five years. The most obvious partner to host the DSS for the next five years was a technology enabler who had been part of RESILIO from the beginning. Hesitant to develop two software applications at the same time, the technology enabler proposed instead to focus on the improvement of their own software The water authority formulates an agreement that ensured the hosting of DSS for at least five years as separate modules to which BG roof owners can be connected within the software of the technology enabler. The final arrangements outlined which new future BG roofs would be serviced through the DSS and that the open-source code would be provided on request through the general RESILIO website The development provoked a new discussion between the water authority and the technology enablers regarding the role of the water authority as a service provider in competition with the technology enablers. Both partners were inspired to continuing brainstorming about new possibilities, such as the application an open standard for protocols for smart BG rooftops as used in smart traffic devices

needed to be arranged by the Micro Water-Management Agreement, the meaning of, and terminology used to refer to, the DSS as a governance instrument was constantly changing as well. With most alternations coming from the water authority and the municipality, our analysis shows how these were often both related to the positions presented in Table 3, and affected the strategies used by the water authority and the municipality to deal with conflicts regarding control of the valves. Table 5 illustrates these main alternations.

This overview highlights not only how the water authority's conception of the role of the DSS changed over the project's three-year period, but also how these changes affected the water authority's position and accountability in the final outcome of RESILIO. Having determined that their wish to control the valves was unfeasible,

Table 5 Main alternations on the role of the DSS in governance of the valves

Dates	Summary of alternate meanings regarding the role of the DSS in controlling the valves
April '19	Role of the DSS according to the water authority: *a smart, open-source governance instrument that balances and translates all partners' stakes into real-time advice on how to control each BG rooftop valve while also incorporating each roof owner's individual preferences for water usage* Consequence of this role: the water authority viewed the DSS as central to BG rooftops governance and the technology enablers as mere service providers for execution of the water-usage agreement between roof owners and the water authority. The municipality followed this representation of the DSS
Sept. '19	Role of the DSS according to the municipality: *just one of the many assets used to inform BG rooftops governance, along with data on the design of each individual rooftop and its: valve, water-retention system and amount of water contained* Consequence of this role: having become aware of the full complexity of BG rooftops governance, the municipality realised that DSS development could not be the only or main vehicle used to negotiate the partners' conflicting positions. While the DSS activity leader initially protested, by the end of 2019 the DSS team had also started downsizing the role of the DSS; although they still felt responsible for formulating the KPIs upon which the rules for control of the valve would be based
March '21	Role of the DSS according to the water authority: *one of multiple water-management instruments, with the definition of its KPIs remaining the most important (unanswered) question. Despite the water authority having already built the DSS, the partners had yet to decide on how to use it: which users need a dashboard and who should be able to control the valves through the dashboard?* Consequence of this role: partners were surprised and confused by this major shift in the water authority's conception of the DSS. While the DSS was technically 'ready to perform', it remained unclear who was actually in charge of controlling the valves. Discussions on the matter moved in circles. The DSS team felt they had done all they could, given the partners' lack of clear instructions regarding which system was dominant: that of the technology enablers or the DSS. The housing associations only wanted a button to empty the rooftops in case of emergency, but were unsure whether the DSS provided that. The technology enablers were hesitant to connect the roofs to the DSS given this lack of clarity
Oct. '21	Role of the DSS according to the water authority: *instead of focusing on development of the best algorithm for valve control, the DSS should focus on supporting the market and stimulating other companies to adopt a DSS model* Consequence of this role: part of a larger conversation between the water authority and the municipality, this change in focus corresponded with the assignment of DSS host to one of the two technology enablers. Although initially triggered by the water authority's temporary suspension of all innovative ICT projects, this switch also resonated with a deeper insight by the water authority that it could not and did not wish to service all future BG roof owners in control of their valves

the water authority changed its position towards the technology enablers. Each alternation can be seen as an attempt by participants to influence the discourse to suit their own position.

4.5 Managing Positions Through Accountability, Identity and Stakes

The main goal of our research was to understand how participants actually engage in, or avoid, conflicts that are inescapable in light of a transition initiative's overall ambition, as was the case with RESILIO. Because the summaries in Table 5 are merely paraphrases of what was actually said in the process of meaning-making regarding control of the valves, they fail to reveal how participants attempted to steer these alternations in personal interactions. For an analysis of the strategies used by participants in personal interactions, following our third marker, we now turn to our conversational data samples.

Data Sample 1 presents a discussion between the DSS team and the governance team about what was needed for RESILIO to achieve its goals—a telling example of how partners made use of accountability management (marker 3) to affect perceptions of responsibility. Occurring during a session organised by the DSS team to discuss the role of each partner in further DSS development, the team expressed its concern about the governance protocols' lack of a clear focus, going so far as to suggest making DSS development dependent on the activities of the governance team. Furthermore, they pointed to an acute need for structural funding in the form of a viable business case and a need for strengthened political support. By asking whether or not these issues were part of the plan of the governance team, the DSS team was essentially attempting to assign these responsibilities to the governance team, whom for their part avoided accountability by suggesting that a feasible business model would be part of the end product—still more than two and half years down the line—and that the 'money conversation' was in fact the responsibility of the project management team.

Although the accountability management in this data sample can easily be explained as responsibility- and conflict-avoidance, the participants' discursive actions also represented their perspectives: willing to develop an instrument but sensing a lack of ownership for future upscaling and its necessary fundraising, the DSS team passed on these responsibilities to the governance team, who felt unable to assume these responsibilities due to their 'neutral' role as researchers, resulting in further passing to the project management team. Participants' constant shifts in accountability management should therefore, we argue, not only be seen as strategies for avoiding conflict or responsibility, but also as offering insight into a deeper issue within the transition initiative of RESILIO: the importance of clear political choices and support as a necessary condition for actual change.

Data Sample 1 Accountability and identity management relating to the success of RESILIO (Nov'19)

Marker 3—accountability management by the DSS team	P1 DSS team: [1.10.58] There's a bit of a feeling that we're always beating around the bush. I get that it's difficult. So that's why I'm actually asking: what is the scope [of the governance model]? Are we going to literally [claps hands] say to the aldermen of the city: 'Look these are the KPIs, this is the plan'? [Short silence] What is the process [of the governance team] to get to results? I'm very curious about that. If those things were clear, DSS development would go so much more smoothly. That's what we're depend… DSS is actually an instrument that shows how a model works. Those facts are the facts
	P2 DSS team: That's why we haven't delivered anything yet. As I said before, we could have delivered already, but that doesn't feel right—as if that changes anything. Is that what you want to deliver—as a group, for five million [euros]? That wouldn't feel fulfilling to us
	P3 Governance team: Yes
	P1 DSS team: I don't see it settled in the governance, that people won't end up being trapped. Something really needs to change. And we need to start talking about the money side of things—like, for example, a conversation needs to be had with the aldermen in which you say: 'Tell me what you need to make this eligible for subsidising.'
	P2 DSS team: Yes, 'on what basis would you provide funding?'
	P1 DSS team: Yes, 'on the basis of what?' Then you have to actually—then the city becomes a stakeholder and yeah—
	P2 DSS team: Is that also in your governance model? Is that in it—in your scalable business model?
Marker 3—accountability and identity management by the governance team	P3 governance team: Yes, it is. We've talked about it with the project management team, they want to be able to go to the aldermen at the end of the project and explain: 'This is working, this is not working, and these are the benefits.' Their goal is in the context of the city. From our role, I think—as far as I'm concerned, as a researcher—it can also include other cities, to be able to say, 'This is what works in this city and here are the context-specific elements that make it different from Copenhagen or Barcelona.'
	P2 DSS team: Yes exactly, I have no idea about the rest of Europe. Do they also have housing corporations? And so on, I don't know
	P3 governance team: And to gain insight into specific pros and what should be done differently. I see that as an important advantage of a project like this: that you can learn and that you can compare different rooftops. [1.13.49]

In Data Sample 2, we present another example of position management, this time by the housing associations. In the context of RESILIO, participating housing associations were reimbursed for 80% of the initial construction costs of BG rooftops. They owned, and were responsible for maintenance of, the valves for a period of five years. During RESILIO meetings, the housing associations repeatedly pointed out a lack of direct interest for them—their top priority being to provide affordable housing. Re-allocation of stakes and responsibilities was then, in their eyes, necessary for the long-term sustainability of the project's ambitions. Data Sample 2 shows how the housing associations tried to change their position both by emphasizing their lack of direct stakes and by highlight those of others, e.g., those of the water authority.

By referring to their primary responsibility of providing affordable housing, the housing associations made a strong argument for shifting their role from funder to 'access giver'. However, the housing associations also repeated such discursive attempts to re-allocate responsibilities over and over throughout the entirety of the project, without substantial impact. This suggests that the initial EU funding had perhaps not only been essential to starting the innovation project, but also that it may have enabled the ongoing neglect of opposing financial positions. Furthermore, the funding conditions left little room to translate the lessons learned throughout the project into changes to the initiative's financial structure. The temporary financial support from the EU, and its strict conditions, may have in fact enabled the other participants to functionally ignore discussions concerning the position of the housing associations.

Data Sample 2 Stake and accountability management by housing association 2 (Sept '20)

Marker 3—stake and accountability management by housing association 2	P1 housing association: [1.02.56] And furthermore, what I'm now seeing is that the maintenance and management costs of the valves are extremely high and that—as a result, if you don't directly benefit from this [the rooftops] yourself—you might ask yourself whether this responsibility would be more appropriately placed on someone else. That maybe, as a roof owner, you need to make your roof accessible—make your roof available, so to speak—to the water authority so that *they* can further arrange for the system to work
	P2 researcher: So you would prefer to be unburdened completely—and preferably at little or no cost?
	P1 housing association: Yes, of course. You have the cost of the roof itself, but extra costs—for optimising water storage while the benefits are for another party—yes, that's not interesting for a roof owner. [1.03.42]

5 Conclusion

The key challenge of RESILIO—to explore new forms of rainwater management instead of just creating a new market for one specific technical innovation—necessitated a search by participants to define the project's collective, and each partner's individual, responsibilities. This shows the ambition of the experiment: to develop a new pathway for (system) change, including several new propositions for governance arrangements. Hence, it may not come as a surprise that our discursive analysis shows a relationship between the potential for conflict and participants' attempts to redefine positions, including these positions' related responsibilities and interdependencies.

Having shown that the main conflicts were caused by incompatible positions between the water authority, technology enablers and housing association, we identified when the conflicts in the collaboration emerged by focusing on the alternations in participants' discursive interactions regarding control of the valves. This resulted in six critical episodes in the discussion on the redistribution of the responsibility for urban water-management, that revealed ongoing position management by participants to handle these conflicts in their joint meaning making process. Participants alternated on central concepts regarding the control of the valve, trying to manage not only their own positions, but also those of others and, of course, of the initiative as a whole.

Additionally, our analysis increased our understanding of how the partners strategically (albeit not necessarily consciously) addressed conflict or avoided overt confrontations. Appearing to be participants' dominant strategy for engaging in conversations that involved incompatible positions, accountability management offered participants a way to both protect their own position, and explore possible new arrangements and the roles of others.

As a case study, the transition initiative of RESILIO serves as a good example both of urban experimentation and of participant learning—in this case about how to increase the resilience of urban systems and mitigate the effects of climate change. Diverse organisations were involved, all of them willing to spend time discussing possibilities for realising the new BG roofs, for how to overcome hindrances and for how to take experimental action. Thus, conditions that are considered key for urban experimentation seemed to be in place [27].

RESILIO was, under the pressure of European subsidy constraints, successful in achieving its direct innovation goals. However, as our discourse analysis showed, participants often found it difficult to work through and transform conflict, though the presence of conflict was sometimes acknowledged, for example in the Hard Talk. As a result, the more fundamental long-term impact of the project, in this case a more future-proof decentralised urban water-management system, remained out of reach. Our results also suggest that a valuable, and perhaps even necessary, first result of any urban experiment may be 'nothing more than' participants gaining a better understanding of their seemingly intangible conflicts and incompatible positions. Discourse monitoring between partners may provide a possible method for making this happen.

6 Discussion

Urban transition initiatives, or urban living labs, as they are also called, have become common in the repertoire of urban experimentation and policy innovation [29, 33]. This concept, in which local governments seek cross-disciplinary collaboration to create shared responsibility for sustainability transitions, is regarded as a positive development that stimulates multi-actor involvement and reflexive governance [29]. At the same time, these urban transition initiatives also face ongoing criticism in terms of their transformative potential. Research shows that participants in transition initiatives often find it difficult to address the political character of sustainability challenges, tending instead to reproduce existing structures of power and dominant perspectives [20, 32]. Scholars have therefore urged that, to increase their transformative potential, research on learning in transition initiatives and transition theory should focus more on the nature and role of conflict in pathways to sustainability transitions [7, 36].

Many of these initiatives seek to change dominant structures, relations and practices by developing alternatives while, at the same time, interacting with these dominant systems [4, 9]. This interaction with the dominant system forms an essential component both of participants' learning processes and of the conflicts they encounter. Our combined discourse analytical perspective helps us identify not just any kind of learning, but especially learning related to structural change: its focus on incompatible positions between partners connects participants' learning to the transformative goals of a transition initiative. In that sense, it presents a more purposeful way of framing learning in transition initiatives compared to other approaches and helps practitioners and researchers reframe conflict as a potential source of learning within the initiative.

Many traditional learning approaches advocated in both transition initiatives and the literature consider joint learning to be a deliberative process in which different perspectives are confronted and united in a shared problem-analysis [13, 34]. Our perspective and focus on conflict affirm that this conceptualisation of participant learning within transition initiatives often presents an overly optimistic view of reality. The results of this chapter illustrate that there was limited space for actual change in the governance over the valves; changes that were crucial in the light of the transformative ambitions of the transition initiative. Our approach can thus provide a more open approach to investigating participants' learning—an approach focused on a process of meaning-making and struggle over defining the hegemonic discourse between an initiative and its environment. In so doing, our discourse analytical perspective also extends dominant definitions of learning, which tend to presuppose a progressive, positive process of increased mutual understanding, to a process of discontinuance of relationships and selective reframing of problems and potential solutions [17].

Although our approach proved useful for examining potential conflict and the strategies participants used to address these conflicts, our results also evoke a need to more precisely analyse whether and how participants' ideas and positions actually change throughout the process of a transition initiative. While alternations in meaning proved to be a reliable indicator for the emergence of conflict, it is unclear whether they could also be used to mark processes of progressive insights among the participants. If so, such an approach, in conjunction with existing learning theories, may help us gather concrete evidence of participant change. For instance, a (re)framing perspective on learning—as often used in theories of policy learning—could help shed light on how initiative participants broaden the problem-definition and whether or not new issues and perspectives were integrated into it [10]? A social learning perspective may help reveal whether new cognitive understanding on the division of responsibilities for water management and new relationships emerged [23]? Such additional research may support transition initiatives both by helping participants realistically reflect on their joint learning processes, and by contributing to a further empirical and theoretical understanding of how to foster participants' learning to support sustainability transitions.

References

1. Andenæs E, Kvande T, Muthanna TM, Lohne J (2018) Performance of blue-green roofs in cold climates: a scoping review. Buildings 8(4):55. https://doi.org/10.3390/BUILDINGS8040055
2. Angermuller J, Maingueneau D, Wodak R (2014) The discourse studies reader: an introduction. In: Angermuller J, Maingueneau D, Wodak R (eds) The discourse studies reader: main currents in theory and analysis. John Benjamins Publishing Company, pp 1–14
3. Baum F, MacDougall C, Smith D (2006) Participatory action research. J Epidemiol Commun Health 60(10):854. https://doi.org/10.1136/JECH.2004.028662
4. Beers PJ, van Mierlo B, Hoes A-C (2016) Toward an integrative perspective on social learning in system innovation initiatives. Ecol Soc 21(1). art33. https://doi.org/10.5751/ES-08148-210133
5. Buizer M, Kurz T (2016) Too hot to handle: Depoliticisation and the discourse of ecological modernisation in fire management debates. Geoforum 68:48–56. https://doi.org/10.1016/j.geoforum.2015.11.011
6. City of Amsterdam (2018) RESILIO application Annex 1-UIA03-093-RESILIO application form
7. Cuppen E (2018) The value of social conflicts. Critiquing invited participation in energy projects.https://doi.org/10.1016/j.erss.2018.01.016
8. Cuppen E, Pesch U, Remmerswaal S, Taanman M (2016) Normative diversity, conflict and transition: shale gas in the Netherlands. Technol Forecast Soc Change. https://doi.org/10.1016/J.TECHFORE.2016.11.004
9. Grin J (2020) 'Doing' system innovations from within the heart of the regime. J Environ Plan Policy Manag 22(5):682–694. https://doi.org/10.1080/1523908X.2020.1776099
10. Grin J, Loeber A (2017) Theories of policy learning: agency, structure, and change. : Handbook of public policy analysis: theory, politics, and methods, pp 201–219. https://doi.org/10.4324/9781315093192-25
11. Grin J (2010) The governance of transitions: an agency perspective. In: Transitions to sustainable development: new directions in the study of long term transformative change, pp 265–284. Routledge. https://doi.org/10.4324/9780203856598
12. Holstein AN, Langewen J (2022) RESILIO final report. https://resilio.amsterdam/wp-content/uploads/2022/03/Final-Report-RESILIO.pdf
13. Hölscher K, Wittmayer JM, Avelino F, Giezen M (2019) Opening up the transition arena: An analysis of (dis)empowerment of civil society actors in transition management in cities. Technol Forecast Soc Change 145:176–185. https://doi.org/10.1016/j.techfore.2017.05.004

14. Jørgensen M, Phillips L (2002) The field of discourse analysis. In: Jørgensen M, Phillips L (eds) Discourse analysis as theory and method. Sage Publications Ltd., pp 1–23
15. Jørgensen M, Phillips L (2012) Laclau and Mouffe's discourse theory. In: Discourse analysis as theory and method. SAGE Publications Ltd., pp 24–59. https://doi.org/10.4135/9781849208871.n2
16. Kemp R, Schot J, Hoogma R (1998) Regime shifts to sustainability through processes of niche formation: the approach of strategic niche management. Technol Anal Strateg Manag 10(2):175–195. https://doi.org/10.1080/09537329808524310
17. Kenis A, Bono F, Mathijs E (2016) Unravelling the (post-)political in transition management: interrogating pathways towards sustainable change. J Environ Plan Policy Manag 18(5):568–584. https://doi.org/10.1080/1523908X.2016.1141672
18. Kouévi TA, van Mierlo B, Leeuwis C, Vodouhê SD (2016) Sensitive issues in natural resource management and discursive strategies addressing them. J Environ Planning Manage 59(7):1168–1185. https://doi.org/10.1080/09640568.2015.1062746
19. Laclau E, Mouffe C (2001) Beyond the positivity of the social: antagonisms and hegemony. In: Hegemony and socialist strategy: towards a radical democratic politics, 2nd ed. Verso, pp 93–148
20. Meadowcroft J (2009) What about the politics? Sustainable development, transition management, and long term energy transitions. Policy Sci 42(4):323–340. https://doi.org/10.1007/s11077-009-9097-z
21. Molder H, Potter J (2005) Conversation and cognition. In: Conversation and cognition. Cambridge University Press. https://doi.org/10.1017/CBO9780511489990
22. Mouffe C (1999) Deliberative democracy or agonistic pluralism? Soc Res 66(3):745–758
23. Pahl-Wostl C, Hare M (2004) Processes of social learning in integrated resources management. J Commun Appl Soc Psychol 14(3):193–206. https://doi.org/10.1002/casp.774
24. van Poeck K, Östman L, Block T (2020) Opening up the black box of learning-by-doing in sustainability transitions. Environ Innov Soc Trans 34:298–310. https://doi.org/10.1016/j.eist.2018.12.006
25. Potter J (2003) Discursive psychology: between method and paradigm. Discourse Soc 14(6):783–794. https://doi.org/10.1177/09579265030146005
26. Potter J (1996) Representing reality: discourse, rhetoric and social construction. Represent Real Discourse Rhetor Soc Constr. https://doi.org/10.4135/9781446222119
27. Puerari E, de Koning JIJC, von Wirth T, Karré PM, Mulder IJ, Loorbach DA (2018) Co-creation dynamics in urban living labs. Sustainability (Switzerland) 10(6):1893. https://doi.org/10.3390/su10061893
28. Sneijder P, te Molder HFM (2005) Disputing taste: foods pleasure as an achievement in interaction. Appetite 45(1):51–61. https://doi.org/10.1016/J.APPET.2005.03.002
29. Torrens J, Schot J, Raven R, Johnstone P (2019) Seedbeds, harbours, and battlegrounds: on the origins of favourable environments for urban experimentation with sustainability. Environ Innov Soc Trans 31:211–232. https://doi.org/10.1016/J.EIST.2018.11.003
30. Torrens J, Johnstone P, Schot J (2018) Unpacking the formation of favourable environments for urban experimentation: the case of the Bristol energy scene. Sustainability 10(3):879. https://doi.org/10.3390/SU10030879
31. Versteeg W, te Molder H, Sneijder P (2018) "Listen to your body": participants' alternative to science in online health discussions. Health (United Kingdom) 22(5):432–450. https://doi.org/10.1177/1363459317695632
32. Voß JP, Smith A, Grin J (2009) Designing long-term policy: rethinking transition management. Policy Sci 42:275–302. https://doi.org/10.1007/s11077-009-9103-5
33. von Wirth T, Fuenfschilling L, Frantzeskaki N, Coenen L (2019) Impacts of urban living labs on sustainability transitions: mechanisms and strategies for systemic change through experimentation. Eur Plan Stud 27(2):229–257. https://doi.org/10.1080/09654313.2018.1504895
34. Wittmayer JM, Loorbach D (2016) Governing transitions in cities: Fostering alternative ideas, practices, and social relations through transition management. Governance of urban sustainability transitions: Euro Asian Exper 13–32

35. Yanow D (2015) Making sense of policy practices: interpretation and meaning. In: Handbook of critical policy studies, pp 401–421.https://doi.org/10.4337/9781783472352.00031
36. Yuana SL, Sengers F, Boon W, Hajer MA, Raven R (2020) A dramaturgy of critical moments in transition: understanding the dynamics of conflict in socio-political change. Environ Innov Soc Trans 37:156–170. https://doi.org/10.1016/J.EIST.2020.08.009
37. van Mierlo B, Beers PJ (1–15 Aug 2018) Understanding and governing learning in sustainability transitions: a review. Environ Innov Soc Transit. https://doi.org/10.1016/j.eist.2018.08.002

Waterfront Revitalization as Opportunity for Sustainability Transitions—The Huangpu River in Shanghai

Harry Den Hartog, Fan Ding, Xiaojing Wang, and Jiang Wu

Abstract This chapter will explain how the Huangpu River plays a key role in shaping the city of Shanghai, and how the revitalization projects along its waterfront can stimulate a sustainable transition. During the years a range of functional transformations took place along the Huangpu River, resulting in changing relations between city and water. After decades of industrialization a new postindustrial phase just started. Inspired by urban planners the city leaders aim to make Shanghai into a frontrunner city for sustainable transition policies in China and beyond. The authors argue why the Huangpu River is so important in today's quest for sustainability transitions, as proclaimed in Shanghai's latest master plan and offer suggestions for improvements to contribute to a more sustainable urban development.

Keywords Globalization · Huangpu river · Port city · Shanghai · Sustainability transitions · Waterfront redevelopment · Urban regeneration

1 Introduction: Multiple Transitions Simultaneously Along the Huangpu River

The classic Chinese painting Qingming Shanghetu—which depicts the former capital Kaifeng during the Song Dynasty early twelfth century—forms the perfect illustration of the importance of water in Chinese urban development. It shows the rich mix of economic activities on and along the water. World port Shanghai owes its existence

H. Den Hartog (✉)
College of Architecture and Urban Planning, Tongji University, Shanghai, China

F. Ding
Academic Development Department, College of Arts and Media, Tongji University, Shanghai, China
e-mail: dingfan_2006@tongji.edu.cn

X. Wang · J. Wu
College of Architecture and Urban Planning, Tongji University, Shanghai, China

Z. Allam (ed.), *Sustainable Urban Transitions*, Urban Sustainability,
https://doi.org/10.1007/978-981-99-2695-4_9

and prosperity to the water, especially the Huangpu River a partly man-made river strategically located within the highly urbanized Yangtze River Delta.

During last three decades the world's economic center of gravity shifted eastward, accompanied by extreme fast and largescale urbanization. China's rapid urban transition has global impact. But there is also serious collateral damage. Decisions on climate actions are substantially influenced by rising powers, especially China. According to the World Economic Forum "China as a relatively poor country has made a conscious choice to shift focus from quantity to quality of growth" [22] and is leading the world toward sustainable development.

In 1992 revolutionary and former statesman Deng Xiaoping named Shanghai as "Head of the Dragon" [13]: China's economic gateway to the world. Nowadays Shanghai is still China's most progressive and innovative city, thanks to its port as "window to the world". The tolerance and openness along the Huangpu River in the early days resulted in a vibrant urban scene with various activities and cultural pluralism where local and foreign cultures coexist. Trading vessels and immigrants brought capital, technology, talents, culture, and art. From this perspective the Huangpu River is the source of Shanghai-style culture, which not only drives the development of the city, but also unites. The river witnessed the evolution of Shanghai and plays a significant role in shaping the city's unique geographical culture and urban landscape. The waterfronts of the Huangpu River have always been and still are the prime symbol of Shanghai, as international cosmopolitan metropolis.

Worldwide especially since the 1960s, due to containerization a process of deindustrialization and urban regeneration of waterfronts is gradually taking place, accompanied by rapid development of the tertiary industry. This process started relatively late in Shanghai, after the millennium, but offers new opportunities. Shanghai's latest master plan (Shanghai Municipal People's Government, 2016) shows that policy makers are very aware of the opportunities that waterfront transformations offer to upgrade the quality of life while reducing emissions in the central city. Unique is that very detailed plans for the waterfronts as a unity [28] have been made in which shanghai aims to guide all promises into realities. Both master plans emphasize that Shanghai wants to be a model city and promises, and according to its municipality, Shanghai will show how the city of the future should be shaped. The Huangpu River forms a backbone for urban regeneration and further internationalization, and also for sustainability transitions [10].

Waterfronts in port cities form an accumulation of conflicting claims for space, living, working, recreation, nature, and water safety. Simultaneously they offer a setting as experimental zones for multi-coloration and sustainable solutions, especially by urban planning and design.

Shanghai's city-water relations are continuously affected by changing conditions. During the last three decades the city experienced dramatic urbanization pressure, socio-economical transitions and simultaneously a range of new environmental challenges. Over the past three decades and ongoing, several other transitions are taking place as well in Shanghai (China). First of all, a political and economic transition into a post-communist system, into a state-controlled global-oriented market economy that started under former Paramount Leader Deng Xiaoping in 1978 with the reform and opening-up, better known under its officially name: Socialism with Chinese characteristics. This transition has been closely connected with a transition from an industrial into a post-industrial society with a fast emerging new socio-economic middle-class—and upper-class—working in the tertiary sector and accompanied by a neo-liberal inspired transformation in urban governance [15]. Increasing urbanization and a new lifestyle of consumption causes increasing pressure on the environment, adding the urgency of sustainable transitions—including e.g., an energy transition and low-carbon transitions [8], and climate change adaptation. In Shanghai all these transitions occur simultaneously in a relative short period of time, amplified in scale and thus also in impact. Moreover, they influence each other and cannot be seen separately. These transitions are ultimately manifested along the Huangpu River waterfronts.

This chapter explains how waterfront transformations and industrial development shaped the city, and how in a postindustrial phase a sustainable transition approach along the Huangpu River reshaped the city again. The waterfront of Shanghai's Huangpu River experienced a similar process as other world ports with a rise as "river for trade and prosperity", "industrial river", "decline into a rust belt", and currently a "revitalization as service belt" [11, 12]. These phases form a basis for understanding current sustainable transition of the Huangpu River into an "urban living room" [27, 28]. Sorting out the development stages of the Huangpu River and analyzing its characteristics can help us better understand and examine the sustainability efforts of the city of Shanghai and learn how to further optimize the city's future public waterfronts. In the next four sections the four phases—Phase 1: River for trade and prosperity (1843–1949), Phase 2: River for industrialization (1949–1990); Phase 3: Deindustrialization and development of Pudong District (1990–2002); Phase 4: Redevelopment of both Huangpu Riversides under globalization (since 2002)—are explained in a microhistory, based on which in the discussion and conclusion findings and recommendations are given to support a sustainable transition (see Fig. 1).

2 (Phase 1) River for Trade and Prosperity: Opening Up and Accumulation of Power Spaces (1843–1949)

After the Treaty of Nanjing in 1842 Shanghai reopened its port for international trade. At that time its urban land area was mainly composed of the walled city of the Shanghai County Town—the nowadays the Old Town—and the former neighborhood and docks of Dongjiadu which was the first international port area of Shanghai outside the wall along the Huangpu River. The port started to expand rapidly, accompanied by accelerated urbanization. Within a century Shanghai grew from a relatively small settlement into the third largest metropolis on earth. After London and New York, it was the world's largest financial center.

Since the Treaty of Nanjing foreign settlements and concessions continued to expand, the British Concession (1846), and the American Concession (1948), and

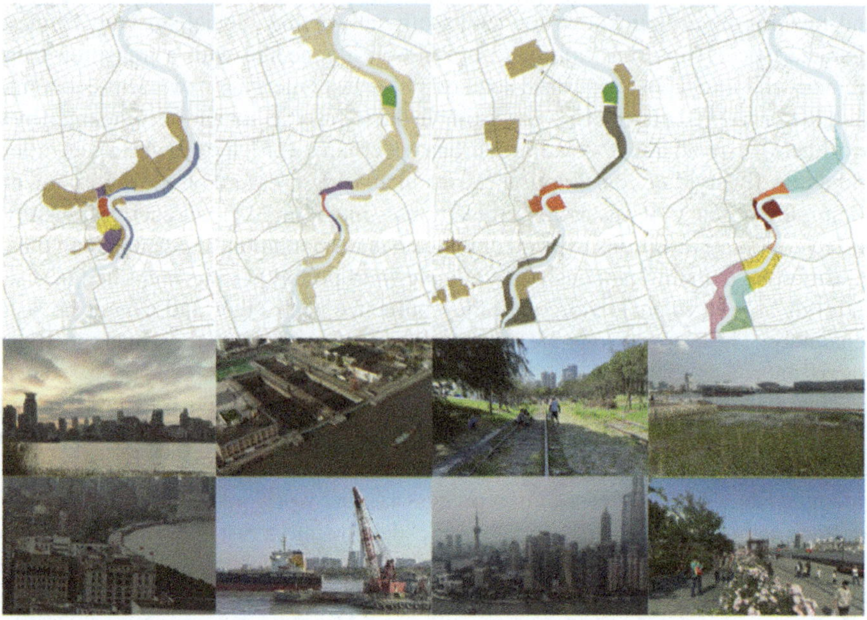

Fig. 1 The four phases: Phase 1: River for trade and prosperity (1843–1949); Phase 2: River for industrialization (1949–1990); Phase 3: Deindustrialization (1990–2002); Phase 4: Redevelopment (since 2002). *Source* by authors

the French Concession (1849). This development resulted in a duo-core structure, where the old town core got further neglected, especially after the creation of the Bund [18]. The concessions structured the urban development of Shanghai until today [26], even during the rapid urbanization with new towns since the millennium [7].

Simultaneously the Huangpu River forms a backbone for spatial-economic and urban redevelopment [10]. With the input of foreign capital and the rise of national capital the Huangpu River became the prime zone in the process of Shanghai's modernization. The riversides are the origin of modern Shanghai's industrial development and a driver for urban development. At the beginning of the twentieth century foreign shipping companies dominated the west bank of Huangpu River with multiple piers along a three kilometers long embankment area starting from the Old Town (South Pier and Shiliupu) to the north (up to the current Beijing Road). The core section of the Bund was dominated by banks and financial offices, and in the side streets there was entertainment and shops. Shanghai's port area continued to extend downstream along Huangpu River with port related industries, textile, docks, and shipbuilding, as well as municipal services such as electricity and water supply facilities [19]. After the Cultural Revolution the urban functions along the Huangpu River waterfront area expanded to the south with the Jiangnan Shipyard and to the north with Fuxing Island Industrial Zone. More and more heavy industries emerged on both sides of the Huangpu River.

Meanwhile since the Opium Wars the original port area of Dongjiadu vanished, and recently this part of Shanghai faces another drastic urban transformation [6] as a new CBD zone.

3 (Phase 2) River for Industrialization: Spatial Claims and Environmental Challenges (1949–1990)

After the foundation of the Peoples Republic of China in 1949 Shanghai's urban planning system gradually adjusted to the new industrial system and economic structure in which the secondary sector started to become dominant with manufacturing and assembly processes [31]. Shanghai became the center of industrial development and engine of China's economic development. Almost 80% of the land in the city center was occupied by factories [1]. During the Great Leap Forward (Second Five-Year Plan) industries started to be decentralized to satellite cities and state farms in rural areas.

During the 1950ies a large number of factories, shipyards and warehouses rose rapidly along Yangshupu Road, in the former American Concession, especially textile mills, spinning mills, paper mills, shipbuilding facilities, waterworks, power plants, etc. [4]. Many innovations were introduced here to China, e.g., the first mechanic paper mill, the first mechanic cotton mill, the first water purifying plant, the first gas plant, also for the first-time reinforced concrete and steel structures were produced here. We could find here the largest shipyard, and the largest thermal power plant in the Far East [4]. In the 1970ies Shanghai still had a shortage of mooring places to load and unload vessels. In response new wharfs were built for 10,000-ton vessels in Wusong, near at river's mouth, including mechanized capacity. In the early 1980s, the construction of container terminals began, and two foreign trade loading and unloading areas were planned and constructed, and the Shiliupu passenger station was reconstructed. In 1984, the port's cargo throughput exceeded 100 million tons, entering the ranks of the world's largest port [30]. In the early 1980s, the Port Authority of Shanghai, together with the Municipal Planning Bureau and relevant departments developed and constructed the Waigaoqiao Free Trade Zone, with new quays for the latest generation container ships [30].

All these developments influenced Shanghai's urban growth, but also resulted in environmental challenges. Industries had made many waterfronts dirty and smelly, while others became repositories of household waste. The areas adjacent to the waterfronts and industries were dominated by working class neighborhoods, often with poor hygienic conditions. There were some merchant houses and offices. Nevertheless, since the industrialization living by the waterside was until recently an unattractive option in Shanghai as it had negative connotations with poverty in working-class accommodations and pollution by industries.

4 (Phase 3) Deindustrialization: Decline into a Rust Belt, and Development of Pudong District (1990–2002)

The reform and opening up in 1978 made Shanghai once again influential and boosted urban development and port expansion. In 1992 the 14th National Congress of the Communist Party of China put forward as strategic goal to shift the national economic focus from secondary industry to tertiary industry. This resulted in further de-concentration and renewal of traditional industries into suburban industrial parks, accompanied by new towns [7]. At the same time, some traditional industries on both sides of Huangpu River started to decline in the 1990s. Textile industry, shipbuilding, chemical industry, steel, machinery, and building materials, they were all in need for expansion and upgrading [32]. Steel, chemical industries, and shipbuilding industries have been officially restricted in their development on both sides of Huangpu River due to environmental pollution. Shanghai's old port area reached a level of saturation and could no longer expand. Because of limitations of water depth and space, and the development of large-scale and containerized ships, most of the port and related industries started being relocated. This provided development space for the upcoming tertiary industry. Since the beginning of the twenty-first century, Shanghai has started to relocate its ports far outside the urban core to accommodate the expected massive urban growth and to allow the new ports to further develop.

This separation of the port from the rest of the city is a trend influenced by globalization due to a modernization of maritime technologies, the introduction of containers, and a decline of port-related employment in port cities [17]. Shanghai is a typical example of the "Anyport Model" [2], due to the increase in marine transportation and vessel size, the old port area needed to expand, and new port areas were built near the coastline of Pudong District, with a deep-sea port in the Hangzhou Bay. The Jiangnan Shipyard moved to Changxing Island. Shanghai's port became a space-and-capital-intensive rather than a labor-intensive port, same as other global port cities.

These relocations made the link between the port and urban development weaker, which previously also happened in other global port cities such as Rotterdam [21]. This offered opportunities to create new functions for the city and to inject new blood into these abandoned old port areas. This transformation from industrial construction to finance and consumption had impact on the planning and development of the city and its overall structure. The relocation of the port also brought opportunities to reconnect both sides of Huangpu River and connect the city with the river, similar to Rotterdam's 'jump' across the river.

In an amplified scale and accelerated way the development of Pudong New Area—especially Lujiazui Financial District—is the first sample in China of the redevelopment of former port-related areas. To guide the redevelopment process of the Pudong New Area, the Shanghai Pudong New Area Management Committee was established in January 1993 [5]. This redevelopment covered initially more than 200 km^2 to accommodate 2.5 million inhabitants. The redevelopment of Pudong was expected to revitalize the whole of Shanghai, even with spinoff beyond into the wider region of the Yangtze River Delta. The waterfront redevelopment provided a unique opportunity to reset the city's image—and of the wider Yangte Delta Region, even the country as a whole—with the photogenic skyline of Lujiazui.

5 (Phase 4) Revitalization as Service Belt and Redevelopment of the Riverbanks Under Globalization (Since 2002)

Since the 1990s the main functions along the Huangpu River changed from logistics, storage, and assemblage to finance, cultural tourism, and—to a lesser extend—residential and commercial functions [10]. This transformation from a production-oriented to a service and consumption-oriented waterfront introduced the post-industrial era, and a reappreciation and perception of the waterfront [9].

In 1998, Shanghai P&K Development Corporation joined forces with the Shanghai Port Authority (owner of most of the waterfront assets) to create a waterfront development framework suitable for investment and marketization. Shanghai Urban Planning and Urban Research Institute and Skidmore Owings and Merrill International Ltd (SOM) prepared a redevelopment plan for seven kilometers waterfront between Yangpu Bridge and Nanpu Bridge [20]. This plan aimed to reconnect the city with the waterfront by opening up much of it as public space [29].

In 2000 the Shanghai Urban Planning and Administration Bureau organized a large-scale international urban design competition, attracting world-renowned experts, including SOM again, Sasaki Association, and the team of Philip Cox [23] for another 13.6-km-long waterfront (22.6 km^2 of redevelopment area) subdivided into four locations and aimed to integrate the development of five municipal districts on both sides of Huangpu River. Especially the main industrial areas of Hongkou District and Yangpu District were largely occupied by abandoned warehouses, docks, and former factories. Most industries already moved to the urban fringes or further in new development zones along the coastline of Pudong.

During the Eleventh Five-Year Plan period, under the guidance of the national policy framework "Four Centers" (international economy, finance, trade, shipping) the modern service industry become a main sector of Shanghai's industrial structure. The Shanghai Municipal Party Committee and Municipal Government promptly put forward as strategic decision to redevelop both sides of Huangpu River with as core idea to reconnect citizens with the water. The new skylines along the waterfront symbolize this reconnection, and historical continuation, by integrating industrial historical assets, and adding financial and cultural clusters in a waterfront landscape with strong visual effects [10]. Since in 2005, Shanghai became the world's busiest cargo and container port, and Shanghai's capital linkages with the world became closer [14].

In 2002, the comprehensive development plan on both sides of Huangpu River was officially launched for about 42.5 km river length (on both sides is about 85 km), involving seven districts (Pudong, Luwan, Huangpu, Hongkou, Yangpu, Baoshan and Xuhui). This created an opportunity for massive urban regeneration of the formerly inaccessible and polluted waterfronts of the Huangpu River. Located central along the Huangpu River, the Expo 2010 played a key role in redefining Shanghai's relationship with its waterfronts under the motto "Better City, Better Life". In order to prepare for the construction of the World Expo site, the Shanghai Urban Planning Bureau organized a series of conceptual design competitions in 2000 and 2001, won by SOM. The plan included space for modern office buildings, commercial buildings, residential buildings, and leisure facilities after the Expo.

Before the expo started a second renovation of Shanghai's prime waterfront public space on the Bund took place in 2008, which was costly, it buried a ten-lane high-intensity motorway underground in favor of pedestrian space along the water with views on Lujiazui's skyline. This provided the ideal starting conditions for the 2010 Shanghai World Expo, and an opportunity to improve Shanghai's modern service functions.

Although Lujiazui's development was the first urban megaproject along the Huangpu River, especially the Expo formed a trigger to start a dozen large redevelopment project, or urban mega projects along the Huangpu [10], accompanied by biennials, e.g., four editions of the Shanghai Urban Space Art Seasons. The Huangpu River became a main display platform for Shanghai's urban culture. The rigor and scale of the already implemented and ongoing projects is unprecedented, both in scale as well as in terms of quality [9]. According to Shanghai's current master plan, the Huangpu River waterfront is earmarked to be a "demonstration zone for the development capability of the global city of Shanghai" [27]. The Huangpu River waterfront space became a showcase for innovations and—in the Chinese context—new forms of urban design [10]. Compared to the usual urban planning practice in contemporary China, the projects along the Huangpu indeed introduced new elements, such as walkable and car-free environments, functional mixing, and the inventive use of underground spaces.

Less than two decades ago ships rolled off the ramps and today, about half of the planned 120 km of regenerated waterfronts has been realized with lush greenery and trails. This landscape forms the backdrop for impressive real estate projects, skyscrapers, and malls, that are in aesthetic quality and appeal surpassing other world ports as New York, London, Tokyo, and Singapore. As global city Shanghai relinks with the global economic system in a strive to compete for global resources, enterprises, and talents [24].

6 Shanghai Reconnected with the River and with the World as Excellent Global City

Worldwide, urban planning and design have been influenced by neoliberal urban policy experiments [3, 25], especially also along waterfronts. Also, in Shanghai since the late 1990s, urban planning practices have been increasingly combined with market-driven developments. The emergence of speculative urban megaprojects along the Huangpu River is characteristic of what Harvey identifies as 'neoliberalism with Chinese characteristics' [16]. This trend in a way represents a departure from the socialism-with-Chinese-characteristics approach and emphasizes an adaptation of Marxism-Leninism ideals to local Chinese conditions. It aims to improve the quality of life for millions by stimulating the national economy.

Shanghai is aiming high to play a leading role in enabling a sustainable transition. Its urban planning and design promises are well intended and of high-quality in their elaboration. Recent (master) plans and practices show that Shanghai's city leaders and planners are aware of environmental challenges and that the city is willing to play a leading role in a green transition. Shanghai is promising to show how ecological values can be combined with other aims, such as flood defense, place-making, and the preservation of industrial heritage, on a large-scale and in the context of high building density, land scarcity, and booming real estate prices.

But what makes these new waterfront related projects sustainable? First of all, the removal of polluting industries and building low-carbon buildings in return reduces local emissions—though the outplaced modernized but enlarged new port areas still cause emissions. This removal of industries not only improved the city's image, but it also created space for a large amount of green and a habitat for species. Innovative in the context of Shanghai and China is also the large amount of green and public space aimed for recreational use. The formerly inaccessible waterfronts have been made more porous and accessible to the public, changing the waterfront into a very appealing urban space.

At first sight the impressive new waterfronts along the Huangpu River are a successful effort to improve quality of living in a congested city, impressive because large-scale, visual attractive, and with high impact. The general quality of the new waterside projects is very high, but yet fails to fully utilize all potentials. There are still mismatches with daily life reality on the ground. Most of the new waterfront spaces still run short on sufficient daily life facilities (benches, places for drinks) and there are functional limitations [9]. A main problem is uncertainty about the actual need for the huge amount of new office spaces, which is still largely unoccupied, and there is an absence of affordable housing—existing neighborhoods even have been removed in favor of speculative real estate [6].

All buildings meet the National Green Building standard; in many cases even international labels such as LEED are obtained, although the labels are questionable [8]. In short, the megaprojects along the Huangpu River appear to be primarily oriented to support the desired image of a global city, and to stimulate the economy with new investments and tourism. Accessibility and usability are not optimal yet:

only a relatively small segment of the new middle- and upper-class profits. The new waterfronts mainly facilitate a new white-collar (upper) middle class and tourists. This is not an inclusive development, as meant in the Sustainable Development Goals, especially in the context of replacing affordable working-class neighborhoods with exclusive—and still for a large share empty—real estate. Nevertheless, a lot of improvements have been achieved.

China's internationalization is a quick process, characterized by experimentalism, adjustability, and flexibility, supported by political vigor (financially strong, quick changes and decisions are possible, all landownership belongs to the government). This is admirable, but vigor and drastic spatial or economic measures don't automatically result into the realization of expectations, despite good intentions. A more inclusive approach is needed, aimed for all people, and not only a selective group. Involving more stakeholders will generate new insights and innovative ideas. This is especially needed in a port city that claims to have an open culture, as window to the world.

The polluting industry is gone, and the waterfront increasingly resembles the classic active image as in the Qingming Shanghetu. The new waterfront generates money as a setting for major real estate projects, and the visual quality of the city is high, capable of attracting tourists and investment. The promises for more urban greenery and improvement of the living environment have been fulfilled. This phase arises from economic motives, which serve a larger purpose, but does not yet fit well with local needs. Hopefully that will come soon in a next phase. Inclusiveness is needed, as promised in the master plan, hopefully with continuing (international) academic and non-academic exchanges, and full social engagement of local citizens. As a result, the waterfront will act as platform to facilitate knowledge exchange and broader involvement of all stakeholders, including citizens, aimed at stimulating real innovation, ultimately leading to an inclusive sustainable transition that goes beyond a beautiful image.

Disclaimer This chapter is based on a compilation of previous articles by the authors, and a further elaboration thereof. Thanks to Xiaojing Wang for help with translation assistance and interpretations.

References

1. Bergère MC (2014) Shanghai history: the road towards modernity. Shanghai Academy of Social Sciences Society, Shanghai. [Translated from Chinese]
2. Bird JH (1971) Seaports and seaport terminals. Hutchinson, London
3. Brenner N, Theodore N (2002) Cities and the geographies of 'actually existing neoliberalism.' Antipode 34(3):349–379
4. Chen H (2010) Shanghai Old industry. Shanghai part 1: Hai People's Fine Arts Publishing House. [Translated from Chinese]
5. Chen Y (2003) Shanghai, a port-city in search of new identity: transformation in the bond between City and Port. In: Carmona M (ed) Globalization And City Ports: The Response Of City Ports In The Southern Hemisphere. Delft University Press

6. Den Hartog H, González Martínez P (2022) Integrating heritage assets in large commercial complexes: de-contextualization and re-signification of memory in Shanghai. Habitat Int 126https://doi.org/10.1016/j.habitatint.2022.102601
7. Den Hartog H (2010) Shanghai new towns: searching for community and identity in a sprawling metropolis. Rotterdam: 010
8. Den Hartog H, Sengers F, Xu Y, Xie L, Jiang P, De Jong M (2018) Low-carbon promises and realities: lessons from three socio-technical experiments in Shanghai. J Clean Prod 181:692–702https://doi.org/10.1016/j.jclepro.2018.02.003
9. Den Hartog H (2019) Re-defining the appreciation and usability of urban watersides in the urban center and Peri-urban fringes of Shanghai. Eur J Creat Pract Cities Landsc 2(1):37–64.https://doi.org/10.6092/issn.2612-0496/8918
10. Den Hartog H (2021) Shanghai's regenerated industrial waterfronts: urban lab for sustainability transitions? Urban Plan 6(3):181–196. https://doi.org/10.17645/up.v6i3.4194
11. Ding F, Wu J (2018) Post-industrial city in the context of globalization research on the mechanism of urban waterfront rejuvenation: taking Shanghai, Huangpu and Jiangxi an example of shore. Modern Urban Res (1):25–34. [Translated from Chinese]
12. Ding F, Wu J (2019) Modern and contemporary history and characteristic analysis of the waterfront development of Shanghai Huangpu river. Housing Technology. [Translated from Chinese]
13. Foster HD, Lai DC, Zhou N (1998) The Dragon's Head: Shanghai. China's emerging megacity. Western Geographical Press
14. Gil I (2008) Shanghai transforming: the changing physical, economic, social and environmental conditions of a global metropolis. Actar, Barcelona
15. Harvey D (1989) From Managerialism to Entrepreneurialism: The Transformation of Urban Governance in Late Capitalism. Geografiska Annaler Ser B Hum Geogr 71(1):3–17. https://doi.org/10.1080/04353684.1989.11879583
16. Harvey DA (2005) A brief history of neoliberalism. Oxford University Press, New York
17. Hoyle BS (1989) The port-city interface: trends, problems, and examples. Geoforum 20(4):429–435
18. Lanning G, Couling S (1921) The history of Shanghai. Shanghai: Kelly & Walsh
19. Li J, Yu Y (2012) Huangpu River waterfront area Cultural strategies for the reuse of industrial remains. Urban Const Archit (3):34–36. [Translated from Chinese]
20. Marshall R (2004) Shanghai's waterfront: presenting a new face to the world. In: Rowe PG, Kuan S (eds): Architecture and Urbanism for Modern China, Shanghai. Prestel, Munich
21. Meyer H, City and Port (1999) Urban planning as a cultural venture in London, Barcelona, New York, and Rotterdam: Changing relations between public space and large-scale infrastructure. International Books, Utrecht
22. Roach SS (2019) China is leading the world in sustainable development. https://www.weforum.org/agenda/2019/09/china-leading-sustainable-development/
23. Rowe PG (2011) Emergent architectural territories in East Asian cities. De Gruyter, Basel
24. Sassen S (2008) Disaggregating the global economy: Shanghai. In: Gil I (ed) Shanghai transforming: the changing physical, economic, social and environmental conditions of a global metropolis. Actar, Barcelona
25. Sassen S (2014) Expulsions, brutality and complexity in the global economy. The Belknap Press of Harvard University
26. Sha Y, Wu J, Ji Y (2014) Shanghai urbanism at the medium scale. Springer, Berlin, Heidelberg. [Translated from Chinese]
27. Shanghai Municipal People's Government (2016) Shanghai master plan 2017–2035: striving for the excellent global city
28. Shanghai Municipal People's Government (2018) Shanghai "One River and One Creek" Waterfront construction plan (2018–2035)-striving for a world-class waterfront area: Shanghai Huangpu river and Suzhou river planning
29. Som SF (1999) Shanghai waterfront redevelopment plan-a vision for the transformation of the Huangpu River. Jianzhu: Dialogue, vol 23, pp 68–73

30. Sun P (1999) Shanghai urban planning chronicle. Shanghai Academy of Social Sciences Press, Shanghai. [Translated from Chinese]
31. Wu J (2009) The 2010 world expo promotes the urbanization of Shanghai planning and construction. Time + Architecture (4):20–23. [Translated from Chinese]
32. Ye G (2007) Following traces and renewing: the city of Shanghai planning evolution. Tongji University Publishing She, Shanghai. [Translated from Chinese]

Policy Advances in Urban Sustainability

How 15-min City, Tactical Urbanism, and Superblock Concepts Are Affecting Major Cities in the Post-Covid-19 Era?

Luca Maria Francesco Fabris, Federico Camerin, Gerardo Semprebon, and Riccardo Maria Balzarotti

Abstract This chapter analyses three strategies proposed to redefine current urban policies to deal with issues inherited from the contemporary city evolution. The case study analysis focuses on applying the concepts of 15-min City, Tactical Urbanism, and Superblock in global cities such as Barcelona, Shanghai, and Milan. Have these cities changed the urban environment and mobility patterns dealing with health, social, and economic inequities? Which have been the impacts of urban regeneration, governance, and inclusion towards achieving the Sustainable Development Goal 11, emphasizing the need for inclusivity and equitability in urban areas? These questions find the answer in three main aspects. First, the regeneration of the existing built environment; second, short-, medium-, and long-term governance issues; third, the concerns about the possible risk of gentrification. An introductive part explains the adopted methodology, follows an analysis of the three case studies, and, eventually, remarks on what we learned. Two are the primary outcomes: a comparison between different global cities and diverse ways to deal with the impacts of people-centered solutions for urban environments and an evaluation of 15-min City, Tactical Urbanism, and Superblocks feasible solutions for sustainable urban transition.

Keywords Global cities · Urban regeneration · Sustainable development · Healthy city · Resilience

---

L. M. F. Fabris (✉) · G. Semprebon · R. M. Balzarotti
Politecnico di Milano, Milan, Italy
e-mail: lucamariafrancesco.fabris@polimi.it

G. Semprebon
e-mail: gerardo.semprebon@polimi.it

R. M. Balzarotti
e-mail: riccardomaria.balzarotti@polimi.it

F. Camerin
Universidad UVA de Valladolid-Universidad Politécnica de Madrid (Grupo de Investigación en Arquitectura, Urbanismo y Sostenibilidad), Madrid, Spain
e-mail: federico.camerin@uva.es

Z. Allam (ed.), *Sustainable Urban Transitions*, Urban Sustainability,
https://doi.org/10.1007/978-981-99-2695-4_10

1 Introduction

In the last decade, numerous urban models have been emerging in response to worldwide environmental and socio-economic urgencies envisioned in need to provide people-centered urban environments [16, 27]. These models, in turn, have been based on a raft of global agreements such as the UN's New Urban Agenda [68] and the 2016 Paris Climate Agreement [36] to foster accessible, healthier, inclusive, safer aimed to prevent climate change effects from becoming worse.

Within the current 'urban era' [34], the main targets are large cities, especially 'global' ones, because they are vigor places that can enable and accelerate flows of capital, goods, and knowledge. Among the solutions to reduce socio-economic and ecological inequalities, the notions of 'Tactical Urbanism,' 'Superblock,' and '15-min City' have gained support to provide a more 'human-needed city' [17]. Tactical Urbanism [43], also known as DIY Urbanism, Planning-by-Doing, Urban Acupuncture, or Urban Prototyping, is a recent US-based urban planning movement that envisages short-term and community-based projects to carry out quick, often low-cost, and creative actions [44, 45]. Starting in the 2010s from American grassroots, Tactical Urbanism has diffused globally thanks to its easy way of producing temporary changes for human-centered public spaces at the street level [67]. Superblocks and 15-min cities are iterations of the original idea of 'neighborhood units,' i.e., residential design models comprising several blocks usually closed to through traffic for a population of about 5,000–9,000 residents, with school, worships, and recreational areas at their center. In 1916 William Drummond [25] introduced what was successively developed by Clarence [56], who, in turn, took inspiration from Ebenezer Howard's 'Garden Cities of Tomorrow.' The idea of neighborhood units has evolved over more than a century according to locally oriented debates [14]. It has recently been implemented as 'Superblocks,' such as in the cases of Barcelona [58], 'Megaplots' in expanding Chinese metropolis [41], and a '15-min City' in Paris [12].

These solutions have quickly gained popularity during the global lockdowns prompted by the Covid-19 pandemic for delivering urban sustainability transition and converting cities into healthier, safer, and more inclusive places [26, 51]. In such a context, it is relevant to understand better this change, its impact on the urban environment, and its implementation limits.

The chapter analyses three strategies proposed to redefine current urban policies, especially urban mobility, to deal with issues inherited from the capitalist city evolution. The research questions are: May these global cities' experiences have changed urban environments and mobility patterns and dealt with health, social, and economic inequities? What impact have urban regeneration, governance, and inclusion on achieving Sustainable Development Goal 11, emphasizing the need for inclusivity and equitability in urban areas?

These questions find answers through a desk research activity, and specific fieldwork carried out between 2017 and 2022 in the cities under analysis. These activities have helped the Authors understand how Barcelona, Shanghai, and Milan have implemented these concepts, dealing with three main aspects of each city. First, the regeneration of the existing built environment; second, short-, medium-, and long-term governance issues; third, the concerns on the possible risk of gentrification as the improvement in the quality of citizens' life may lead to the rise of real estate values. The research's structure is the following: an introductive part with the explanation of the methodology adopted, an analysis of the three case studies, and, eventually, the conclusions remarking on what we learned. In particular, the analysis results in two primary outcomes: first, a comparison between different global cities and diverse ways to deal with the impacts of people-centered solutions for urban environments; second, an evaluation of how 15-min City, Tactical Urbanism, and Superblocks feasible solutions for sustainable urban transition, although with limits, could be applied in other contexts through a correct assessment of these three variables.

2 Barcelona: Extending the Neighborhood Unit's Principles to the Whole City

2.1 Built Environment Transformation

The implementation of Superblocks in Barcelona developed in the frame of the so-called 'Ecological Urbanism' movement fostered by the local entity BCNEcologia [59]. The idea is to divide the city into 503 neighborhood units to adapt according to the urban morphology, such as the standardized blocks of Cerda's Extension that would form 400-x-400-m urban cells, including interior and exterior streets. This model is expected to be fully applied in 2030 and would strongly impact the citizens' lives by diminishing private motorized transport by 19.2% and speed limit to 10 or 20 km/h in the interior streets, and expanding green surfaces from 2.7 m^2/inhabitant to 6.3 m^2/inhabitant in the Extension district. These changes at the street level would, in turn, lower NO_2 pollution by 24% and heat waves by 35.9% [53]. Doing so is expected to prevent 667 premature deaths annually and increase life expectancy by almost 200 days.

The launch of the local government document "Let's fill the streets with life" [6] was the ground zero for two pilot projects in Poblenou and Sant Antoni neighborhoods in the pre-pandemic period (2016–2019). Poblenou's Superblocks (September 2016–January 2017) is a small-scaled project located on the border between "El Parc i la Llacuna del Poblenou" and "El Poblenou" neighborhoods, covering 1.6 ha in which are living 5,800 inhabitants for a total density of 348 inhab/ha,[1] while Sant Antoni urban cell covers the entire neighborhood (80.10 ha, 38,182 residents, and a total density of 479 inhab/ha).[2]

[1] https://ajuntament.barcelona.cat/superilles/ca/content/poblenou.

[2] https://ajuntament.barcelona.cat/superilles/ca/content/sant-antoni.

Despite showing different intervention scales, the two interventions radically transformed the public spaces by cutting vehicular traffic to create human-scale communities [7: 104, 126]. Poblenou neighborhood involved the conversion of 13,350 m^2 of streets for pedestrians and cutting car-parking spaces from 575 to 316 for creating human-centered facilities and equipment. They were the planting of 212 trees and 349 benches, the increase of green areas and tree-lined streets from 9,722 to 18,632 m^2, the creation of 1,000 m^2 of cycle path, a new 2,483-m^2 playground and interactive games area for both elderly and young (Fig. 1) that included an athletics track (Fig. 2), 37 new premises for activities at street level; and an electric vehicle charging point. The implementation of Sant Antoni Superblock occurred in two phases. The first one regarded rehabilitating the 53,388-m^2 Sant Antoni market (Fig. 3) and its immediate surroundings' open spaces. The works ended in May 2018 for a total cost of 80 million euros and provided a new public square (1,800 m^2) plus 3,200 m^2 of green areas. The second phase regarded the redevelopment of 38,328 m^2 of streets that resulted in the pedestrianization of 23,709 m^2 of public spaces, the realization of three tactical urbanism actions (Fig. 4) for a total of 17,947 m^2, the improvements of 1,272 m^2 of tree pits, and the cut of 838 parking lots.

These initiatives have demonstrated the capacity to reduce environmental factors that affect human health [2]. Vehicular traffic decreased by 58% within the Poblenou unit, with a consequent fall in daytime noise levels of five decibels. Although no data on the changes in air quality has been available, Sant Antoni registered a decrease

Fig. 1 A playground area for children. *Source* F. Camerin (April 2022)

Fig. 2 The Poblenou's athletics track. *Source* F. Camerin (April 2022)

in concentrations of air pollutants NO_2 (25%, $-14{,}6$ g/m^3) and PM10 (17%, $-4{,}1$ μg/m^3) since the implementation of the Superblocks system. The success of these experiments leveraged the implementation of Superblock-related measures in the whole Extension district with the slogan "green axes" (*eixos verds* in Catalan). The completion launched by the City Council in late 2020 aimed to convert the area with the most daily vehicle traffic flow (350,000 cars/day) and high levels of pollution (50 μg/m^3 on overage in 2019, 10 μg/m^3 more than the WHO recommended 40 μg/m^3) into a healthier place. The aim is to turn 21 streets into green hubs and create 21 new squares at the crossroads, devoting 33.4 hectares to pedestrians and 6.6 hectares to urban green space filled with 4,000 new trees [9].

2.2 Urban Governance

Two are the main aspects related to governance. The first one regards the high cost and time of implementation. While Salvador Rueda urgently claimed to convert Barcelona into a great Superblock by 2030 for a total expenditure of approximately 300 million euros [47], the City Council still does not offer any information. Poblenou's and Sant Antoni's Superblocks cost 3.8 and 10.5 million euros, respectively, while the expenditure for the green axis project in Eixample, finished by 2030,

Fig. 3 The Sant Antoni market's surroundings. *Source* F. Camerin (April 2022)

rises to 37.8 million euros. Up to date, the City Council financed all Superblocks-related interventions, but the latter has still not provided an official plan or scenario for completing the task.

The second is the need for a shared vision with residents. Poblenou's Superblock was a tactical urbanism experiment that introduced removable items to show how the urban cell may work but was not the result of previous public consultation with the local population. The loss of driving privileges benefitted slow mobility: for instance, no underground parking replaced the removed parking lots in Poblenou and Sant Antoni neighborhoods, apart from the Sant Antoni market's 390 underground parking. In addition, the unclear potential of the Superblock caused citizens to set up the "Association of People Affected by the Poblenou Superblock" (*Plataforma d'Afectats per la Superilla del Poblenou*). The association lobbied the administration to reverse the intervention, drawing intense media attention [54]. However, a closer consultation with residents resulted in more structural changes and resolved conflicts. The lessons learned for successfully implementing Superblocks have been mutual for both public authorities and citizens. This first experience showed that the move from tactical to structural cannot be too quickly or slow, but this action needs the community to figure out what it wants. Poblenou Superblock was unconsciously conceived as a low-level shock therapy to kick-start the process. Residents could not express whether they needed or wanted people-centered public space; they were instead faced with it and successively asked what they like to make of it. As a result,

Fig. 4 A Tactical Urbanism intervention at a crossroad in Sant Antoni neighborhood. *Source* F. Camerin (April 2022)

the following Superblock in Sant Antoni was solidly developed through open debates with citizens and monitoring activities.

Third, the pandemic acted as an accelerator of an already more comprehensive policy committed to providing an accessible, resilient, safer, and sustainable environment in the last ten years. A pivotal document enabling this challenge was the "Citizen Commitment to Sustainability 2013–2022" [3], followed by ad hoc tools such as the "green infrastructure and biodiversity plan towards 2020" [4] and the "2013–2018 urban mobility plan" [5], along with scientific research provided by Salvador Rueda's team [60]. This forward-thinking policy demonstrated the capacity to elaborate short- and medium-term strategies resulting in the gradual reorganization of Barcelona's urban grid towards a greener and more human-centered environment with less pollution and noise [8].

2.3 Risk of Gentrification

Since the celebration of the 1992 Olympics, Barcelona has been experiencing a relevant growth in the cost of living, tourist gentrification flows, and socio-economic inequalities [62]. Superblocks demonstrated to improve community green and social infrastructure and provide a healthier environment so that they can be at risk of rising rents with a consequent displacement of long-term residents. Flourishing literature demonstrates how human-centered programs like Superblocks are catalysts for the so-called 'environmental gentrification' and how the resilient and sustainability framework enables this process [13]. Housing displacement and gentrification are currently concerns in both Poblenou [55: 780] and Sant Antoni [15]. However, real estate market value trends have appeared in line with the average prices, not suffering from the Superblock-related intervention. The housing and rental costs data prove this claim [10]. El Parc i la Llacuna del Poblenou and El Poblenou neighborhoods' price €/m^2 of second-hand houses for sale increased respectively from 3,761 and 4,223 €/m^2 in 2016 to 4,268 and 4,469 €/m^2 in 2021, thus exceeding the average price of Sant Martí district (from 3,382 €/m^2 in 2016 to 3,563 €/m^2 in 2021). Sant Antoni neighborhood's price €/m^2 of second-hand houses for sale deflected from 4,591 €/m^2 in 2017 to 4,170 €/m^2 in 2021, is in line with the slight decrease of the average price of Eixample district (from 5,005 €/m^2 in 2017 to 4,538 €/m^2 in 2021).

In addition, recent government measures demonstrated policymakers' awareness of limiting the social risks related to eviction, displacement, and inequitable urban development at the regional and local levels. On the one hand, in September 2020, the Catalan Parliament passed a law regulating rent prices on new housing contracts (approved in December 2020) to guarantee accessibility in sixty Catalan municipalities, including Barcelona [24]. On the other hand, the City Council's Barcelona Superblocks Government Measure [11: 8] declared that Superblock-related urban regeneration strategies are eager to avoid gentrification.

3 Superblocks and the 15-min City in China: Insights from Shanghai

3.1 Built Environment Transformation

Superblocks, or 'Megaplots,' constitute the central spatial units of contemporary Chinese developmental cities [41: 10]. These fenced compounds usually lay on parcels of land measuring between 300 and 500 m in length and width. They physically separate the housing communities enclosed in semiprivate neighborhoods from the external urban reality, often consisting of fast circulation roads. Repeated standard condominiums dot the area inside the gates indifferently, leaving space at the ground level for greeneries, car parks, and various amenities for children and elderlies (Fig. 5). The Superblock model, whose density and covered area ratio vary according to planners' or developers' desires, represents an efficient mechanism for optimizing land values and creates a point of encounter between state and entrepreneurial interests [29]. As a result of *tabula rasa* land clearings, Superblocks usually follow so-called rationale planning and scientific method based on generic zoning plans,

Fig. 5 Typical Chinese Superblock as seen from inside a housing unit. Shanghai, Minhang District. *Source* G. Semprebon, September 2017

which leave little space for the definition of cohesive, diversified, and inclusive urban environments [19].

The Megaplots' size requires the developers to provide «large capital reserves and high political standing and must also possess the operational and financial capacity to produce a megaproject» [50: 47]. Consequently, the bureaucratic process is fastened by the reduced number of transactions that put the top choices in the developers' hands. The market forces are then predominant in shaping expanding Chinese cities, which are territories often lacking a sense of urbanity, spatially and socially segregated from the serendipitous vitality of mixed used-and-shaped urban fabrics (Fig. 6).

Whether the Superblock is the result of a contamination history with exogenous models or is the evolution of a distinctive one, the Chinese version has been widely debated. For instance, Miao [49] saw in Shanghai's new gated communities a copy of the US suburban model based on social and spatial segregation and questioned whether Chinese cities would have «indiscriminately model themselves» after foreign settlement patterns. Successively, Xu and Yang [74] argued that the idea of gating dwelling spaces belongs to a rooted local tradition that has been able to adapt to historical and socio-political situations.

Criticalities on the Superblocks model have also emerged recently. Cheshmehzangi and Butters [18] identified housing density applied to urban morphology as a key to pragmatically unpack possible alternative settlement principles more sensitive to social and ecological dynamics, such as climate change and neighborhood gentrification. At the same time, recent research by design [40, 50] suggested how redeveloping gated communities could have reshaped public–private relationships by maintaining high densities. These issues appear today emphasized by the sequence of

Fig. 6 Typical Chinese Superblock as seen from the outside. Shanghai, Minhang District. *Source* G. Semprebon (September 2018)

unexpected shocks that have perturbated geopolitical equilibriums in the last years. One of these was the slump in China's housing market, epitomized by the Evergrande debt crisis [31]. Observers are following how the story will unfold and whether the situation will be absorbed by the system or will reverberate globally. What is sure is that the episode has thrown shadows on the current real estate machine, substantially based on urbanization via Superblock development.

3.2 *Urban Governance*

Since the formulation of the Shanghai Master Plan 2017–2035, "Striving for the Excellent Global City" [64], some key concepts and implementation formulas have been enucleated to move forward with new eco-oriented forms of urban living. Under the aegis of strengthening its domestic and international attractivity, Shanghai has been the first megalopolis in China to have identified the 15-min City concept, also named 'Community life circle,' as a core developmental strategy [66]. In 2016, the urban planning and land resources Administration Bureau released a guidelines manual. The book contains the first comprehensive research on the 15-min City concept applied to Shanghai and collects data from interviews, questionnaires, and practical implementations. The most critical aspects emerging from this document were two.

On the one hand, the attempt to define a benchmark for the 15-min life circle construction as part of a broader urban renewal action. On the other hand, the

Fig. 7 The Columbia Circle has hosted the Shanghai Urban Space Art Season 2021, focusing on the 15-min city. *Source* Gerardo Semprebon (April 2019)

government will engage communities from below, promoting grassroots participation [65]. Both factors entwine inextricably with Chinese Superblocks' strategic approach to city-making, questioning, for instance, the relations materializing between neighborhoods supersizing and communities' accommodation.

The Shanghai Urban Space Art Season 2021 can be framed as a cultural response to the issues that materialized in the last decades by enforcing the Superblock model. The exhibition has opened a reflection on the possible implications of the 15-min City, sharing practices of local neighborhoods to create community life circles and calling for exemplary cases of urban renewal, such as Chaoyang and Xinhua Communities.[3] The critical exhibition site itself, the Columbia Circle, is the result of a process of urban renewal focused on the community's reactivation which saw the participation of renowned architectural offices like OMA and West8 (Fig. 7).[4] In parallel, the government is fostering the life circle community concept also in the suburbs [30] as part of a broader plan to use fifteen pilot projects to monitor the 15-min city impacts on residential communities [66].

The 15-min City strategy is still in its experimental phase, and some criticalities have already emerged. For instance, in Guangzhou, Zhou found that the implementation is imposed with excessive uniformity regardless of local differences (2019). The principal reason resides in the spatial characteristic of housing compounds that

[3] https://www.susas.com.cn/en/home.html.

[4] https://www.oma.com/projects/columbia-circle and https://www.west8.com/historic_columbia_circle_revitalization_project_in_shanghai_opened/.

substantially repeat ubiquitously and generate homogeneous urban environments. According to Hou and Liu, the 15-min City concept describes an urban portion that should cover between 3 and 5 km^2 hosting 50–100 thousand people [37]. Apprehensions of social inequalities have already been pinpointed as well. For instance, in some cases, the concept of "walkable neighborhoods" targeted groups differing in socioeconomic status or age, reducing urban inclusivity [70]. Moreover, Wu et al. [72] denounced that the 15-min concept was not adherent enough to a supply and demand analysis.

3.3 Risk of Gentrification

Alongside the mass housing bubble, residential segregation and gentrification issues have emerged. For instance, the impressive numbers of the real estate industry [20, 71] coupled with housing problems for a large slice of the population, mainly the migrant workers. Shanghai has a relatively low degree of segregation compared to Beijing or Guangzhou [42]. Nevertheless, such issues remain unsolved and generate striking frictions, i.e., unfair compensation, people relocations, and even homelessness [45, 73].

Gentrification and speculation are the two reasons why the Shanghai Master Plan 2017–2035 set the goal of realizing at least 8–10% of housing dedicated to low-rent housing, public rental housing, common-property rights housing, and resettlement housing [64]. The Master Plan for 2035 portrays a scenario in which Shanghai will drastically increase the presence of public space across the city: providing parks and squares covering over 400 m^2 within five minutes walking distance, furnishing inhabitants with at least 4 m^2 of community public space per capita; distributing 99% of public facilities within fifteen minutes walking distance in communities; and reducing to 2,5 km the average travel distance for the daily life needs [64]. Today, Shanghai's developmental agenda sets urban renewal as a necessary and complementary strategy to current estate-driven development to improve living quality by moving toward a human-scale built environment.

Unfortunately, the COVID-19 outbreak and the subsequent *zero-COVID policy* have delayed the roadmap. The harsh restrictions imposed to prevent contagion have exerted tremendous socio-political pressures on Superblocks. They became de facto large-scale trial samples unveiling communities' reactions in concomitance to unexpected and extraordinary situations. In Spring 2022, the images of locked Shanghai showcased the effects of a megalopolis sealed for the second time, with high-rise buildings and gated communities assuming more of the characteristics of reclusion structures rather than family places (Fig. 8). On one side, Superblocks proved the capacity to contain the population, and consequently, the virus spread, but, on the other side, their inhabitants paid a steep price in terms of human experience. Even after the lockdown, reports warned about the side effects of such a prolonged measure on people's mental health, encapsulating an unprecedented condition of fragility in urban residents. Among the impacts on local communities, the dismantling of close

Fig. 8 Distribution of primary goods inside a Shanghainese Superblock in Minhang District. *Source* Sergei Khasikov (April 2022)

social ties and the rebuilding of new ones have taken place, not only dictated by the need to access primary goods but also being subject to authority's control, reshaping neighborhood relationships [69]. Even after the relief from Spring 2022's restriction, the detection of new positive cases has determined new local confinements, creating frustration among many residents who fear new enclosures.

These recent experiences prove that the fate of mass housing hosted in the Superblocks still represents a crucial challenge to reshape the pact between citizens and places, especially in an economic system fueled by the real estate industry like the Chinese one.

4 Milan and the Attempts for a 15-min City and Tactical Urbanism Experiments

4.1 Built Environment Transformation

Covid-19 has extensively highlighted Milan's issues regarding poor connections, demanding access to local services, and public open areas. However, the city leveraged this situation to give a fast programmatic response based on the 15-min City strategic document 'Milan 2020. Adaptation strategy' that fostered an innovative urban and mobility plan named *Strade Aperte* (Open Streets) [21, 22] to answer the pre- and post-pandemic problems [1]. Tactical Urbanism tools and examples were a consistent part of this document.

This rapid response based its *raison d'être* on an ongoing renewal process that started with the 2012 PGT *Piano di Governo del Territorio* (the new city Master Plan). The PGT identified the so-called NIL *Nuclei di Identità Local* (Local Identity Cores), a scientific mapping of 88 neighborhoods characterized by geographical, social, and morphological affinities, much smaller than the administrative division of Milan into nine sub-municipalities. Their goal was to underline local problems, identify specific needs, and discover the eventual scarcity of life-quality improving elements, such as public services, green areas, safety, and schools, in a very capillary way.

A second key element in the urban renewal process of Milan appeared in 2018. The City Council launched the *Piano Quartieri* (Neighborhoods Plan), an annual program that defines and finances urban interventions to solve some of the issues underlined by the NIL mapping. This plan included an essential strategy for the future development of the post-pandemic Milan: the *Piazze Aperte* (Open Squares) program. It was an urban project [35] that sought to recover and create new public spaces. The primary tool to implement it was Tactical Urbanism, which allows fast implementation and low-budget interventions that modify public spaces by re-organizing vehicular traffic and parking spaces using temporary and movable urban furniture and ground paintings.

After the first two interventions in two peripheral areas (*Piazza Dergano and Piazzale Corvetto*), 14 more followed in 2019 (Fig. 9). In late 2019, the second edition of *Piazze Aperte* was open to ideas from citizens and associations, but the pandemic outbreak caused the open call to stop temporarily.

The Milanese administration converted the pandemic into an occasion to experiment even more with the Tactical Urbanism strategies. The programmatic document *Strade Aperte* introduced the 15-min City objective for Milan to be gradually achieved through two actions. The first was the creation of cycling paths and public open spaces for pedestrians using Tactical Urbanism tools, i.e., ground painting, temporary separations and urban furniture, and new road sections that accommodate cycling paths just by reshuffling the vehicular and parking lanes order. The second was the possibility for bars and restaurant owners to occupy public surfaces free of

Fig. 9 Image of Piazza Spoleto, one of Milan's first Tactical Urbanism interventions. *Photo* Riccardo M. Balzarotti (2021)

charge, even subtracting parking and part of the road, following a specific rule called *Delibera Tavolini* (Dehors Tables Act), issued in May 2020 (Fig. 10).

In late 2021, *Piazze Aperte* and *Strade Aperte* spread across Milan with 67 km of new cycling paths [48], and, since 2018, 38 new public squares have been created by subtracting redundant vehicular space. These 38 local interventions, displaced mainly in peripheral areas, led to the re-organization of 22,000 m^2 of pedestrian open space and the installation of 250 benches, 310 vases with plants, 380 bike stalls, and 35 public tables, and 32 table-tennis units. Despite the quantitative data, it seems too early to have a significant scientific output of indicators, such as traffic and air pollution. Still, it is possible to assess from the on-field surveys that the new squares enormously improve the perception of public space quality. Also, these programs are not the only urban strategy used by Milan to increase life quality standards. Alongside Tactical Urbanism, more traditional urban and strategic planning is making progress in creating bigger-scale interventions, such as public parks and soft mobility infrastructures, as planned in the updated version of the PGT [23].

Fig. 10 Via Borsieri parking occupied by bars and restaurants dehors. *Photo* Riccardo M. Balzarotti (2021)

4.2 Governance

Tactical Urbanism interventions may insignificantly increase services, availability of public open spaces, and cycling paths. The City Council reported a 22.000 m^2 growth of pedestrian space, which, compared to the size of the sole Piazza Duomo—39.000 m^2—tends to mislead the assessment of the value of the Tactical Urbanism program.

Compared to the total number, size, and the many different typologies of public spaces in the metropolitan area [57], the *Piazze Aperte* and *Strade Aperte* initiatives cover a minimal surface. Nevertheless, Tactical Urbanism interventions may have a relevant impact in the following way. First, there is the possibility of constructing high-speed and very low-budget solutions, according to the so-called LQC approach: Lighter, Quicker, and Cheaper [46]. The flexibility of light and not-expensive interventions allows more radical choices that can be adjusted after a period of observation and accordingly be more responsive to the variations of social needs and public perception of the public space [61]. By working on modifying existing situations, Tactical Urbanism provides the addition of public space while reclaiming it from other previous use, mainly vehicular. Second, the small scale and fast activation allow the capillary diffusion in different areas of the city, which is an excellent strategy to fulfill the 15-min City goal. The City Council mapped the 38 built interventions marking around an 800 m radius, considered the 15-min influence area.

According to this map, they could state that 1 out of 2 citizens benefited from the availability of a new public open space.

However, negative outputs occurred. First, using low-budget solutions made Tactical Urbanism interventions subject to fast aging and decaying, vanishing the improvement in terms of quality perception of the public space. The short-term modifications risk becoming ineffective when they lose their original appearance. For long-term results, they must become permanent solutions. The mistake was to consider Tactical Urbanism as the end of a process when it should be the first step due to the ease and rapidity of its implementation. At the end of 2021, three Tactical Urbanism squares became permanent (*Piazza Dergano, Piazza Angilberto II, and Piazza San Luigi*) (Fig. 11), realizing a conversion project able to preserve the original concept while using more traditional and durable materials.

Another element that recently showed the necessity of transmutation to permanent solutions is using plants in standalone vases. The current and exceptional drought hitting Northern Italy during spring–summer 2022 led to the death of many plants in vases because they lacked irrigation and a consistent amount of soil (Fig. 12). This situation vanished the effect of using plants as a quality-increase element in terms of aesthetics, air pollution, and ground heat mitigation [28]. Areas reclaimed from traffic with Tactical Urbanism may become new squares with permanent green spaces and plants. This solution can fulfill the goal of other extensive strategic plans for greenery,

Fig. 11 Piazza Dergano, one of the first two Tactical Urbanism interventions in Milan, become soon after a permanent solution. *Photo* Riccardo M. Balzarotti (2022)

Fig. 12 Via Sardegna, the plants in vases suffer extreme drought. *Photo* Riccardo M. Balzarotti (2022)

such as *ForestaMI* [32], launched in 2018, which aims to urban reforestation by planting 3,000,000 new trees in Greater Milan Metropolitan area by 2030.

A second issue is related to *Delibera Tavolini*. From the public order point of view, the space occupied by the 2,500 new outdoor areas sometimes led to congestion of the sidewalks, noise during the nighttime, and sudden scarcity of parking, generating complaints among the residents. The temporary outdoor facilities, whose realization was the responsibility of the single owners, also had poor quality and aesthetic standards. Other criticisms emerged about the rule's balance between a public utility and private benefits, considering the free-of-charge of the space occupancy a form of relief no longer necessary in the pandemic "normalization." The local administration began charging the occupation fee again from April 2022 with a 20% discount and plans to maintain approximately 80% of the temporary outdoor areas created from 2020, asking for an onerous permanent occupation permit [63].

Critics raised the safety and utility of paths built by just using paint and reducing the space for car lanes, especially the first and most significant intervention of the *Strade Aperte* program, the cycling path in Corso Buenos Aires the main northern access point to the city center.

Despite the accident increase, which rose from 36 in 2019 to 182 in 2021, the cycling traffic experienced a significant increase. The share of bikes, cars, and motorbikes used in 2019 was 5%, 75%, and 20%, respectively, while it became 27%, 50%, and 23% in 2022. Today the Tactical Urbanism bike lane of Corso Buenos Aires has been adapted in its most dangerous intersection and will become permanent, with safer and stronger concrete separation from the car lanes [38].

4.3 Risk of Gentrification

The pre-and post-pandemic planning strategies are experimental and fast adapting systems that are an innovative part of urban renewal; they are just a tiny part of it. Other large-scale masterplans and guidelines, mainly developed through public–private cooperation processes, generated vast recent urban transformations, which include, alongside private buildings, "public" green areas and facilities, such as real estate (*CityLife Area, Garibaldi-Porta Nuova Area*) and cultural-led operations (*Prada Foundation Area*) [52].

The cooperation between the public and the private in creating public spaces will become even stronger in the future with the updated Milan 2030 PGT Master Plan. The most significant challenge is the upcoming development of large urban voids [39] left by the abandoned commercial railway yards.

The benefits of public infrastructures provided by private investors may have benefits in the short term, but in the long term, they may provoke threats of gentrification on the local cores of the NIL. In this idea of maximum diffusion of services and public spaces over the territory, even though market prices usually follow the improvements in citizens' life quality, Tactical Urbanism actions may not significantly influence the real estate market values. The average cost for m^2 in the *Corvetto* area, where one of the first two Tactical squares appeared, shows a minimal variation between 2014 and 2020, passing from a range of 2,700–3,900 €/m^2 for a second-hand house in excellent conditions to a range of 2,800–3,600 €/m^2. In the same period, the average price of the *Farini-Isola* district raised from 3,500–4,400 to 4,300–5,600 €/m^2. This semi-central residential area surrounds the new luxury district of *Garibaldi-Porta Nuova* (whose average prices were almost constant in a range of 7,000–9,500 €/m^2), suggesting a correlation with the proximity of the new public (and fancy) infrastructure built and maintained by private developers.

5 Conclusion

This research is a first approximation to monitor, assess, and compare the implementation modalities of urban strategies such as 15-min City, Tactical Urbanism, and Superblocks in three first-tier global cities. We questioned whether such experiences had changed urban environments and mobility patterns when dealing with health, social, and economic inequities.

The case of Barcelona shows how the Superblocks and 15-min City strategies have coped to reorganize the urban environment more sustainably and efficiently.

The gradual implementation of Superblock-related measures has been strictly related to several tools launched by the City Council over the last decade. While evidence of improvements has been detected, governance issues and risks of gentrification have risen. The path towards the conversion of the whole city into a great Superblock is still at the beginning, but the local administration has challenged the way of organizing the city, influencing a new way of living. In Shanghai, the 'Community Life Circle' concept—the Chinese version of the 15-min City—has been proposed pioneeringly to strengthen neighborhood relationships in an archipelago-like environment, in which urban expansion follows the unquestioned formula of Superblocks construction. Superblocks proved to work as micro-communities regulated by both endogenous and exogenous forces according to the political will, but also as coercive compounds where authoritarian rules have been relentlessly exerted. Allegedly, Tactical Urbanism found little application mainly because of the confined possibilities of ground rent.

Conversely, it has been vigorously implemented in Milan, giving a significant impulse to the capacity of turning pandemic-related difficulties into opportunities. Tactical Urbanism's extensive use contributed to the redesign of some peripheral areas, increasing the availability of public space, sometimes becoming the natural starting point of episodes of urban regeneration. At the same time, Tactical Urbanism interventions had a low impact on housing market prices, mainly because of their small-scale, often requested by local communities, their attractiveness limited to the neighborhood scale, and unable to generate competition between districts. However, their fast and cheap implementation allowed the development of many projects simultaneously, reducing polarization effects.

We identified strengths and weaknesses entangled with the 15-min City, Tactical Urbanism, and Superblocks strategies by framing how:

- they transformed the urban environment;
- they were stewarded during the implementation phases;
- they interacted with gentrification issues.

Each cultural, economic, environmental, and political dimension has profoundly influenced urban development, coagulating distinctive forms and implementation modalities that contribute to understanding ongoing urban transitions.

This research has also been an occasion to reflect upon the possible transferability of actions undertaken under the umbrella of the concepts of 15-min City, Tactical Urbanism, and Superblocks as feasible approaches and strategies for governing urban transitions across geographical contexts.

Interventions related to the 15-min City, Tactical Urbanism, and Superblocks were planned in the pre-pandemic period to sort out inequalities, climate change, and pollution. After the COVID-19 outbreak, air pollution, social vulnerability, and extreme weather conditions were vectors and multipliers of the transmission rate of the virus [75]. The pandemic performed as an accelerator, inducing several administrations to invigorate implementing these actions. Many strategies were implemented as temporary solutions, successfully converting into permanent changes for their public health or social benefits.

Our current era imposes to build more resilient cities, adapt, and respond to present and future challenges [33]. In front of us are no more only scenarios of risk reasonably determined by past and present experiences. We will face uncertain scenarios whose connotations cannot always be identified by current forecasts. More efforts must be made to imagine, design, and shape our environment toward more sustainable urban transitions.

Acknowledgements The whole paper is the product of joint research work by the four Authors. However, the various parts are to be attributed as follows: Prof. Fabris had the role of proposal and research coordinator and contributed to editing the final version of the article and wrote the Introduction and the Conclusions; Dr. Camerin wrote the parts dedicated to Barcelona; Dr. Semprebon wrote the parts dedicated to Shanghai, and Riccardo Maria Balzarotti wrote the parts illustrating the case of Milan. Federico Camerin has participated as co-author within the research project "*La Regeneración Urbana como una nueva versión de los Programas de Renovación Urbana. Logros y fracasos*". This project is co-funded by the Spanish Ministry of Universities in the Recovery, Transformation, and Resilience Plan framework, the European Union—NextGenerationEU, and the Universidad de Valladolid.

References

1. Abdelfattah L, Deponte D, Fossa G (2022) The 15-minute city as a hybrid model for Milan. TEMA J Land Use Mobil Environ 1:71–86. https://doi.org/10.6093/1970-9870/8653
2. Agència de Salut Pública de Barcelona (2021) Result report "Salut als carrers" (Health in streets). https://www.aspb.cat/wp-content/uploads/2021/10/English-ASPB_salut-carrers-resultsreport-Superblocks.pdf. Accessed 4 Jul 2022
3. Ajuntament de Barcelona (2012) Compromiso Ciudadano por la Sostenibilidad 2013–2022. https://ajuntament.barcelona.cat/ecologiaurbana/sites/default/files/CompromisoCiudadanoSostenibilidad.pdf. Accessed 4 Jul 2022
4. Ajuntament de Barcelona (2013) Barcelona Green Infrastructure and biodiversity plan 2020. https://ajuntament.barcelona.cat/ecologiaurbana/sites/default/files/Barcelona%20green%20infrastructure%20and%20biodiversity%20plan%202020.pdf. Accessed 4 Jul 2022
5. Ajuntament de Barcelona (2014) Pla de mobilitat urbana 2013–2018. https://ajuntament.barcelona.cat/ecologiaurbana/sites/default/files/PMU_Sintesi_Catala.pdf. Accessed 4 Jul 2022
6. Ajuntament de Barcelona (2016) Omplim de vida els carrers. La implantació de les Superilles a Barcelona. https://bcnroc.ajuntament.barcelona.cat/jspui/bitstream/11703/97356/1/mesuradegovernomplimdevidaelscarrerslaimplantacidelessuperi-160518111846.pdf. Accessed 4 Jul 2022
7. Ajuntament de Barcelona (2020) Pla de Movilitat urbana 2024. https://www.barcelona.cat/mobilitat/sites/default/files/documentacio/pmu_bcn_2024_per_ceuim_20201214_compressed.pdf. Accessed 4 Jul 2022
8. Ajuntament de Barcelona (2021) Superblock Barcelona. Towards the city we want. https://ajuntament.barcelona.cat/superilles/sites/default/files/20210202_Superblock_Barcelona_web.pdf. Accessed 4 Jul 2022
9. Ajuntament de Barcelona (2021) Resolución de los concursos de ideas de Superilla Barcelona. https://ajuntament.barcelona.cat/superilles/es/content/resolucion-de-los-concursos-de-ideas-de-superilla-barcelona. Accessed 4 Jul 2022

10. Ajuntament de Barcelona (2021) Precio medio de oferta en los barrios (€/m^2). Datos anuales 2013–2021. https://ajuntament.barcelona.cat/estadistica/castella/Estadistiques_per_temes/Habitatge_i_mercat_immobiliari/Mercat_immobiliari/Preu_oferta_habitatge_segona_ma/evo/t2mab.htm. Accessed 4 Jul 2022
11. Ajuntament de Barcelona (2021) Mesura de govern Superilla Barcelona per regenerar Barcelona i els seus barris. https://bcnroc.ajuntament.barcelona.cat/jspui/bitstream/11703/123221/1/Llibret_SUPERILLA_MdG_A4_web.pdf. Accessed 4 Jul 2022
12. Allam Z, Bibri SE, Chabaud D, Moreno C (2022) The '15-Minute City' concept can shape a net-zero urban future. Hum Soc Sci Commun 9(126). https://doi.org/10.1057/s41599-022-01145-0
13. Anguelovski I, Connolly JJ (eds) (2021) The Green City and Social Injustice. 21 tales from North America and Europe. Routledge
14. Brody J (2016) How ideas work: memes and institutional material in the first 100 years of the neighborhood unit. J Urban Int Res Placemak Urban Sustain 9(4):329–352. https://doi.org/10.1080/17549175.2015.1074602
15. Caballero L (2018) Procesos de desplazamiento y desposesión en contextos urbanos. El caso de Sant Antoni (Barcelona). Perifèria. Revista de Recerca i Formació en Antropologia, 23(1):31–51. https://doi.org/10.5565/rev/periferia.634
16. Caprotti F (2018) Future cities: moving from technical to human needs. Palgrave Commun 4(35). https://doi.org/10.1057/s41599-018-0089-5
17. Cardoso R, Sobhani A, Meijers E (2021) The cities we need: towards an urbanism guided by human needs satisfaction. Urban Stud. Advance Online Publication.https://doi.org/10.1177/00420980211045571
18. Cheshmehzangi A, Butters C (2017) Chinese urban residential blocks: Towards improved environmental and living qualities. Urban Des Int 22:219–235. https://doi.org/10.1057/s41289-016-0013-9
19. Chow RY (2015) Changing Chinese cities. The potential of field urbanism. National University of Singapore Press
20. China Development Research Foundation (2013) China's new urbanization strategy. Routledge
21. Comune di Milano (Apr 2020) Milan 2020. Adaptation strategy. Open document to the city's contribution. https://www.comune.milano.it/documents/20126/7117896/Milano+2020.+Adaptation+strategy.pdf/d11a0983-6ce5-5385-d173-efcc28b45413?t=1589366192908. Accessed 20 Jul 2022
22. Comune di Milano (Apr 2020) Open Streets. Strategies, actions, and tools for cycling and walking, ensuring distancing measures within the urban travel and towards a sustainable mobility. https://www.comune.milano.it/documents/20126/7117896/Open+streets.pdf/d9be0547-1eb0-5abf-410b-a8ca97945136?t=1589195741171. Accessed 20 Jul 2022
23. Comune di Milano (Feb 2020). Milano 2030. Piano di Governo del Territorio. Documento di Sintesi. https://www.pgt.comune.milano.it/sites/default/files/allegati%20home/SintesiPGT_Milano2030.pdf. Accessed 20 Jul 2022
24. Comunidad Autónoma de Cataluña (2020) Ley 11/2020, de 18 de septiembre, de medidas urgentes en materia de contención de rentas en los contratos de arrendamiento de vivienda y de modificación de la Ley 18/2007, de la Ley 24/2015 y de la Ley 4/2016, relativas a la protección del derecho a la vivienda. https://www.boe.es/buscar/doc.php?id=BOE-A-2020-11363. Accessed 4 Jul 2022
25. Drummond W (1916) Plan by William Drummond developed from a sketch submitted in competition. In: Yeomans AB (ed) City residential land development. Competitive plans for subdividing a typical quarter section of land in the outskirts of Chicago. The University of Chicago Press, Chicago, pp 37–44
26. Eggimann S (2022) The potential of implementing Superblocks for multifunctional street use in cities. Nat Sustain 5:406–414. https://doi.org/10.1038/s41893-022-00855-2
27. European Commission (2019) The human-centred city: opportunities for citizens through research and innovation. https://doi.org/10.2777/859158. Accessed 4 Jul 2022

28. European Space Agency (7 Jul 2022) Land-surface temperature in Milan on 18 June 2022. https://www.esa.int/ESA_Multimedia/Images/2022/07/Land-surface_temperature_in_Milan_on_18_June_2022. Accessed 20 Jul 2022
29. Fabris LMF, Semprebon G (2019) The Chinese 'high and slender' condominium. Techne 17:100–109. https://doi.org/10.13128/Techne-23892
30. Fan Y (2021) Shanghai expands '15-Minute Community' to the Suburbs. Sixth Tone. https://www.sixthtone.com/news/1008493/shanghai-expands-15-minute-community-to-the-suburbs. Accessed 14 Sep 2021
31. Farrer M (2021) Evergrande: 'Everyone bet on inexorably rising Chinese property prices'. The Guardian. https://www.theguardian.com/business/2021/dec/31/evergrande-chinese-property-sectors-debt-crisis-to-intensify-in-2022. Accessed 31 Dec 2021
32. Forestami (Feb 2021) Report 2020. https://forestami.org/wp-content/uploads/2021/03/report_2020.pdf. Accessed 20 Jul 2022
33. Frantzeskaki N, Castán Broto V, Coenen L, Loorbach D (eds) (2017) Urban sustainability transitions. Routledge
34. Gleeson B (2012) Critical commentary. The urban age: paradox and prospect. Urban Stud 49(5):931–943. https://doi.org/10.1177/0042098011435846
35. Global Designing Cities Initiative & NACTO (2016) Global street design guide. Island Press
36. Gordon DJ, Johnson CA (2017) The orchestration of global urban climate governance: conducting power in the post-Paris climate regime. Environ Polit 26(4):694–714. https://doi.org/10.1080/09644016.2017.1320829
37. Hou L, Liu Y (2017) Life circle construction in china under the idea of collaborative governance: a comparative study of Beijing, Shanghai and Guangzhou. Geogr Rev Jpn Ser B 90:2–16
38. Implicito F (16 Mar 2022). Corso Buenos Aires, la ciclabile che non piace è tra le più usate di Milano. MItomorrow. https://www.mitomorrow.it/verde/corso-buenos-aires-ciclabile-dati/. Accessed 20 Jul 2022
39. Infussi F, Montedoro L, Pasqui G (2019) Scali a Milano. Prove di innovazione? J Sustain Des 20:25–62
40. Kan H, Forsyth A, Rowe P (2017) Redesigning China's superblock neighbourhoods: policies, opportunities and challenges. J Urban Des 22(6):757–777. https://doi.org/10.1080/13574809.2017.1337493
41. Lee CMC (ed) (2013) Xiamen. The Megaplot. Harvard University Graduate School of Design
42. Li Z, Wu F (2010) Post-reform residential segregation in three Chinese Cities: Beijing, Shanghai and Guangzhou. In: Wu F, Webster C (eds) Marginalization in urban China comparative perspectives. Palgrave Macmillan
43. Lydon M, Garcia A (2012) Tactical urbanism. Short-term action for long-term change, vols 1 and 2. The street plans collaborative
44. Lydon M, Garcia A (2015) Tactical urbanism: short-term action for long-term change. Island Press
45. Ma X (2018) Housing challenges in Chinese cities under urbanization. In: Ye L (ed) Urbanization and urban governance in China. Palgrave Macmillan, pp 161–181
46. Maciver M (Aug 8 2010). Eric Reynolds, master of low-cost, high-return public space interventions in London and NYC. Project for public space. https://www.pps.org/article/eric-reynolds-master-of-low-cost-high-return-public-space-interventions-in-london-and-nyc. Accessed 20 Jul 2022
47. Manchón M (Apr 2022) Salvador Rueda, sobre las superillas de Colau: "Son ineficientes". https://www.metropoliabierta.com/informacion-municipal/movilidad/salvador-rueda-sobre-superillas-colau-son-ineficientes_53919_102.html. Accessed 4 Jul 2022
48. Maran P (2022) Le città visibili. Solferino ~~Editore.~~
49. Miao P (2003) Deserted streets in a jammed town: the gated community in Chinese cities and its solution. J Urban Des 8(1):45–66. https://doi.org/10.1080/1357480032000064764
50. Monson K (2008) String block vs Superblock patterns of dispersal in China. Archit Des 7(1):46–53. https://doi.org/10.1002/ad.608

51. Moreno C (2020) Vie urbaine et proximité à l'heure du Covid-19. Éditions de l'observatoire/ Humensis
52. Mosciaro M (2021) Selling Milan in pieces: the finance-led production of urban spaces. Eur Plan Stud 29(1):201–218. https://doi.org/10.1080/09654313.2020.1735309
53. Mueller N et al (2020) Changing the urban design of cities for health: the Superblock model. Environ Int 134:105132
54. Riart M, Soro S (Jan 2017) Veïns del Poblenou tallen el trànsit contra la superilla. https://www.ara.cat/societat/veins-poblenou-transit-contra-superilla_1_1367779.html. Accessed 4 Jul 2022
55. Oscilowicz E, Honey-Rosés J, Anguelovski I, Triguero-Mas M, Cole H (2020) Young families and children in gentrifying neighbourhoods: how gentrification reshapes use and perception of green play spaces. Local Environ 25(10):765–786. https://doi.org/10.1080/13549839.2020.1835849
56. Perry CA (1929) The neighborhood unit, a scheme of arrangement for the family-life community. Monograph one in neighborhood and community planning, regional plan of New York and its environs. Committee on regional plan of New York and its environs
57. Quinzii C, Terna D (2022) Milan public space. An in-progress atlas of everyone's space. LetteraVentidue
58. Rueda S (2019) Superblocks for the design of new cities and renovation of existing ones: Barcelona's case. In: Nieuwenhuijsen M, Khreis H (eds) Integrating human health into urban and transport planning. Springer, pp 135–153
59. Rueda S (2020) Regenerando el Plan Cerdà: de la manzana de Cerdà a la supermanzana del urbanismo ecosistémico/Regenerating the Cerdà Plan. From Cerdà's Block to the Ecosystemic Urbanism Superblock. Agbar
60. Rueda S (2022) La complejidad urbana y su relación con la morfología de los tejidos urbanos y la proximidad. Ciudad y Territorio Estudios Territoriales 54(M):227–250. https://doi.org/10.37230/CyTET.2022.M22.10
61. Sadik-Khan J (25 Jun 2020) Streets for pandemic. Response & recovery. NACTO. https://nacto.org/wp-content/uploads/2020/09/Streets_for_Pandemic_Response_Recovery_Full_20-09-24.pdf. Accessed 20 Jul 2022
62. Sánchez-Aguilera D, González-Pérez JM (2021) Geographies of gentrification in Barcelona. Tourism as a driver of social change. In: Dominguez-Mujica J, McGarrigle J, Parreño-Castellano JM (eds) International residential mobilities. geographies of tourism and global change. Springer, pp 243–268
63. Senesi A (2022) Bar e ristoranti a Milano, cosa cambia per 2.500 dehors: dal 1° aprile torna la tassa sui tavolini (che resteranno). Il Corriere della Sera, March 9
64. Shanghai Urban Planning and Land Resource Administration Bureau (2018) Shanghai master plan 2017–2035. Striving for the excellent global city. http://www.shanghai.gov.cn/newshanghai/xxgkfj/2035004.pdf. Accessed 4 Jul 2022
65. Shanghai Urban Planning and Land Resources Administration Bureau (2016) Planning guidance of 15-minute community-life circle: planning and construction guidance, action guidelines. Shanghai Government Printing Office (in Chinese)
66. Shanghai Municipal People's Government Information Office (2021) Shanghai's construction of a 15-minute community life circle. https://ghzyj.sh.gov.cn/ldft/20210715/cc2adc4ffff244418d2a99fd6d7756ad.html. Accessed 4 Jul 2022
67. The Street Plans Collaborative (2016) Tactical urbanist's guide to materials and design. John S. and James L. Knight Foundation
68. United Nations (2017) New urban agenda. https://habitat3.org/wp-content/uploads/NUA-English.pdf. Accessed 4 Jul 2022
69. Wang V (2022) 'Very Fragile': Shanghai wrestles with psychological scars of lockdown. The New York Times, June 29. https://www.nytimes.com/2022/06/29/world/asia/shanghai-lockdown-china.html
70. Weng M, Ding N, Li J, Jin X, Xiao H, He Z, Su S (2019) The 15-minute walkable neighborhoods: measurement, social inequalities and implications for building healthy communities in urban china. J Transp Health 13:259–273. https://doi.org/10.1016/j.jth.2019.05.005

71. World Bank (2014) Urban China: toward efficient, inclusive, and sustainable urbanization. The World Bank
72. Wu H, Wang L, Zhang Z, Gao J (2021) Analysis and optimization of 15-minute community life circle based on supply and demand matching: a case study of Shanghai. PLoS One 16(8). https://doi.org/10.1371/journal.pone.0256904
73. Wu F, Webster C, He S, Liu Y (eds) (2010) Urban poverty in China. Edward Elgar Publishing
74. Xu M, Yang Z (2009) Design history of China's gated cities and neighbourhoods: Prototype and evolution. Urban Des Int 14:99–117. https://doi.org/10.1057/udi.2009.12
75. Zhang Z, Xue T, Jin X (2020) Effects of meteorological conditions and air pollution on COVID-19 transmission: evidence from 219 Chinese cities. Sci Total Environ 727:138704. https://doi.org/10.1016/j.scitotenv.2020.140244

Urban Mobility and Time Geography Within the Smart City Narrative

Aditya Dixit and Zaheer Allam

Abstract The subject of mobility in cities in gaining increasing interest in view of rapid urbanisation, coupled with urban sprawled, being the expansion of city size, highlighting the need for people to move to access various urban amenities. With cities transforming to data-driven urban centers, this chapter explores how the emerging Smart City narrative can aid in rendering more efficient mobility options and can aid in increasing the quality of life of urban dwellers. Of interest to this chapter is the inclusion of the concept of time geography within the urban sphere, while integrating the smart city narrative, and building the motive for transitioning towards more connected urban services, hence increased proximity, such as the celebrated' 15-min' City model.

Keywords Smart mobility · Smart cities · Transportation · Urban policy · Sustainability · Intelligent transport systems

A. Dixit (✉)
Department of Mechanical Engineering, Friedrich Alexander University Erlangen-Nuremberg, Schloßplatz 4, 91054 Erlangen, Germany
e-mail: aditya.dixit@fau.de

Z. Allam
Panthéon-Sorbonne, Chaire Entrepreneuriat Territoire Innovation (ETI), IAE Paris Sorbonne Business School, Université Paris, 75013 Paris, France

Live+Smart Research Laboratory, School of Architecture and Built Environment, Deakin University, Geelong, VIC 3220, Australia

Network for Education and Research On Peace and Sustainability, Hiroshima University, Higashihiroshima 739-8530, Japan

Z. Allam
e-mail: chabaud.iae@univ-paris1.fr; carlos.moreno@univ-paris1.fr

Z. Allam (ed.), *Sustainable Urban Transitions*, Urban Sustainability,
https://doi.org/10.1007/978-981-99-2695-4_11

1 Introduction

The need and growth of automobiles in cities have been growing, especially after the Second World War 2 (WW2), due to a number of factors. Among such include the rapid economic growth in most cities, especially in North America and in Western countries. Such prompted unprecedented rise in the standard of living, improved purchasing power, and other factors like low inflation, enabling people to not only afford private cars, but also access low-cost mortgages [1]. Consequently, such factors brought about the aspect of urban sprawl triggered by demand for housing (prompted by vast urban population growth), growth of suburban areas and accessibility made possible by availability of private cars [2, 3]. On this, the automobiles were seen to facilitate the last mile connectivity between urban areas and suburban areas. As such, the dynamic changes in the transportation sector have had significant impacts on urban planning, with most decisions, especially placement of different urban components being greatly influenced by availability and accessibility of automobiles [4]. The dynamics have further influenced other urban elements like urbanisation, demographic compositions, and economic growth.

In the study of time geography—mainly focusing on how humans order different activities within their pace relative to time, the aspect of urban mobility, especially focusing on emerging technologies is becoming very critical. With time geography, changes within a society in a given environment can be viewed to resonate with the amount of time that people have for particular activities. Therefore, in the case of urban growth, modern trends of increased urban population growth (over 55% of population now lives in cities) and urbanisation can be viewed on how people spend time moving from different nodes in cities, as well as accessing different basic services within cities. This then qualifies why the emergence and subsequent increase of vehicles have had significant impacts on the nature of urban planning models, relative to time and space resources.

The importance of time, as a resource, in urban areas has thus influenced the emergence of new ventures, especially in the transport industry, where focus is shifting to adoption of autonomous vehicles like the robo-taxis [5], which allow data-driven urban planning approaches. On this, it is worthwhile to note that conventional automobiles have led to numerous urban challenges, including increased pollution (emissions—approximately 28% of global CO_2, noise, and solid wastes—like from tires), traffic congestions and accidents et cetera [6]. As such, the emergence of autonomous vehicles is gaining traction in urban areas due to their potential in helping address the above issues. That is, they have been reported to provide added advantages like reduced crashes and fatalities, environmental friendliness, traffic efficiency and societal cost-saving, especially in terms of time, money and resource consumption in urban areas [7].

In addition to influencing direct innovations in the mobility sector, the aspect of time geography is further argued to be playing critical roles in helping shape other urban spheres related to human-oriented dimensions. Such include the liveability status (especially in terms of housing and neigborhoods) influenced by emergence of new planning models such as the 15-min City concept that advocates for proximity, diversity, digitalization and density [8, 9]. On these, the availability of data, as noted by Rowe [1], is critical in making available information that would allow for sound decision making on placement of different urban amenities, on application of diverse technologies et cetera. Such have become especially important post COVID-19 pandemic, where the need to strengthen different human aspects in relation to time and spaces have become an urgent matter. Essentially, the pandemic prompted the emergence of new trends like work-from-home policies, new communication models, new learning approaches (virtual learning) [10], et cetera, that could not be implemented successfully without relying on available and emerging data, especially in relation to urban mobility. As such, this chapter dwells on how the aspects of urban mobility and time geography are related within the smart city narrative.

2 Surveying the Field of Smart Urban Mobility

Urban mobility is an important component and area of concern in urban planning, especially in view of the diverse impacts the transport sector has on urban morphology, urban residents, the environment, economy and politics. From a wide range of literature [11–13], it is documented that most cities across the globe have been fashioned and planned to align with road transport networks. As such, the sector has been a major target amongst urban planners as they seek to address diverse urban issues such as traffic congestion, climate change, housing, sprawl and others. It is also a target in the pursuit of emerging urban planning models such as compact cities; especially with the intention of reducing over-reliance on vehicular use and instead, adopting alternative mobility options (e.g. soft mobility solution) [14]. This explains why a substantial number of emerging technologies, such as Internet of Things (IoT), Artificial Intelligence, Big Data and others have been widely adopted in the manufacturing of devices and tools with direct impacts on the mobility sectors.

The quest to transform the mobility sector has been congruent with growing urban concerns such as population growth, sprawl and other; hence, finding amicable solution in the mobility sector is seen as a step toward addressing those different urban challenges [11, 15]. For this reason, it is reported that the adoption of technology to influence urban mobility (by making it 'smart') began as early as technologies deployment in urban areas began to become popular among planners [16]. In particular, as from the 2004s, the quest to establish alternative mobility options, together with other smart solutions for cities become inevitable following advancement in technologies. For instance, as from that period, new innovations emerged such as autonomous vehicles, ride sharing and car-pooling services et cetera that are seen to widen the scope of the mobility sector. Further, there has also been an increased agitation by different urban players for the need to reduce conventional automobiles in urban areas, by replacing them with alternative more sustainable and safer options [11].

Those alternatives include creating infrastructure and providing conducive environments for bicycle riders, pedestrians, and other non-motorized options [12, 17]. These are gaining popularity mostly in cities planned to comply with tenets of compact cities and sustainability.

Smart urban mobility options have also gained popularity due to commitments that different cities are taking in the pursuit of SDG 11 [18], which promote the need for inclusivity and safety in urban areas. On this it is evident that urban mobility has played a significant role in promoting exclusion of different demographics in cities, especially in regard to areas like housing, and accessibility to basic amenities and services. For instance, there are clear disparities in most cities in terms of infrastructure development between different neigborhoods, all influenced by mobility options [19]. That is, most neighbourhoods hosting middle and upper income groups are associated with improved and advanced mobility infrastructures unlike those areas occupied by lower income groups [20]. Further, noting that the most popular mobility option available in most cities entails the use of automobiles (whether private or public means), where the posh and lavish neighborhoods are occupied by individuals and households with means to access such conveniently. However, low income earners are concentrated in neighborhoods near public transportation routes, hence, experience overcrowding and excessive traffic congestions [21]; and further highlighting a disparity of travel modes between the rich and the poor.

The myriad of urban challenges, especially pointing to the human dimensions could therefore be resolved by correctly predicting, anticipating and solving mobility challenges. In particular, by adopting and implementing smart technologies in the urban sector as well as aligning mobility infrastructure development plans with emerging urban planning models. As such, it would become possible to make provision for the integration of infrastructures such as footpaths and cycling lanes in urban land use plans. Further, it would become possible to provide for other amenities such as the end-of-journey facilities such as bathrooms and locker rooms, which are essential incentives for people to transition to adopting alternative mobility behaviours [19]. With technology, it would further be possible to deploy smart devices such as sensors and cameras to facilitate security, efficiency and convenience for those using the smart urban mobility options – especially for cycling [17]. Technology would further be instrumental in helping implement other services like smart parking that not only help drivers save time in identifying available parking spots within cities, but also help in reducing emissions from vehicles. The technologies would help in incentivising the adoption of autonomous vehicles [22]; hence, prompting a natural decline on fossil-fuel reliant vehicles.

The quest to implement smart urban mobility cannot only be viewed in the lens of available technologies and subsequent emergence and gradual increase in autonomous vehicles. It should also be viewed on the scope of emerging urban challenges like the recent case of COVID-19 outbreak that prompted unprecedented grounding of different mobility options [23]. On this, it has been affirmed that urban residents in cities with existing infrastructures such as cycling lanes had some reprieve. Following this, it became apparent that other cities adopted this approach and converted some road networks into bicycle lanes to provide residents with alternative mobility options to access different urban facilities. Such strategies have been further seen to give some impetus to the adoption of the 15-min City concept that emphasizes on the adoption of soft mobility options as the solution to numerous urban human scale challenges [9, 24]. In the context of sustainability, climate mitigation and improved liveability status, the adoption of smart urban mobility is seen to be highly revered, especially in respect to the reduction of emissions.

3 Time Geography and Smart Urbanism

In the recent days, there has been an increased attention on urban areas and their impacts on sustainability agenda. The focus on cities is influenced by a myriad of reports that showcases that cities contributes approximately 70% of the global emissions, as well as consuming over 78% of the global natural resources [9]. By delving deeper into understanding the underlying factors that leads to these outcomes, it is evident that the aspect of time and space—advanced in the concept of time geography are very critical. On this, it is rightly argued that the concept of time geography entails understanding how humans activities within a given geographical spaces are influenced or ordered relative to time [25–27]. In other words, the concept help explore and examine how human activities over a given period prompt different change processes, that in most cases have significant impacts on the environment, the society, economies, politics and other components that are encompassed in human spaces.

This concept is therefore very critical in understanding processes for instance in cities, and how such could be ordered to embrace approaches that are congruent with sustainability agendas. Within this dispensation when global attention is focus on ordering different geographical spaces (especially cities) such that they promote human scale outcomes such as improved liveability, increased inclusivity and safety in urban areas, transitioning to sustainable practices like use of alternative energy and conscious consumption and production et cetera, the study of time geography is very paramount. Relating with the arguments by Hägerstrand [28], the aspect of time and space in urban areas is further important, especially in planning the placement of different urban components such as basic amenities and services, residential areas and public spaces. It is further profound in understanding and planning for urban infrastructures such as education facilities, transportation networks, health institutions et cetera that directly impact on human activities.

In regard to smart urban mobility, the concept of time geography is paramount in view of trends such as increasing overdependence on automobiles, population increase and the rapid rate at which the world is urbanizing. All those occurrences have been argued to have significant spatial impacts on urban areas, prompting urban challenges such as congestion, overconsumption of resources, sprawl, pollution and social inequalities among others [29]. Understanding these relative to time therefore would help explore diverse approaches such as application of technology in urban planning that would essentially help in optimizing available space within a given time period. For instance, this concept would help in appreciating the need for smart urban mobility concepts that promote the adoption of alternative transportation modes that fosters increased human interaction, increase accessibility within cities, promotes reduces negative ecological impacts and improve inclusivity. Understanding this concept would further help appreciate emerging urban planning models such as the Compact City approach where concepts such as the 15-min City falls. In particular, it will elucidate the need to have basic amenities and services within neighborhoods placed in such a way that they are accessible within a given time frame, especially post-pandemic.

4 Autonomous Vehicles, Robo-Taxis and Digital Urban Networks

The quest for using data in urban planning began in 1979 in Los Angeles [30], and since then, cities have grown to become the largest avenues for big data. This was particularly cemented as from 2000s, during the fourth industrial revolution, making advanced technologies cheaper, accessible and ubiquitous [31]. As a consequence, numerous new opportunities and frontiers have emerged in cities, especially in the mobility sector; hence, allowing for a departure from the traditional transportation options. In perspective, due to availability of technologies (especially 'smart' oriented), diverse cities have now managed to plan, install and build infrastructures with capacities to allow for autonomous vehicles, cycling, robo-taxis, and walkability as alternatives to the traditional road and/or rail systems.

In particular, the availability of diverse infrastructures have made it possible for automobiles to manufacture and install smart devices (Internet of Things enabled) such as sensors and cameras that make it possible for deployment of autonomous vehicles [22]. With those devices, Alonso-Mora et al. [1] note that vehicles have resulted with impressive capabilities to communicate with each other, and with their surroundings; hence, increasing road safety. Further, such technologies have been argued to allow for new opportunities for start-ups such as those dealing with car sharing; thus diversifying mobility options as well as new income streams for a majority of urban population [32]. For instance, in 2022 alone, the car sharing segment revenue will hit approximately $11.88 billion globally. This is due to a compounded growth rate of approximately 8.13% which is anticipated to remain or increase up to 2026 [32]. In terms of users, it is anticipated that by 2026, over 60.5 million people will have used car sharing from diverse platforms [33].

The communication aspect in the autonomous vehicles is enhanced greatly by smart digital technologies, and the underlying digital urban networks. For instance, it is reported that due to the presence of 'basic forward-collision warning systems', around 7% of reduction in crashes were reduced in 2014. Further, for those vehicles with automatic braking options, crashes were reduced by approximately 15% [34].

Such technologies have been hailed as a majority of crashes as in over 90% of cases associated with human errors; hence, their presence help complement human efforts [35].

A leading innovation in the autonomous vehicle market is the Robotaxi—self-driving (driverless), complex cars that are integrated with modern technologies, to allow among other things, car-sharing and ride hailing. The popularity of these cars have seen them start to gain traction in markets in regions like Asia and at global scale. To put this in perspective, their market is estimated at $10 billion, with an anticipation that their valuation would rise to approximately $38.61 billion in 2030 [36]. Currently, these AVs are popular with human transportation, but anticipation is that they will be hugely deployed in goods transportation such that this segment would contribute at least $21.71 billion of the market valuation [36]. The steadily growing market for the robo-taxi, as highlighted in a report by McKinsey [37], is due to the increased attention on data in most cities, where modern technologies are making it possible to garner data from diverse spheres. With data, especially from the mobility sector, it is becoming possible for urban managers to make plans for alternatives like AVs, which not only reduce accidents as noted above, but also open opportunities to focus on new ideas like reduction of private cars in cities through trends such as car-sharing. Further, as noted by Hutson [38], such has helped in reducing traffic. For instance, it is noted that in Manhattan, the reduction was approximately 7% in 2018, while Salter [39] observe that it is possible for AVs to help reduce the traffic by over 35%.

5 On Compact Urbanism and Soft Mobility

During the twentieth century, cities across the globe experienced notable morphological transformations—triggered by a number of factors. Such include unprecedented population growth, changes in income levels, demographic compositions, advancement in technologies, and substantial growth in the number of vehicles [40]. Among the most notable occurrence is the urban sprawl, which became evident as from the 70 s (after WW2), due to rising income levels that led to improvement in the purchasing powers of residents, especially those based in urban areas. The increased purchasing capacity prompted the emergence of a number of things, but most important; an exponential increase in the number of private cars and secondly, an outward expansion of existing cities, as people sought to settle in residential areas outside cities. According to Gross [1], the increase private cars allowed people to travel quickly, and more conveniently; hence, influencing their decision to settle in areas relatively cheaper than cities.

Further, in the literature on urban planning, it is also evident that the growing economic status of cities made it possible for the extension of basic infrastructures like transport networks (especially roads) beyond urban perimeters; hence, promoting sprawl. On these, while urban sprawl has its advantages, it has been associated with critical disadvantages regarding environmental pollution, especially from vehicular mobility [13, 41]. It is also associated with tampering and the contraction of urban buffers such as forests, wetlands et cetera. Further, in relation to this study, sprawl has been very significant in prompting even further demand for automobile dependency, as most of the sprawled areas are in most cases not covered by public transportation systems.

In a bid to counter urban sprawl; hence address underlying challenges, a number of proposals have been fronted. Among these include the need for new planning agendas and models that would guarantee that the growing urban population is catered for, while at the same time reducing the need for vehicular mobility. As such, planning models that promote the creation of compact cities have been argued to be potent solutions, especially for their potential to help incorporate human-oriented aspects in cities as well as promote sustainability agendas. Particularly, the proposals encapsulated in compact city models such as the 15-min City concept focus on increasing accessibility of different urban amenities and services to pedestrians without them necessary relying on automobiles [8]. On this, the emphasis is on expanding and promoting infrastructures and other elements that would promote soft mobility solutions such as walkability and cycling within cities. Further, to promote the adoption of those solutions, it is becoming clear that the aspect of time taken between different nodes need to be factored in planning and placement of the different urban amenities as well as infrastructures.

Within the 15-min City concept, which is currently booming globally, especially after the reality of the COVID-19 pandemic, there is clear emphasis on the need for paradigm shift in the mobility options. According to the proponents of this model [8, 9, 42], the benefits derived from its adoption and implementation would not only increase in the mobility sector, but would have significant impacts on the human dimensions. For instance, soft mobility options are expected to increase inter-human relations through increased interactions as people interact with each other. Such would also be achieved as people share diverse public spaces such as multi-use parking, recreation centres and green spaces that would make part of the compact cities. With soft mobility options, it is also argued that the liveability status of urban areas will further improve as resources consumed in conventional urban areas would be utilized in improving the different nodes of the compact cities.

6 Discussions and Conclusions

The role of cities at the global scale is indisputable, especially in regard to human population, economic contribution, and politics and in environmental discourse. This has been affirmed in a wide range of global agreement, accords and policy documents (e.g. Paris Agreement [43], New Urban Agenda [44], the SDGs [45], the 2030 Agenda for Sustainable Development [46] et cetera). The common denominator in these accords being the need to address various issues such as resource consumption, adoption of alternative energy and solving inclusivity challenges. From the literature, it has been widely argued that cities are the epicentres of all those challenges, and re-thinking operations and planning in cities would have positive impacts in addressing those issues. It is on this premise that the aspect of smart city narrative has been gaining substantial focus, as amicable solutions hinged on technologies are being sort.

It has been established above that urban areas and the activities therein are as a result of the aspect of time and space. As such, in the thinking of how to plan and reconstruct the cities to make them responsive to emerging demands such as reduced carbon footprint, increased attention on sustainable, inclusive and safe cities, and improved liveability status, the aspect of time geography cannot be ignored. However, pertinent to this chapter, the focus has been on urban mobility and how this could be seamlessly integrated into the narrative of smart city, guided by the understanding that cities are constrained in regard to space and time resource. Appreciating those constraints, and also acknowledging the contribution of the transport sector into global emissions, it becomes paramount to focus on how emerging trends such as adoption of 'smart' technologies could be integrated into the urban mobility to help increase efficiency, as well as reduce the carbon footprint from the sector. In light to this, it has been established those notable achievements have been achieved in many cities, including focusing on establishing soft mobility infrastructure that would promote and encourage a paradigm shift from conventional automobiles to alternative mobility options.

From the literature discussed above, it has been established that substantial steps have been made in aligning the urban mobility option with the smart city agenda. For instance, a sizeable number of cities across the globe are reported to have made provision for the adoption of autonomous vehicles, by installing relevant smart technologies. Further, the emphasis on the use of data as tools for planning have made it possible for different cities to prioritize alternative mobility options, especially those that promote safety, reduce emissions, promote societal benefits (such as human interactions and human health) among others [47, 48]. However, it is worth noting that despite the availability of diverse technologies, and goodwill in cities to promote smart urban mobility, there still lingers outstanding issues that need to be addressed. For instance, the aspect of privacy occasioned by data generated and collected from the 'smart' devices has been found to remain an obstacle that make a substantial number of urban residents apprehensive of these emerging technologies. As such, there is need for proactive public participation and inclusion of stakeholders in decision making. This is particularly important in the understanding of the concept of time geography, where human beings (urban residents in this case) are the target of those changes, and how such would influence their activities within their spaces. The challenge of cost is also paramount and need to be addressed, as this has a higher probability of promoting exclusion, especially if alternative mobility options are made robust such that they extend to urban neigborhoods.

References

1. Carnes MC (2007) The columbia history of post-world war II America. Columbia University Press, New York, NY
2. Petit S World vehicle population rose 4.6% in 2016. Available online https://wardsintelligence.informa.com/WI058630/World-Vehicle-Population-Rose-46-in-2016. Accessed on 10 Dec 2021
3. De Janosi PE (1959) Factors influencing the demand for new automobiles. J Mark 23:412–418. https://doi.org/10.2307/1247410
4. Moss S End of car age: how cities are outgrowing the automobile. Available online https://www.theguardian.com/cities/2015/apr/28/end-of-the-car-age-how-cities-outgrew-the-automobile. Accessed on 11 Dec 2021
5. Liu M, Wu J, Zhu C, Hu K (2022) Factors influencing the acceptance of robo-taxi services in china: an extended technology acceptance model analysis. J Adv Transp 2022:1–11. https://doi.org/10.1155/2022/8461212
6. Magill B Urban Sprawl, Cars hamper cities' best efforts on CO2. Available online https://www.climatecentral.org/news/city-tailpipe-emissions-18861. Accessed on 10 Aug 2022
7. Cugurullo F (2020) Urban artificial intelligence: From automation to autonomy in the smart city. Frontiers Sustain Cities 2
8. Moreno C The 15 minutes-city: for a new chrono-urbanism! Available online http://www.moreno-web.net/the-15-minutes-city-for-a-new-chrono-urbanism-pr-carlos-moreno/. Accessed on 10 Nov 2020
9. Allam Z, Nieuwenhuijsen M, Chabaud D, Moreno CJTLPH The 15-minute city offers a new framework for sustainability, liveability, and health. 6:e181-e183
10. Wang B, Liu Y, Qian J, Parker SK (2020) Achieving effective remote working during the COVID-19 pandemic: A work design perspective. Appl Psychol. https://doi.org/10.1111/apps.12290,doi:10.1111/apps.12290
11. Nieuwenhuijsen MJ, Khreis H (2016) Car free cities: Pathway to healthy urban living. Environ Int 94:251–262. https://doi.org/10.1016/j.envint.2016.05.032
12. Glazebrook G, Newman P (2018) The city of the future. Urban Planning 3:1–20. https://doi.org/10.17645/up.v3i2.1247
13. Newman P, Kenworthy J (1999) Costs of automobile dependence: global survey of cities. Transp Res Rec 1670:17–26. https://doi.org/10.3141/1670-04
14. La Rocca RA (2010) Soft mobility and urban transformation. TeMA-J Land Use, Mobil Environ 2. https://doi.org/10.6092/1970-9870/125
15. Newman P, Kenworthy J (2015) Automobility and resilience: A global perspective. In: The Urban Political Economoy and Ecology of Automobility Driving Cities, Driving Inequality, Driving Politics Walks. Routledge, pp 28–58
16. Jones P (2014) The evolution of urban mobility: The interplay of academic and policy perspectives. IATSS Research 38:7–13. https://doi.org/10.1016/j.iatssr.2014.06.001
17. Stamatiadis N, Pappalardo G, Cafiso S (2017) Use of technology to improve bicycle mobility in smart cities. In: Proceedings of the 2017 5th IEEE International conference on models and technologies for intelligent transportation systems (MT-ITS). Napoli, pp 86–91
18. United Nations Department of Economic and Social Affairs (UNDESA) Goal 11: Make cities and human setttlements inclusive, safe, resilient and sustainable. Available online https://sdgs.un.org/goals/goal11. Accessed on 4 Feb 2022
19. Cavoli C (2021) Accelerating sustainable mobility and land-use transitions in rapidly growing cities: Identifying common patterns and enabling factors. J Transp Geogr 94:103093. https://doi.org/10.1016/j.jtrangeo.2021.103093
20. Lee J, Arts J, Vanclay F, Ward J (2020) Examining the social outcomes from urban transport infrastructure: Long-term consequences of spatial changes and varied interests at multiple levels. 12:5907. https://doi.org/10.3390/su12155907

21. Fraser MS, Wachira BW, Flaxman AD, Lee AY, Duber HC (2020) Impact of traffic, poverty and facility ownership on travel time to emergency care in Nairobi, Kenya. Afr J Emerg Med: Rev Afr Med D'Urgence 10:40–45. https://doi.org/10.1016/j.afjem.2019.12.003
22. Paiva S, Ahad MA, Tripathi G, Feroz N, Casalino G (2021) Enabling Technologies for Urban Smart Mobility: Recent Trends, Opportunities and Challenges. Sensors 21. https://doi.org/10.3390/s21062143
23. Transport Scotland (2020) COVID-19 Transport and trend data–20–26 April 2020. Transport Scotland: Edinburgh, UK
24. Deloitte 15-Minute city. Available online https://www2.deloitte.com/global/en/pages/public-sector/articles/urban-future-with-a-purpose/15-minute-city.html. Accessed on 20 Mar 2022
25. Neutens T, Schwanen T, Witlox F (2010) The Prism of Everyday Life: Towards a New Research Agenda for Time Geography. Transport Review 1:1–23
26. Miller HJ (2018) Chapter 5: Time geography. In Handbook of Behavioral and Cognitive Geography; Edward Elgar Publishing, Cheltenham, UK
27. Ellegård K, Svedin U (2012) Torsten Hägerstrand's time-geography as the cradle of the activity approach in transport geography. J Transp Geogr 23:17–25. https://doi.org/10.1016/j.jtrangeo.2012.03.023
28. Hägerstrand T (1970) What about people in regional science? Papers of the Regional Science Association 24:6–21
29. Van Mierlo J, Messagie M, Rangaraju S (2017) Comparative environmental assessment of alternative fueled vehicles using a life cycle assessment. Transp Res Procedia 25:3435–3445. https://doi.org/10.1016/j.trpro.2017.05.244
30. Soja E, Morales R, Wolff G (1983) Urban restructuring: an analysis of social and spatial change in Los Angeles. Econ Geogr 59:195–230. https://doi.org/10.2307/143613
31. Ghobakhloo M (2020) Industry 4.0, digitization, and opportunities for sustainability. J Clean Prod 252:119869. https://doi.org/10.1016/j.jclepro.2019.119869
32. Livingston M Smart infrastructure for smart cars: a conversation with dr. stilgoe on the hidden dependencies shaping automated vehicles. Available online https://www.wilsoncenter.org/blog-post/smart-infrastructure-smart-cars-conversation-dr-stilgoe-hidden-dependencies-shaping#:~:text=These%20technologies%20have%20the%20potential,including%20through%20increased%20fuel%20efficiency. Accessed on 10 Aug 2022
33. Statista Car-Sharing. Available online https://www.statista.com/outlook/mmo/shared-mobility/shared-rides/car-sharing/worldwide. Accessed on 10 Aug 2022
34. Consumer Reports Avoiding crashes with self-driving cars. Available online https://www.consumerreports.org/cro/magazine/2014/04/the-road-to-self-driving-cars/index.htm#:~:text=The%20Insurance%20Institute%20for%20Highway,for%20those%20with%20automatic%20braking. Accessed on 10 Aug 2022
35. Haghi A, Ketabi D, Ghanbari M, Rajabi-Vardanjani H (2014) Assessment of human errors in driving accidents; analysis of the causes based on aberrant behaviors. Life Sci J 11
36. Singh A, Katare L, Mutreja S Robi taxi market. Available online https://www.alliedmarketresearch.com/robo-taxi-market. Accessed on 10 Aug 2022
37. Ambadipudi A, Heineke K, Kampshoff P, Shao AE (2017) Gauging the disruptive power of robo-taxis in autonomous driving. McKinsey, p 10
38. Hutson M Watch just a few self-driving cars stop traffic jams. Available online https://www.science.org/content/article/watch-just-few-self-driving-cars-stop-traffic-jams#:~:text=The%20self%2Ddriving%20cars%20sped,cars%20passing%20through%20by%207%25. Accessed on 13 Aug 2022
39. Salter A Will autonomous vehicles really put an end to traffic congestion? Available online: https://www.2025ad.com/will-autonomous-vehicles-really-put-an-end-to-traffic-congestion. Accessed on 13 Aug 2022
40. Gross M (2016) A planet with two billion cars. Curr Biol 26:R307–R310. https://doi.org/10.1016/j.cub.2016.04.019
41. Brown LR (2003) Redesigning cities for people: car-centered urban sprawl. In: Eco-economy: Building an Economy for the Earth, Earthscan: Washington DC

42. Reimer J The 15-minute infrastructure trend that could change public transit as we know it. Available online: https://www.departures.com/travel/maze-like-park-vienna-social-distancing. Accessed on 5 Nov 2020
43. UNFCCC The paris agreement. Available online https://unfccc.int/process-and-meetings/the-paris-agreement/the-paris-agreement#:~:text=The%20Paris%20Agreement%20is%20a,compared%20to%20pre%2Dindustrial%20levels. Accessed on 2 Feb 2022
44. United Nations (2017) New Urban Agenda. Habitat III:1–66
45. United Nations Development Programme (UNDP) (2015) Sustainable development goals. UNDP: Geneva, p 24
46. United Nations. *Transforming our World: The 2030 Agenda for Sustainable Development*; A/RES/70/1; ustainable Development Branch, UN: 2011.
47. Dixit A, Kumar Chidambaram R, Allam Z (2021) Safety and risk analysis of autonomous vehicles using computer vision and neural networks. Vehicles 3:595-617. https://doi.org/10.3390/vehicles3030036
48. Allam Z, Sharifi A (2022) Research structure and trends of smart urban mobility. Smart Cities 5:539–561

Is Gamification an Enabler of the Sustainable Urban Transition Process?

Maria Giovina Pasca, Laura Di Pietro, and Roberta Guglielmetti Mugion

Abstract The chapter aims to understand the role of gamification in sustainable urban transitions. In several European countries, environmental issues and the consequent adverse effects on the planet have attracted policymakers, companies, consumers and academics (McKenzie in Comput Environ Urban Syst 79, 2020). Cities must transform into increasingly sustainable and intelligent smart cities, generating synergies between different sectors and companies and providing citizens with more efficient and easily usable products/services. In this regard, transport is a crucial service industry in the transition towards urban sustainability. This chapter will focus on the role of gamification in sustainable mobility and the urban transition process. The study adopted a systematic literature review approach to summarize existing knowledge on gamification in the transport field. The review synthesized 40 studies highlighting the bibliometric characteristics, the main gamification outputs' and research gaps detecting potential implications and relevant insights for service literature. The research has evidenced how gamification can change consumer behavior by creating and increasing loyalty and engagement, raising awareness of environmental issues or developing an eco-friendly lifestyle. The study provides a systematic literature review investigating gamification's role in sustainable urban mobility. The findings present managerial implications and some insights for companies and policymakers into implementing gamification to pursue the sustainable urban transition.

Keywords Gamification · Sustainable mobility · Urban transition · Smart cities

M. G. Pasca (✉)
Niccolò Cusano University, Via Don Carlo Gnocchi, 3–00166 Rome, Italy
e-mail: mariagiovina.pasca@unicusano.it

L. Di Pietro · R. Guglielmetti Mugion
Department of Business Studies, Roma Tre University, Via Silvio D'Amico, 77–00145 Rome, Italy
e-mail: laura.dipietro@uniroma3.it

R. Guglielmetti Mugion
e-mail: roberta.guglielmettimugion@uniroma3.it

Z. Allam (ed.), *Sustainable Urban Transitions*, Urban Sustainability,
https://doi.org/10.1007/978-981-99-2695-4_12

1 Introduction

In several European countries, environmental issues and the consequent adverse effects on the planet have attracted policymakers, companies, consumers and academics [47]. Cities must transform into increasingly sustainable and intelligent smart cities, generating synergies between different sectors and companies and providing citizens with more efficient and easily usable products/services. In this regard, transport is a crucial service industry in the transition towards urban sustainability. Sustainable mobility is recognized as an EU priority by the Agenda 2030, and it is aimed at developing Quality, reliable, sustainable and resilient infrastructures (SDG-9.1),improving the safety and quality of the service provided (SDG-11.2), and promoting an accessible and flexible solution that reduces emissions (SDG-13). For instance, sharing mobility services promotes sustainable cities' development [32, 33]. Indeed, shared mobility services reduce traffic and polluting carbon dioxide (CO2) emissions as most shared vehicles are electric [29]. Electric vehicles in intelligent mobility are another strategic key to achieving the circular economic model [22, 44].

However, it is necessary to promote integrated and innovative infrastructures to ensure resilience to climate change, involving all actors in the urban context [30]. In this perspective, U.N. member states have set a specific goal in the Agenda 2030 to make cities inclusive, safe, resilient and sustainable. "By 2030, access to safe, affordable, accessible and sustainable transport systems must be provided for all" (SDG-11.2. In addition, to making cities sustainable, sustainable mobility improves citizens' well-being and quality of life by pursuing sustainable development. Recently, many European cities have witnessed the spread of shared micro-mobility services (e.g. Bike-sharing docked and dockless, scooter sharing, kick scooter [61, 51, 60]. These services facilitated reducing traffic congestion and polluting emissions by improving the quality of life [47]. Micro-mobility services were an effective solution for short journeys or to reach other services such as public transport. Indeed, it had the potential to solve the last-mile problem (passenger journeys of less than 5 miles) [6, 26]. However, despite the convenience and usefulness of these means for short journeys, cities face several issues such as parking spaces, efficient use, means' micro-mobility coverage, and vehicle safety [42, 60]. However, the outbreak of the Covid-19 pandemic has been changing worldwide the social and economic context by generating impacts on mobility field. First, it emerged how pollution could contribute to the spread of the Covid-19 virus, generating damage to public health, considering that cities are responsible for 60–80% of energy consumption and 75% of emissions of harmful substances [75]. Moreover, due to the pandemic restrictions, the people movement was drastically reduced, causing a strong decrease in service usage, namely shared mobility services. For instance, during the Italian lockdown in the first 2020 semester, the mobility rate dropped from 85% in 2019 to 32%, the effects were also evident in sharing mobility, whose services recorded an 80% drop in average demand. Anyway, post quarantine, shared mobility services usage has started to grow again; in particular, the bike and scooter sharing services recorded a recovery of 60% and 70%. For the car-sharing service, a recovery of 30% was recorded (Sharing Mobility Observatory 2021. In the last months, shared micro-mobility services were overgrowing across the United States and abroad [41, 52] as they were considered safer than public transport closed vehicles. Besides, the proliferation of shared mobility services has been supported by technology (peer-to-peer platforms, social networks, GPS, etc.) which has encouraged these new green transport solutions [3, 12], changing the citizens' lifestyles [59]. However, in the service literature, there is a need to empower and educate users on sharing and, in general, to understand how to motivate users in sustainable behaviours [47, 51, 73, 74]. Thus, although sustainable mobility is a crucial aspect of promoting sustainable development, it is necessary to investigate these issues by analysing how user behaviour has been changed [15]. To change users' behaviour, technological innovations included persuasive strategies integrated into mobile applications, such as gamification mechanics, which guided the user in choosing the most sustainable transport mode [1, 21]. The study conducted by Anagnostopolou et al. [1] highlighted that gamification mechanics are widely applied in the mobility sector, enhancing the usage of sustainable means of transportation and motiving users to change their travel behaviour to the most environmentally friendly habits.

Gamification is a new tool applied in non-gaming contexts to create and increase user engagement and to support the customers' value creation [16]. Huotari and Hamari [31] defined gamification as a procedure of delivering game-like experiences to users, usually to influence the users' behaviour. Gamification can be defined as the set of three elements, namely: affordances, psychological and behavioural outcomes. The offers are the mechanics behind the game that involve users (such as point and leadboard..). These affect the psychology of users, generating involvement, knowledge, social recognition or enjoyment [11]. Instead, behavioural outcomes refer to activities supported through gamification, such as increased user participation in the game [23]. Gamification generates three types of benefits: utilitarian (utility and ease of use), hedonic (fun and playfulness) and social (recognition and social influence) [24]. Gamification promotes interaction between the user and the system by creating and increasing loyalty (Clanton 1988), raising health awareness, or developing eco-friendly. Most of the existing literature is focused on the domains of education and learning, crowdsourcing and health [39]. Gamification is a powerful tool that can change consumer behaviour by creating and increasing their loyalty, and engagement, raising awareness of environmental issues or developing an eco-friendly lifestyle [61, 14 13, 16].

The chapter aims to investigate the role of gamification in sustainable urban mobility by analysing the main results and further research opportunities. The study adopted a systematic literature review (SLR) approach to summarize knowledge on gamification in the transport field.

In particular, this study aims to answer the following questions.

What are the bibliometric characteristics (publication year, document type, methodology, research context, means promoted and affordances implemented) of existing papers?

What are the most frequent affordances and benefits generated in the mobility sector?

What are the main themes, results and further research?

The review highlighted the bibliometric characteristics, the main gamification outputs' and research gaps evidencing potential implications and relevant insights for service literature. The research evidenced how gamification can change consumer behaviour by creating and increasing loyalty, and engagement, raising awareness of environmental issues or developing an eco-friendly lifestyle. The chapter is organized as follows. Paragraph 2 illustrates the methodological approach and describes the results and further research from the literature review. In paragraph 3 some conclusions are drawn.

2 Systematic Literature Review

2.1 Methodological Approach

The study adopted a systematic review approach [66] to investigate the existing literature [58] through a structured, repeatable and transparent process. After the screening phase, the papers selected for the eligibility were analyzed, firstly, author-centrically and then concept centrically to extract the relevant information [71].

To resume the literature, the authors conducted a descriptive and interpretative analysis [17]. The chapter synthesizes research and provides suggestions for future research [66].

The SLR is structured in three phases:

(i) *Planning the review;*
(ii) *Conducting the review;*
(iii) *Reporting and dissemination*

.

(i) *Planning the review*

The chapter aims to understand the role of gamification in sustainable urban transitions. The literature search was conducted in the Scopus database in July 2022.

The literature search was conducted only on one database to ensure the process's clarity, rigour and replicability, avoiding redundancy and overlapping [55]. Scopus database indexes contents of all other databases with potentially relevant content. The search of the literature in the Scopus database was conducted using the following search query:

Keyword (TITLE-ABS-KEY) Gamification AND urban mobility OR urban sustainability OR sustainable mobility OR sustainable urban mobility.

The literature search resulted in a total of 40 records.

(b) *Conducting the review*

The authors included the papers identified in database considering these criteria [49]:

(1) The research included conference papers, articles in journals, reviews, book and book chapters;
(2) The research has no time limits;
(3) The research papers must be written in English;
(4) Papers must be mainly focused on gamification in mobility context from a managerial perspective. Technical and engineering papers are excluded.
(5) Articles must be available in full text and must be already published;

Figure 1 reported records of the systematic review through the PRISMA flow diagram [49].

Two authors independently evaluated the studies, for inclusion and methodological quality, by reviewing the title and abstract to ensure quality and relevance. The other researcher reviewed the whole process [27]. The number of articles selected for eligibility for full-text analysis was 44 hits. 22 articles were excluded: two paper

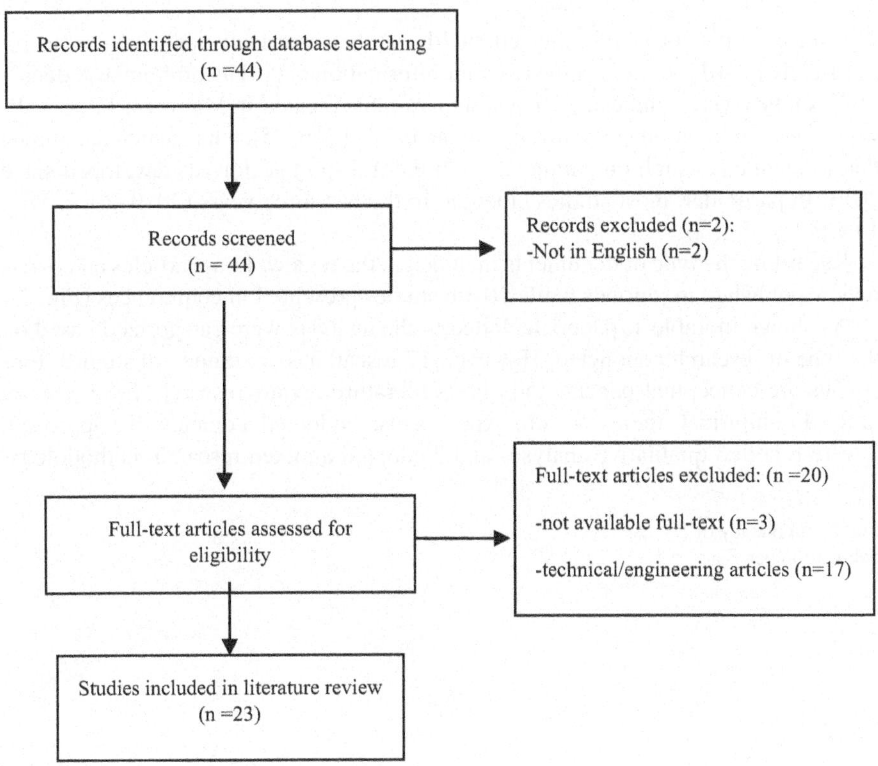

Fig. 1 PRISMA flow-chart

were not in English,3 articles were not available in the full text; 17 studies were analyzed from a technical/engineering perspective.

The final body of literature consists of 22 papers.

(iii) *Reporting and dissemination*

The full texts of the final body of literature were analyzed and the authors collected information such as: Study approach (conceptual or empirical), Methodology (qualitative, quantitative, or mixed method), research country, Aim, Findings, and Further research. These information were extracted from each study, collected in an excel file, and summarized through tables [71]. The reviewed studies were analyzed in order to provide the bibliometric characteristics of existing literature, to identify the main results of the gamification's use in sustainable urban mobility and suggestions for further research.

2.2 Main Results

2.2.1 Bibliometric Characteristics

The selected papers were analyzed, firstly, author-centrically and then concept-centrically in order to extract the relevant information [71]. The bibliometric details of 22 studies were synthesized in a summary table created in Microsoft Excel software. The gamification phenomenon spread in 2011 [16, 25], but research has shown that academic research on gamification in the transport sector has developed since 2013. In particular, most studies emerged in the last three years (2019–2022) with 14 articles (Fig. 2).

Regarding the type of documents included in the review, 11 are articles and one is review published in journals while 10 are studies presented in conferences (Fig. 3).

As shown in Table 1, papers included in the analysis were categorized based on the type of research conducted. However, 17 researches are empirical studies, four studies are conceptual papers, and one is literature review research. Furthermore, of the 17 empirical studies, 9 were experiments, 4 adopted a quantitative approach, 2 were reported qualitative analyses and 2 adopted a mixed research methodology.

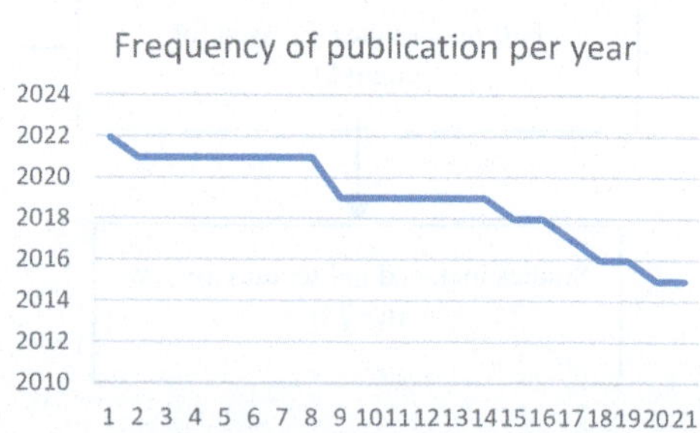

Fig. 2 Frequency of publication per year

Fig. 3 Documents by type

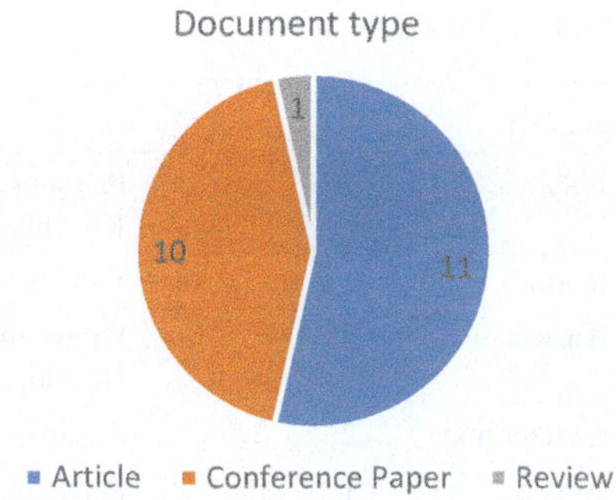

From our study emerges that a majority of empirical research are experiments that tested game mechanics to promote sustainable urban mobility and to involve, retain and modify user behaviour.

From the research context, it emerges that 7 empirical studies were conducted in Italy; 2 in Spain, Austria and Switzerland (Table 2).

In the 22 reviewed papers, it emerges that the affordances major utilized in the mobility field are points, badges, challenges and feedback. The choice of the affordances to be implemented is a strategic aspect and varies according to the objective to be achieved. The use of affordances creates competition among users. Receiving points creates involvement and dependence on the application, generating a sense of accomplishment in the user [64]. The ranking lists allow the comparison of performances between users by creating competition to encourage the use of sustainable means of transport. Rankings also allow participants to compare their scores

Table 1 Study approach

Study approach	Frequency	Study
Empirical	17	
Qualitative methodology	2	Cardoso et al. [8], Sipone et al. [63]
Quantitative methodology	4	Cellina et al. [10], Ferron et al. [18], Ebermann and Brauer [20], Kazhamiakin et al. [37]
Mixed methodology	2	Pasca et al. [56], Marconi et al. [46]
Experiment	9	Paraschivoiu and Layer-Wagner [54], Sipone et al. [35], Kazhamiakin et al. [43], Luger-Bazinger and Hornung-Prähauser [64], Cellina et al. [9]; Petersen et al. [57], Khoshkangini et al. [38], Kazhamiakin et al. [34], Jylhä et al. [36]
Conceptual	4	Wallius et al. [68], Maltese et al. [45], Bucchiarone et al. [7], Wernbacher et al. [72]
Literature review	1	Andersson et al. [2]

Table 2 Research context

Research context	Frequency	Study
Italy	7	Pasca et al. [56], Kazhamiakin et al. [20, 34–37, 46], Khoshkangini et al. [38]
Spain	2	Sipone et al. [63, 64]
Austria	2	Paraschivoiu and Layer-Wagner [54], Luger-Bazinger and Hornung-Prähauser [43]
Switzerland	2	Cellina et al. [9, 9]
Norway	1	Petersen et al. [57]
Germany	1	Ebermann and Brauer [18]
Finland	1	Jylhä et al. [34]
Portugal	1	Cardoso et al. [8]

against others to motivate them to reach a higher score. The rankings promote social confrontation, an essential antecedent of motivation [20] and competence satisfaction. Feedback can support user behaviour change in the long run as it allows people to monitor their behaviour and control their choices [9]. In general, the design of the incentives must be adapted to the corporate culture context and meet the recipients' needs [65].

Our research analysed which benefits are generated by using gamification in the transport sector. To engage users, game mechanics need to afford extrinsic and/or intrinsic motivations [25]: utilitarian (usefulness, ease of use), hedonic (enjoyment, playfulness) and social (recognition, social influence). In particular, from the reviewed articles, the game mechanics implemented generate utilitarian and social benefits. Users used the game mechanics for usefulness, but at the same time, social motivations prevailed, such as the need for recognition in the community.

2.2.2 Review Analysis

To extract, synthesize and analyse the results of studies, we used the content analysis in order to classify the reviewed papers. In particular, the title, abstract, author's keywords context, aim and research gap of the 22 studies were reviewed and coded. The authors extracted, summarized and analyzed the results of the studies in order to identify the main themes, results and further research for each theme in transport sector. The analysis of the body of literature shows that there are 5 key themes. To identify the themes we considered the context of the study, the research gap and the results.

Table 3 outlines the main themes of studies.

The need to change citizens' behaviour has emerged in most of the reviewed papers.

Table 3 Themes in reviewed papers

Themes	Study
Behavioural Change	Paraschivoiu and Layer-Wagner [54], Kazhamiakin et al [36, 37], 2021, Luger-Bazinger and Hornung-Prähauser [8], Cardoso et al. [20], Ferron et al. [43], Cellina et al. [10], Andersson et al. [2], Khoshkangini et al. [38], Ebermann and Brauer [18], Wernbacher et al. 2015
Environmental awareness	Maltese et al. [7], Bucchiarone et al. [45], Cellina et al. [9], Petersen et al. [57], Jylhä et al. [34]
Game edutainment	Sipone et al. [63, 64], Marconi et al. [46]
Urban gamification	Wallius et al. [68]
Service quality	Pasca et al. [56]

Technology is not enough to achieve environmental sustainability goals. It is necessary to change one's behaviour, decreasing the use of private car that generates traffic and pollution [2]. Negative environmental impacts can be mitigated or eliminated by changing human behaviour. Several studies have highlighted how human behaviour is the strategic factor in achieving environmental sustainability, raising awareness and educating users [8]. Green Information Systems [19], which use persuasive strategies such as gamification, require a behavioural change to accept and adopt gamified systems to raise user awareness of environmental issues [35].

The theme *Environmental Awareness* includes studies that encourage green transportation behaviour. Game mechanics are used to sensitizing users on environmental issues and the consequences of their behaviour. Numerous studies highlighted how persuasive strategies aim to increase citizens' awareness and promote their behavioural change. The game educates, sensitizes and informs the player [7, 44, 9, 57, 34].

Game edutainment is a hybrid form that combines the need to educate users and provide entertainment through IT systems. The studies included in this theme show that game mechanics provide an interactive tool for informing, involving, educating and entertaining citizens [46, 63, 64]. The study conducted by [63] introduces a learning game to improve players' knowledge, attitude and willingness to use mobility systems. In addition, to sensitize users, it is necessary to provide information on how to use alternative means to your private car. Through the game, it is possible to integrate competition, cooperation and acquisition of skills and therefore have an educational purpose as well as entertainment [46, 64]. Several studies have highlighted how serious games facilitate learning by being engaging, based on real problems, providing continuous feedback, and helping to change attitudes [4].

The theme of Urban Gamification includes a study that analyses the contemporary forms of playful urban activity that allows to reinterpret and rediscover the city. The case of the e-scooter illustrates how playful urban mobility marks new connections between civic concerns of data security, physical safety, inclusivity, and urban sustainability [68].

The study included in the Service Quality theme highlight how gamification can be a strategic tool to improve the quality of the service and the users' loyalty [56]. The gamification improves users' engagement, transferring rules, facilitates the achievement of goals and quality standards and enhances bike sharing usage [56]. To encourage sustainable mobility, it is necessary to improve the quality of services and therefore to design services considering the needs of the user. The gamification applied to the transport context improve the quality of the interaction between the user and the shared mobility company by providing information.

2.3 Discussion

The study highlighted that gamification is a crucial strategy in the services sector to remain competitive in the market and pursue sustainability. The research provides an analysis of the effects of gamification use' in the mobility sector. In the results section, the main themes of the research on gamification have been identified and presented: behavioural change; environmental awareness; game edutainment; urban gamification; service quality. The systematic literature review confirms the contribution of gamification strategies in transmitting knowledge and increasing awareness among users generating engagement on sustainability issues. Several studies have highlighted the potential of gamification mechanics for motivating users to change behaviour and make more sustainable mobility choices by making them more aware of environmental issues [8, 20, 72, 36]. It is crucial to explore the participants' preferences before designing a game [10, 8]. Customization to the user, relevant and contextualised information and feedback, commitment, appealing design, personalized information and mobile applications with intuitive designs are aspects that influence behavioural change [7, 34, 2, 36].

It emerges as users who would be interested in sustainability had a positive attitude toward using the Smart Citizen application long-term [54]. Indeed, Paraschivoiu and Layer-Wagner [54] found a positive attitude to be more active and acting sustainably locally.

Gamification, through the use of challenges, promotes sustainable mobility and supports the interaction of the many stakeholders involved [35, 36]. Participants showed that citizens become part of a community by using a personal mobility app that incorporates gamified elements and nudging. They are motivated to cycle more and explore sustainable city services through tours and visiting points of interest [43]. Competition and environmental attitude affect the implementation of sustainable behaviour. In fact, the results of Ebermann and Brauer [18] study' reveal that a combination of the hedonic goal "competition" and the normative goal "climate protection" leads a positive impact on sustainable behaviour.

Gamification strategies provide tourists an enjoyable and authentic experience, engaging them in public transportation services [8]. Gamification emerged as an essential asset for promoting sustainable behaviours and city-wide policies related to the services made available within a complex sociotechnical system like a Smart City [37].

Persuasive strategies, such as gamification, have the potential to raise awareness of environmental issues and develop green mindsets [9]. However, a comprehensive approach (e.g. citizen involvement, spatial planning, gamification, data collection, etc.) is lacking to promote sustainable mobility [45]. For instance, through the Mobility-as-a-Service apps, users can plan, pay and use any mode of transport or service in an integrated and seamless way [7]. Through personalised challenges, users become aware of environmental issues and receive rewards that change their transport habits [34]. The study by [57] highlighted how the Smiling Earth application helps to raise awareness among users on environmental issues. The gamified application motivates users to change their behaviour towards environmentally sustainable habits [57].

The study conducted by [46] shows a behavioural change generated by educational value. Kids-Go-Green improves children's mobility habits through gamification mechanics. Riding a bicycle with other children, being part of an active community and seeing the impact on society and the environment of one's choices helps shape the character of children as functional members of society. Through these mechanics, children have acquired new concepts highlighting the importance of the social and economic components linked to sustainable mobility and generating awareness of environmental issues [63]. Through the Classcraft gamified platform, pupils acquired new concepts associated with social and economic components and became aware of how to become an active part of behavioural change [64].

Through the game mechanics, it is possible to discover and interact with the city and its services. Playful urban mobility marks new connections between civic concerns of data security, physical safety, inclusivity, and urban sustainability [68].

Using gamification mechanics in mobility shared services improves users' loyalty and influences service quality. The gamification tool improves users' engagement, transferring rules, facilitates the achievement of goals and quality standards and enhances bike sharing usage [56]. In the context of mobility, the use of Information Systems (IS) can support operational processes by improving the quality of the service.

2.4 *Future Perspective of Research*

Our systematic literature review frames the current state of gamification knowledge in the sustainable urban mobility sector, identifying future research opportunities. The authors propose research suggestions on gamification in services by comparing the study results with existing knowledge.

Regarding the Behavioural change theme, further longitudinal research is needed to investigate how users' willingness to change modes of transport varies to different persuasive strategies and different cultural contexts and how the use of the app influences public engagement in the field of sustainability [54, 2, 36, 19]. It is necessary to carry out qualitative and quantitative studies with larger samples and check whether the "green" behaviour does not change over time [72. Future research will have to examine the implementation of different game mechanics in other countries and in other contexts to improve mobile applications with greater customization [2, 19, 34, 36, 9] evaluating if the gamified systems respond to different interests/needs of users [37, 36]. These studies are useful for policymakers and transport operators to identify how best to implement mobile applications to engage and sensitize citizens on environmental issues.

Studies on environmental awareness have shown that it is necessary to conduct longitudinal and quantitative studies to bring out the effects of the long-term interaction between users and the game system considering different cultural contexts [48, 62], benefits and risks, as well as environmental attitudes [9]. Future studies will have to investigate the link between gamification and other means of transport and the impact of these mechanics, considering the challenges regarding ethics, safety and protection of autonomous vehicles [7, 45]. In addition, the study conducted by [34] highlighted the need to include social features and personalization of challenges based on user characteristics and evaluate the effectiveness of different types of challenges [57].

From the Game edutainment theme emerged that further studies need to be conducted to assess user needs and consider personal concerns [40]. In general, it is necessary to evaluate whether the acquired concepts were maintained after the game activity to generalize and validate the results obtained, considering other cultural contexts [63, 64]. [46] showed that through the Kids-Go-Green game, children are educated to use alternative means, such as bikes, to go to school, and future research prospects plan to update the game with new functionality to promote safe road habits. In the future, a quantitative analysis can be administered to parents to verify whether the sustainable behaviour learned by children has spread to other contexts, such as leisure time and family travel [64].

Regarding the urban gamification theme, the need has emerged to investigate how gamification mechanics influence behaviour differently, considering different cultural contexts [53]. Furthermore, it is necessary to consider in future research how factors such as disability, race, gender and sexuality shape the experience of participants in urban recreational mobility, also by comparing different means of transport (bicycles, cars, self-driving cars, drones and Segways) [68].

The literature review has highlighted a few studies that analyse the role of service quality as a driver in the usage of shared means [50, 56, 67]. However, few studies have emerged focusing on the effects of technological innovations in the sustainable mobility field [5, 28]. To evaluate the efficiency of gamification in improving the quality of services, it is necessary to investigate the gamification tool for other shared services in other Italian cities, in different cultural contexts and after the Covid-19 pandemic [56].

3 Conclusions

The research shows the state-of-the-art of existing literature on gamification use in the sustainable urban transition to identify the main results and future research opportunities. Our research highlights that interest in gamification in the mobility field is growing, and most studies test, through experiments, mobile applications with game mechanics.

Research conducted by the National Sharing Mobility Observatory (2021) showed that Italians changed their mobility habits after the pandemic. Pedestrian, bicycle, and shared mobility services are the solutions that emerge most. Users seek safety and flexibility in means of transport and infrastructure. Through a holistic approach, governments must ensure easy access to local mobility services to efficiently exploit existing means of transport and promote sustainable means of transportation. With the emergency Covid-19, many governments, through mobility incentives, are implementing interventions at the level of infrastructure, services and green mobility policies by promoting sharing mobility services such as using bicycles and scooters. In support of these interventions, it is necessary to sensitize and educate the user on environmental issues to influence their transport habits gradually. It is required to create services that meet users' needs as personalization can allow users' interests to remain constant over time. Existing applications where game mechanics are installed must use persuasive strategies tailored to the needs of individual users, not a "one size fits all" approach. As highlighted by the reviewed papers, gamification is crucial for changing transport habits and generating awareness of environmental issues by educating and raising awareness among citizens. Using these mechanics helps explore citizens' needs by highlighting their travel habits. The information collected, through the use of mobile applications, can be used by governments and private mobility companies as Information and Communication Technologies (ICT) collect helpful data to improve urban strategies and infrastructures to implement a digital transition that favours the achievement of the Agenda 2030 goals.

The chapter has some limitations. The authors have conducted a systematic literature review on one database, following [55] guidelines, to ensure its quality in terms of both rigour and relevance. Although Scopus is the largest, complete online database of peer-reviewed literature and is particularly relevant for social sciences research, some studies may have been missed by not including other databases [70]. Although the papers were reviewed systematically, we may have missed certain studies that did not use search terms included in our search string. In order to minimize coding mistakes, two reviewers created and discussed the coding tables for the papers.

In addition, further research is required to understand how innovative tools may represent an opportunity for the mobility field to reformulate its strategy through technological innovation by evaluating the new needs of users.

References

1. Anagnostopoulou E, Bothos E, Magoutas B, Schrammel J, Mentzas G (2018) Persuasive technologies for sustainable mobility: State of the art and emerging trends. Sustainability 10(7):2128
2. Andersson A, Hiselius LW, Adell E (2018) Promoting sustainable travel behaviour through the use of smartphone applications: A review and development of a conceptual model. Travel Behav Soc 11:52–61
3. Banister D (2008) The sustainable mobility paradigm. Transport policy 15(2):73–80
4. Boyle E, Connolly TM, Hainey T (2011) The role of psychology in understanding the impact of computer games. Entertain Comp 2(2):69–74
5. Brendel AB, Mandrella M (2016) Information systems in the context of sustainable mobility services: A literature review and directions for future research
6. Brunner H, Hirz M, Hirschberg W, Fallast K (2018) Evaluation of various means of transport for urban areas. Energy, Sustain Soc 8(1):1–11
7. Bucchiarone A, Battisti S, Dias TG, Feldman P (2021) Guest editorial diversification in urban transportation systems and beyond: Integrating people and goods for the future of mobility. IEEE Trans Intell Transp Syst 22(4):2008–2012
8. Cardoso B, Ribeiro M, Prandi C, Nunes N (2019) When gamification meets sustainability: A pervasive approach to foster sustainable mobility in madeira. In: Proceedings of the 1st ACM workshop on emerging smart technologies and infrastructures for smart mobility and sustainability, pp 3–8
9. Cellina F, Bucher D, Mangili F, Veiga Simão J, Rudel R, Raubal M (2019) A large scale, app-based behaviour change experiment persuading sustainable mobility patterns: Methods, results and lessons learnt. Sustainability 11(9):2674
10. Cellina F, Bucher D, Veiga Simão J, Rudel R, Raubal M (2019) Beyond limitations of current behaviour change apps for sustainable mobility: insights from a user-centered design and evaluation process. Sustainability 11(8):2281
11. Cheong C, Filippou J, Cheong F (2014) Towards the gamification of learning: Investigating student perceptions of game elements. J Inf Syst Educ 25(3):233
12. Chourabi H, Nam T, Walker S, Gil-Garcia JR, Mellouli S, Nahon K, Scholl HJ (2012) Understanding smart cities: An integrative framework. In 2012 45th Hawaii international conference on system sciences, pp 2289–2297. IEEE
13. Clanton C (1998) An interpreted demonstration of computer game design. In: CHI 98 conference summary on Human factors in computing systems, pp 1–2. ACM
14. Crawford C (2011) The art of computer game design (Kindle version), Seattle from Amazon. Com
15. Cruz IS, Katz-Gerro T (2016) Urban public transport companies and strategies to promote sustainable consumption practices. J Clean Prod 123:28–33
16. Deterding S, Dixon D, Khaled R, Nacke L (2011) From game design elements to gamefulness: defining" gamification". In: Proceedings of the 15th international academic MindTrek conference: Envisioning future media environments, pp 9–15
17. Dixon-Woods M, Bonas S, Booth A, Jones DR, Miller T, Sutton AJ, Young B (2006) How can systematic reviews incorporate qualitwative research? A critical perspective. Qualitative research 6(1):27–44
18. Ebermann C, Brauer B (2016) The role of goal frames regarding the impact of gamified persuasive systems on sustainable mobility behavior
19. Elliot S (2011) Transdisciplinary perspectives on environmental sustainability: a resource base and framework for IT-enabled business transformation. Mis quarterly:197–236
20. Ferron M, Loria E, Marconi A, Massa P (2019) Play&Go, an urban game promoting behaviour change for sustainable mobility". Interaction Des. Archit. J. 40:24–25
21. Fogg BJ (2003) Prominence-interpretation theory: Explaining how people assess credibility online. In: CHI'03 extended abstracts on human factors in computing systems, pp 722–723

22. Garau C, Masala F, Pinna F (2016) Cagliari and smart urban mobility: Analysis and comparison. Cities 56:35–46
23. Hamari J (2013) Transforming homo economicus into homo ludens: A field experiment on gamification in a utilitarian peer-to-peer trading service. Electron Commer Res Appl 12(4):236–245
24. Hamari J, Koivisto J (2015) Why do people use gamification services? Int J Inf Manage 35(4):419–431
25. Hamari J, Koivisto J, Sarsa H (2014) Does gamification work?--a literature review of empirical studies on gamification. In: 2014 47th Hawaii international conference on system sciences, pp 3025–3034. IEEE
26. Heineke K, Kloss B, Scurtu D, Weig F (2019) Micromobility's 15,000-mile checkup
27. Heinonen K, Jaakkola E, Neganova I (2018) Drivers, types and value outcomes of customer-to-customer interaction: An integrative review and research agenda. J Serv Theory Practice
28. Herrenkind B, Brendel AB, Nastjuk I, Greve M, Kolbe LM (2019) Investigating end-user acceptance of autonomous electric buses to accelerate diffusion. Transp Res Part D: Transp Environ 74:255–276
29. Hickman R, Banister D (2014) Transport, climate change and the city. Routledge
30. Hoornweg D (2016) Cities and Sustainability: A new approach. Routledge
31. Huotari K, Hamari J (2012) Defining gamification: a service marketing perspective. In: Proceeding of the 16th international academic MindTrek conference, pp 17–22
32. Jankal R, Jankalova M (2019) 2030 Agenda and sustainable smart city
33. Jaremen DE, Nawrocka E, Żemła M (2019) Sharing the economy in tourism and sustainable city development in the light of agenda 2030. Economies 7(4):109
34. Jylhä A, Nurmi P, Sirén M, Hemminki S, Jacucci G (2013) Matkahupi: a persuasive mobile application for sustainable mobility. In: Proceedings of the 2013 ACM conference on Pervasive and ubiquitous computing adjunct publication, pp 227–230
35. Kazhamiakin R, Loria E, Marconi A, Scanagatta M (2021) A gamification platform to analyze and influence citizens' daily transportation choices. IEEE Trans Intell Transp Syst 22(4):2153–2167
36. Kazhamiakin R, Marconi A, Martinelli A, Pistore M, Valetto G (2016) A gamification framework for the long-term engagement of smart citizens. In: 2016 IEEE International Smart Cities Conference (ISC2), pp 1–7. IEEE
37. Kazhamiakin R, Marconi A, Perillo M, Pistore M, Valetto G, Piras L, Perri N (2015). Using gamification to incentivize sustainable urban mobility. In: 2015 IEEE first international smart cities conference (ISC2), pp 1–6. IEEE
38. Khoshkangini R, Marconi A, Valetto G (2017) Machine learning for personalized challenges in a gamified sustainable mobility scenario. In: Extended abstracts publication of the annual symposium on computer-human interaction in play, pp 361–368
39. Koivisto J, Hamari J (2019) The rise of motivational information systems: A review of gamification research. Int J Inf Manage 45:191–210
40. König A, Kowala N, Wegener J, Grippenkoven J (2019) Introducing a mobility on demand system to prospective users with the help of a serious game. Transp Res Interdiscip Perspect 3:100079
41. Lazarus J, Pourquier JC, Feng F, Hammel H, Shaheen S (2020) Micromobility evolution and expansion: Understanding how docked and dockless bikesharing models complement and compete–A case study of San Francisco. J Transp Geogr 84:102620
42. Li A, Zhao P, He H, Axhausen KW (2020) Understanding the variations of micro-mobility behavior before and during COVID-19 pandemic period. Arbeitsberichte Verkehrs-und Raumplanung, 1547.
43. Luger–Bazinger C, Hornung–Prähauser V (2021) Innovation for sustainable cities: the effects of nudging and gamification methods on urban mobility and sustainability behaviour. GI_Forum 9:251–258
44. Lyons G (2018) Getting smart about urban mobility–aligning the paradigms of smart and sustainable. Transp Res Part A: Policy Pract 115:4–14

45. Maltese I, Gatta V, Marcucci E (2021) Active travel in sustainable urban mobility plans. An italian overview. Res Transp Bus & Manag, 40:100621
46. Marconi A, Schiavo G, Zancanaro M, Valetto G, Pistore M (2018) Exploring the world through small green steps: improving sustainable school transportation with a game-based learning interface. In: Proceedings of the 2018 International Conference on Advanced Visual Interfaces, pp 1–9
47. McKenzie G (2020) Urban mobility in the sharing economy: A spatiotemporal comparison of shared mobility services. Comput Environ Urban Syst 79:101418
48. Millonig A, Wunsch M, Stibe A, Seer S, Dai C, Schechtner K, Chin RC (2016) Gamification and social dynamics behind corporate cycling campaigns. Trans Res procedia 19:33–39
49. Moher D, Liberati A, Tetzlaff J, Altman DG, TP Group (2009) Linee guida per il reporting di revisioni sistematiche e meta-analisi: il PRISMA Statement. PLoS Med 6(7):e1000097
50. Mugion RG, Toni M, Di Pietro L, Pasca MG, Renzi MF (2019) Understanding the antecedents of car sharing usage: an empirical study in Italy. Int J Qual Serv Sciences
51. Mulcahy RF, McAndrew R, Russell-Bennett R, Iacobucci D (2021) Game on! Pushing consumer buttons to change sustainable behavior: a gamification field study. Eur J Marketing
52. National Association of City Transportation Officials (NACTO) (2019) Shared Micro mobility in the US: 2018
53. Pang C, Pan R, Wong S, Neustaedter C, Wu Y (2017) City explorer: Gamifying public transit trips while exploring the City. In: Proceedings of the 2017 CHI conference extended abstracts on human factors in computing systems, pp 2825–2832
54. Paraschivoiu I, Layer-Wagner T (2021) Placemaking for urban sustainability: designing a gamified app for long-term, pro-environmental participation. In: Extended abstracts of the 2021 annual symposium on computer-human interaction in play, pp 186–191
55. Paré G, Trudel MC, Jaana M, Kitsiou S (2015) Synthesizing information systems knowledge: A typology of literature reviews. Inf Man 52(2):183–199
56. Pasca MG, Mugion RG, Toni M, Di Pietro L, Renzi MF (2021) Gamification and service quality in bike sharing: an empirical study in Italy. The TQM J
57. Petersen SA, Ahcin P, Petersen I (2019) Smiling earth-citizens' awareness on environmental sustainability using energy and transport data. In: Joint international conference on entertainment computing and serious games, pp 459–465. Springer, Cham
58. Pittaway L, Robertson M, Munir K, Denyer D, Neely A (2004) Networking and innovation: a systematic review of the evidence. Int J Manag Rev 5(3–4):137–168
59. Scavarda A, Daú GL, Scavarda LF, Korzenowski AL (2019) A proposed healthcare supply chain management framework in the emerging economies with the sustainable lenses: The theory, the practice, and the policy. Resour Conserv Recycl 141:418–430
60. Shaheen S, Cohen A (2019) Shared micromoblity policy toolkit: Docked and dockless bike and scooter sharing
61. Sharing Mobility Observatory Report (2021) Rapporto Nazionale Sulla Sharing Mobility, Osservatorio Nazionale sulla Sharing Mobility
62. Shevchuk N, Degirmenci K, Oinas-Kukkonen H (2019) Adoption of gamified persuasive systems to encourage sustainable behaviors: Interplay between perceived persuasiveness and cognitive absorption. In: Proceedings of the 40th International conference on information systems, ICIS 2019, pp 1–17. Association for Information Systems
63. Sipone S, Abella-García V, Barreda R, Rojo M (2019) Learning about sustainable mobility in primary schools from a playful perspective: A focus group approach. Sustainability 11(8):2387
64. Sipone S, Abella-García V, Rojo M, dell'Olio L (2021) Using ClassCraft to improve primary school students' knowledge and interest in sustainable mobility. Sustainability 13(17):9939
65. Suh A, Wagner C (2017) How gamification of an enterprise collaboration system increases knowledge contribution: an affordance approach. J Knowledge Manage
66. Tranfield D, Denyer D, Smart P (2003) Towards a methodology for developing evidence-informed management knowledge by means of systematic review. Br J Manag 14(3):207–222
67. Tussyadiah IP (2015) An exploratory study on drivers and deterrents of collaborative consumption in travel. In: Information and communication technologies in tourism 2015, pp 817–830. Springer, Cham.

68. Wallius E, Thibault M, Apperley T, Hamari J (2022) Gamifying the city: E-scooters and the critical tensions of playful urban mobility. Mobilities 17(1):85–101
69. Wang Y, Han JH, Beynon-Davies P (2018) Understanding blockchain technology for future supply chains: a systematic literature review and research agenda. Supply Chain Manag: Int Journal
70. Wang Y, Han JH, Beynon-Davies P (2019) Understanding blockchain technology for future supply chains: a systematic literature review and research agenda. Sup Chain Manage: Int J 24(1):62–84
71. Webster J, Watson RT (2002) Analyzing the past to prepare for the future: Writing a literature review. MIS quarterly:13-23
72. Wernbacher T, Pfeiffer A, Platzer M, Berger M, Krautsack D (2015) Traces: a pervasive app for changing behavioural patterns. In: European conference on games based learning, vol 589
73. White K, Habib R, Hardisty DJ (2019) How to SHIFT consumer behaviors to be more sustainable: A literature review and guiding framework. J Mark 83(3):22–49
74. Winslow J, Mont O (2019) Bicycle sharing: Sustainable value creation and institutionalisation strategies in Barcelona. Sustainability 11(3):728
75. Zambrano-Monserrate MA, Ruano MA, Sanchez-Alcalde L (2020) Indirect effects of COVID-19 on the environment. Sci Total Environ 728:138813

Regional and Project Management in the Context of Shaping Sustainable Urban Futures: Theoretical and Practical Context of the EU Programming Period 2021–2027 in the Czech Republic

Martin Pelucha and Jana Kourilova

Abstract The chapter discusses the terminological foundations of cohesion policy, which has a long-term emphasis on urban regions, as well as key megatrends that are reflected in the formation of strategic goals and sub-policies of the EU. The EU's current long-term goals are related to the European Green Deal, which affects the context of cohesion policy, rural development policy, the transformation of coal-fired regions into a green economy, etc. Climate change has become the main leitmotif of the second decade of the new millennium, which is currently reflected in its implementation in policy-making (i.e. mainstreaming of climate change). These global and EU challenges have a major impact, and will continue to influence the formation of urban structures. Although preparations for the 2021–2027 programme period have been significantly delayed, some fundamental new elements that will influence the future sustainability of urban centres are already visible over the long-term, including the design of individual infrastructure projects, and their philosophy and management, e.g. with reference to the "do no significant harm" (DNSH) principle.

The chapter reflects these elements, using the example of the Czech Republic, which has three major urban regions (i.e. the Prague agglomeration and two coal regions—Ustecky region and Moravian-Silesian region), and regions that have a significant representation of small and medium-sized towns. It is the dense network of these towns that creates very specific preconditions for policy-making.

The last section of the chapter compiles critical reflections on the theoretical postulates of traditional approaches to theories of regional and urban development, and their confrontation with broader aspects of socioeconomic development, and the functioning of regional economies and urban agglomerations in a globalised environment.

M. Pelucha (✉) · J. Kourilova
Faculty of Economics, Department of Regional Management and Law, University of South Bohemia in České Budějovice, České Budějovice, Czech Republic
e-mail: Martinpelucha@gmail.com

Z. Allam (ed.), *Sustainable Urban Transitions*, Urban Sustainability,
https://doi.org/10.1007/978-981-99-2695-4_13

1 Terminological Foundations of Cohesion Policy, Key Global Megatrends, and the Context of EU Policy-Making in the 2021–2027 Programming Period

Historically, the emergence of regional policy is closely linked to the problem of significant interregional disparities in key socio-economic indicators, especially in the level of unemployment. An example is the emergence of this policy in Great Britain in the 1930s, as a result of the Great Depression. Also, other countries of Western Europe continued, in different ways, to design various instruments of regional policy during the 1950s. For example, France dealt with significant differences in the performance of particular cities and smaller towns, compared to the socio-economically overwhelmed Paris agglomeration. Italy dealt with problems linked to the differences between the wealthy north and the poorer south, or the Benelux countries dealt with infrastructural problems due to strong urbanisation. Over time, however, external influences, and various factors that influenced the shaping of regional policy instruments, were constantly increasing in the individual stages of the development of the European integration process.

The increasing importance of external influences can be demonstrated in the development of the relevant terminology. There is a fundamental difference between the terms of regional policy, European regional policy, structural policy, and cohesion policy. It should be noted that many different factors and much contextual information can be assigned to individual terminological titles of regional politics and, basically, a separate book could be compiled for such a description. In this section, the key connections that represent the main differences between the aforementioned titles of regional politics are briefly presented.

While the early years (i.e. the 1930s and 1950s) of **regional policy** are briefly discussed above, European regional policy is closely related to the first round of EEC enlargement to involve the United Kingdom (UK) in 1973. The UK negotiated the creation of a new specific fund, which would be a certain counterweight to the dominant Common Agricultural Policy. In other circumstances, the UK would be a net contributor to the European budget, which was not politically acceptable for the British public and, therefore, for British representatives. That is why the European Regional Development Fund (ERDF) was created in 1975, and was presented as a key stimulus for the creation of **European regional policy**. In the 1980s, the southern enlargement of the EEC by Greece (1981), Spain (1986), and Portugal (1986) took place, which also had an impact on discussions about financing regarding the common budget. On the one hand, the British, under the leadership of Prime Minister Margaret Thatcher, exchanged the so-called British rebate (i.e. a discount from contributions to the common budget, which only ended with Brexit), and the negative consequences of human activity on the environment (especially in industrial regions), were also more noticeable. For this reason, the intention was to coordinate more, the previously separate financial flows supporting regional development, i.e. support for economic development in the form of ERDF (hard infrastructure projects), and support for the social dimension of regional development (soft projects focused on human resources), through the European Social Fund (hereinafter ESF), while the environmental dimension of development was newly emphasised in a coordinated manner, in both hard and soft regional development projects. In other words, **structural policy** represents an umbrella for regional, social, and environmental policy. That is why the relevant funding sources were collectively called Structural Funds.

Cohesion policy is the last in a series of terms that need to be explained. Literally translated, it is originally a policy of economic and social cohesion, and after the approval of the Treaty of Lisbon in 2009, territorial cohesion was also added. The essence of the concept of coherence is closely related to the efforts of policy makers and interest groups, to coordinate the implementation of the tools used, and also to place more emphasis on their evaluations [33. In relation to the topic of this chapter, and the book as a whole, territorial cohesion is key, which represents a long-term effort to conceptually solve urban–rural relations. According to the literature review, the debate is about functional regions (ESPON Monitoring Committee, 2007; [26, 36], which usually represent strong urban centres and their catchment areas, which are functionally linked by commuting to work, services, culture, and infrastructure load, while being administratively different regions. In other words, functional regions solve real spatial links, needs, and challenges in the development of localities and regions. This is a real challenge for regional development in the twenty-first century. However, critical remarks can be made about the design of the relevant instruments in the first decade of the new millennium, when for the 2007–2013 programme period, the instruments of the rural development policy were completely removed from the cohesion policy, and transferred under the second pillar of the Common Agricultural Policy, while the cohesion policy was only supposed to focus on urban regions [11, 34]. This strict separation in funding has not been entirely successful, and has been the subject of many discussions, as the socio-economic development of rural areas has only very limited possibilities in the second pillar, with respect to the emphasis on rural "agricultural" development. A clear exception are the long-proven tools related to the LEADER method, and its implementation through local action groups (hereinafter, referred to as LAGs). Therefore, even in the subsequent programming period of 2014—2020, greater emphasis began to be placed on spatial links and territorial development in a more complex concept, especially through a more significant expansion, and the use of integrated territorial investments (hereinafter, referred to as ITI; see also European [8] and integrated territorial development plans, through which it was, and is, possible to access real functional socio-economic ties in the regions, and to support urban–rural relations in a comprehensive way.

From the terminological analysis, a strong emphasis on the endogenous conditions of the development of regions and localities, with an emphasis on strengthening the internal potential of the given territory, is evident. Especially in the 1990s, this was reflected in the expansion of strategic planning of regions, cities, and smaller municipalities [13, 42]. On the contrary, since the beginning of the new millennium, the growing importance of external context and related factors in published critical papers has been evident. In this connection, for example, [33] noted the importance of extralocal factors, which actors of local and regional development may not always be able to identify well, and then adequately counter them. Agarwal et al. [1] pointed out that particular approaches applied in regional development may not always be completely successful, because there is a dominant influence of other economic policies of states and transnational institutions (the European Union and its institutions, e.g. the European Central Bank), or multinational global companies. Woods [38] also noted the diverse abilities of local and regional development actors to adequately perceive and solve relevant extra-local influences, which was also confirmed in the OECD study (2014), in noting that long-term socio-economic trends at the local and regional level are primarily influenced by the external context, while regional or urban policy instruments can only moderate the negative consequences of global trends.

In the case of some global trends, it is really complicated to deal with the respective impacts, and that is why there is often talk of actors adapting to new conditions, or mitigating negative effects. What trends are we actually talking about? Terminologically, the term **megatrend** began to be used frequently in the academic literature and in strategic documents, but it originally arose at the end of the 1980s and was first discussed by Naisbitt [23]. Then, it was more prominently implemented in the practice of policy-making by the European Environment Agency, which published in 2015 a comprehensive report and outlook [7] regarding the European environmental state, trends and prospects, in a global context. Specifically, 11 global megatrends were identified, which should be reflected in the design of policies and strategic documents. Megatrends relate to demographic contexts (including urbanisation), the grave state of pandemics, accelerating technological changes, economic contexts (economic growth, global disparities, competition, resources), environmental challenges (sustainability of ecosystems, climate change, pollution), and also diversifying approaches to governance. Carayannis et al. [3] paid long-term attention to the conceptualisation of this topic, distinguishing between three levels of contextual influences, and naming them as **the helix system** (Triple Helix, Quadruple Helix, and Quintuple Helix). The Triple Helix represents the local microeconomic level at which specific approaches and forms of solutions are created by the relevant actors. The Quadruple Helix represents the national level of influence, and includes the context of society, legislation, culture, traditions. The Quintuple Helix represents the transnational, and especially the global level of factors, that are represented precisely by megatrends (e.g. climate change, technological progress).

From the perspective of policy-making, there is an evident dominant emphasis on the environmental part of megatrends, which is most closely associated with mitigation of, and adaptation to, climate change. During the second decade of the new millennium, the word **greening**, which covers the process of solving relevant environmental problems, as a result of the megatrend of climate change, began to be commonly used at the international level. An example is the OECD study (2013), whose title is absolutely apt, and demonstrates a paradigmatic shift in policy-making, i.e. "*Putting Green Growth at the Heart of Development*". In the EU, The European Green Deal has become an essential strategic document, the goal of which is to achieve zero net emissions by 2050, while interim goals are also set. This strategy represents the main document influencing the setting of all EU policies, and fundamentally sets the parameters for the functioning of EU member states. In the practice of policy-making, this was manifested in the way that member states had to reflect the percentage of actions, which were aimed at climate change in the draft budgets of operational programs, for the periods of 2014–2020 and 2021–2027. The European Commission controlled the mainstreaming of climate change in operational programmes, as part of negotiations and comment procedures. For coal regions, the European Commission has prepared new financial mechanisms that should contribute to speeding up the process of decarbonisation of regional economies, and the long-term strengthening of their competitiveness—the so-called transition process, that was financed through the Just Transition Mechanism. The environmental dimension of megatrends has a fundamental influence on the new generation of cohesion policy instruments.

2 Project Management Within the Green Programming of Urban Sustainability in the EU

Although the EU organises cohesion policy through seven-year programming periods, in recent periods it has become a certain firm regularity that there have been delays in the approval of the EU budget or relevant legislation. However, for the programming period 2021–2027, the delay took on an enormous extent, which was caused by the unprecedented year 2020, when the Covid-19 pandemic broke out and the functionality of states, administrations, and economies was paralysed. Hand in hand with this process, parallel new sources of funding aimed at restoring the economies of EU member states (Recovery and Resilience Fund–RRF) were, therefore, prepared after this pandemic, while these sources are procured on debt, for the first time in the history of the European integration process. Until 2020, the EU (and before that the EEC) always formed its budgets solely from its own revenues. Despite this major delay due to the pandemic, however, the details of programming and, subsequently, the projection of regional development in the EU countries in the diction of the environmental impacts of climate change were still conceived. These new elements will influence long-term urban sustainability, including the conception of the philosophy of individual projects and their management.

Project management is a set of models, methods, procedures, tools, and techniques for planning and implementing projects, which in themselves should be unique and unrepeatable. The basis of this field is project cycle management, which by default includes the key phases of the project. The starting point is programming, then the idea of the project must be identified (usually a one-page description–project fiche), which is further elaborated in greater detail (project proposal). If it is accepted by the management of the given institution, financial resources (own, loan, subsidy, etc.) are secured. Then, the project is implemented with on-going evaluation. After the completion of the project, an ex-post evaluation of outputs and results must take place. This partly concerns the planning of other activities or projects of the given institution, and at the same time, in an aggregated form, for all projects of the relevant intervention, a wider impact evaluation and discussion about setting up new programmes for funding (especially in the case of subsidy programmes) takes place.

However, these basic process steps, which can be found in any project management publication, are continuously under increasingly greater pressure from environmental megatrends, which are discussed in the previous section. In past programme periods, these were primarily major projects that were financed from the Cohesion Fund, and they were focused mainly on transport and environmental infrastructure. As a rule, it was about ensuring a mandatory *Environmental Impact Assessment* (hereinafter, referred to as EIA) for every major project that could have a significant impact on the environment, among other things, as a result of its nature, scale, or location. EIA, therefore, meant a mandatory condition for the launch of real activities of large infrastructure investments. At the level of infrastructure-oriented programmes, a similar process called *Strategic Environmental Assessment* (SEA) was, and is, ongoing. These environmental elements of project management were supplemented for the first time in the 2014–2020 programme period, by another process called **climate-proofing major projects**, which is further strengthened in the period of 2021–2027. The European Commission (2021, p. 7) issued technical guidance for this, and defined climate proofing as "*a process that integrates climate change mitigation and adaptation measures into the development of infrastructure projects. It enables European institutional and private investors to make informed decisions on projects that qualify as compatible with the Paris Agreement. The process is divided into two pillars (mitigation, adaptation) and two phases (screening, detailed analysis). The detailed analysis is subject to the outcome of the screening phase, which helps reduce the administrative burden.*"

The climate proofing process applies to all funds that finance infrastructure investments, i.e. ERDF, Cohesion Fund (CF), Just Transition Fund (JTF), InvestEU, and Connecting Europe Facility (CEF). The basis is then the verification of whether the given project intentions are consistent with the Paris Agreement and EU climate objectives, the **Energy Efficiency First** (EE1st) principle (its goal is to reduce fossil fuel consumption and increase supply independence, including reducing energy production), and with the **Do No Significant Harm principle** (hereinafter, referred to as DNSH), which is aimed at ensuring that investment actions may not cause any significant harm to the environment (it was formulated primarily for RRF). These new requirements, that were incorporated into the project management of infrastructure projects, will be essential for the green programming of urban sustainability. According to the European Commission (2021, p. 20–22), there is a list of different types of interventions and projects, which can be divided into three types:

- Environmental infrastructure and facilities—i.e. wastewater and sludge treatment, waste management facilities, municipal solid waste landfills,
- Transport infrastructure and facilities—road transport, rail transport, urban transport, ports and logistics platforms, airports,
- Renovation of buildings and the heating industry (heating plants and power plants, district heating networks).

It follows from the above that the programming and projecting of investments, not only in urban regions, is and will continue to be under greater pressure from the greening process, which is further continuously refined by partial elements of environmental megatrends (especially climate change). This general trend must be carefully reflected in the process of policy-making and project management, as projects that are not compatible with climate goals, or the broader goals of the greening process will not be eligible for support. It should be noted that this trend does not only concern public spending programmes, but also the banking industry, which has also begun to place great emphasis on green investments, and is gradually beginning to consider providing loans for investments, with potentially negative consequences for the environment.

3 The Specifics of the Czech Urban Structure and the Programming Context of the Respective Challenges in the Period 2021–2027

From the previous two sections, some fundamental shifts in the programming and projecting of urban challenges within the EU cohesion policy, and the wider context of external factors, especially the influence of megatrends, emerge. These theoretical, paradigmatic, and strategic starting points can be presented on the example of the Czech Republic, and the process of its preparation for these key challenges. Attention is first given to the basic description of the specifics of the Czech settlement structure and, thus, the uniqueness of the urban structure, which affects the relevant urban needs and challenges. Attention is also paid to the definition and hierarchy of urban areas in Czech strategic documents, and finally, there is a discussion on the programming of territorial and global challenges in urban areas in the period of 2021–2027.

Specifics of the Czech urban and settlement structure

The settlement structure of the Czech Republic reflects the historical and socio-economic development of the territory. Urbanisation and suburbanisation processes are constantly taking place in the territory, and influence the development of the settlement structure, and thus also the administrative structure (number and size of local administrative units–municipalities). After 1990, there was an increase in the number of municipalities in the Czech Republic and, currently, there are 6,258 municipalities and approximately 15,000 settlements [4], which are relatively high numbers, if we consider the total area of the Czech Republic, and the number of inhabitants. A comparison with other EU member countries is presented in Table 1. The administrative structure of France and Slovakia appears to be the closest, in terms of the mentioned statistical indicators.

Significant regional differences in the administrative and settlement structure can be identified within the Czech Republic (see Table 2). These differences are influenced historically, either by the period of industrialisation and related urbanisation

Table 1 Administrative structure of the EU countries

Country	Number of municipalities, 2018	Average municipality (No. of inhabitants), 2018	Urban population (% of total population, 2021)	Share of municipalities with less than 2,000 inhabitants (%), 2016
Austria	2098	4205	59	55
Belgium	589	19352	98	1
Bulgaria	265	26604	76	4
Croatia	556	7384	58	32
Cyprus	380	2274	67	83
Czech Republic	6258	1695	74	89
Denmark	98	58992	88	1
Estonia	79	16698	69	5
Finland	311	17727	86	14
France	35357	1893	81	86
Germany	11054	7490	78	54
Greece	325	33050	80	7
Hungary	3178	3077	72	76
Ireland	31	155819	64	0
Italy	7960	7598	71	44
Latvia	119	16255	68	5
Lithuania	60	46815	68	0
Luxembourg	102	5902	92	37
Malta	68	6996	95	19
Netherlands	380	45213	93	1
Poland	2478	15326	60	1
Portugal	308	33412	67	2
Romania	3181	6140	54	22
Slovakia	2930	1858	54	85
Slovenia	212	9749	55	12
Spain	8124	5743	81	72
Sweden	290	3497	88	0

Source Own table based on OECD (2018a) and data of World Bank

processes in the nineteenth century, or by developments after the Second World War. This was mainly the removal of Germans from the borderlands and the resettlement of this territory [2], and the following period of building socialism characterised by the collectivisation of the countryside. At the same time, great emphasis was placed on industrial production, which led to the movement of the population to cities, especially to traditional industrial regions (Prague and its surroundings, north-eastern Moravia, north-western Bohemia) and cities (Brno, Pilsen, and others). According to Hampl and Müller [24], the significant political and socioeconomic changes of the 1990s manifested themselves in the distribution of the population only gradually, and their dynamism only occurred after 2000. Hampl and Müller [24] also identified that the most significant changes after 2000 brought about the process of suburbanisation which, however, began to manifest gradually only from the end of the 1990s [28]. Currently, the consequences of these events and processes are manifested in the territory, and are reflected in the relatively high fragmentation of the settlement structure [29], which entails certain problems in terms of transport and technical infrastructure, and ensuring that the territory is served by public transport and other public or commercial services. Small and medium-sized towns play an important role in the Czech settlement structure, which, according to [35], enable the stabilisation of the Czech countryside. However, the dense network of small and medium-sized towns creates very specific preconditions for policy-making.

Urbanisation and suburbanisation processes, in the territory of the Czech Republic, continue even today. Suburbanisation, which is most pronounced in Prague and its surroundings and in Brno and its surroundings, often leads to the unwanted

Table 2 Local administrative units–municipalities in Czech regions, 2020

Region	Number of municipalities	Average municipality (no of inhabitants)	Urban population (% of total population)	Share of municipalities with less than 2,000 inhabitants (%)	Population density
Praha	1	1335084	100.0	0.0	2 690.6
Středočeský	1144	1222	51.4	89.2	127.9
Jihočeský	624	1031	63.6	91.5	64.0
Plzeňský	501	1180	66.7	91.2	77.3
Karlovarský	134	2189	81.7	79.9	88.6
Ústecký	354	2308	79.1	85.3	153.0
Liberecký	215	2058	76.9	84.7	139.9
Královéhradecký	448	1229	66.0	90.4	115.7
Pardubický	451	1159	61.3	92.0	115.7
Vysočina	704	723	56.1	95.6	74.9
Jihomoravský	673	1776	61.5	87.1	166.3
Olomoucký	402	1568	56.3	88.8	119.6
Zlínský	307	1890	58.5	81.8	146.4
Moravskoslezský	300	3976	73.6	72.7	219.7
Czech Republic	6258	1710	68.8	88.6	135.7

Source Own table based on [4]

occupation of agricultural land, creates strong pressure on transport and technical infrastructure and the availability of necessary, especially public, services (e.g. problems with capacities in education and healthcare), but according to [30], it is also often associated with negative impacts on the social environment of suburbanised municipalities.

According to CZSO data (2022), there were only five cities in the Czech Republic with a population of more than 100,000 as of 31 December 2021, namely Prague, Brno, Ostrava, Pilsen, and Liberec. Zipf's law (rank size rule) can be applied to a settlement structure that is characterised by one prominent centre. The Czech Republic is a suitable example because it has the capital city of Prague, where 12.1% of the total population of the country lived (according to CZSO data as of 31. 12. 2021). As can be seen from Fig. 1, the Czech Republic lacks large cities of several hundred thousand people, that would be important socio-economic centres with a significant impact on the territory. For comparison, Fig. 2 shows the situation in Belgium, which has about 0.8 million more inhabitants than the Czech Republic. While there are only five cities with a population of over 100,000 in the Czech Republic, there are 10 such cities in Belgium, and the Zipf curve almost copies the real situation.

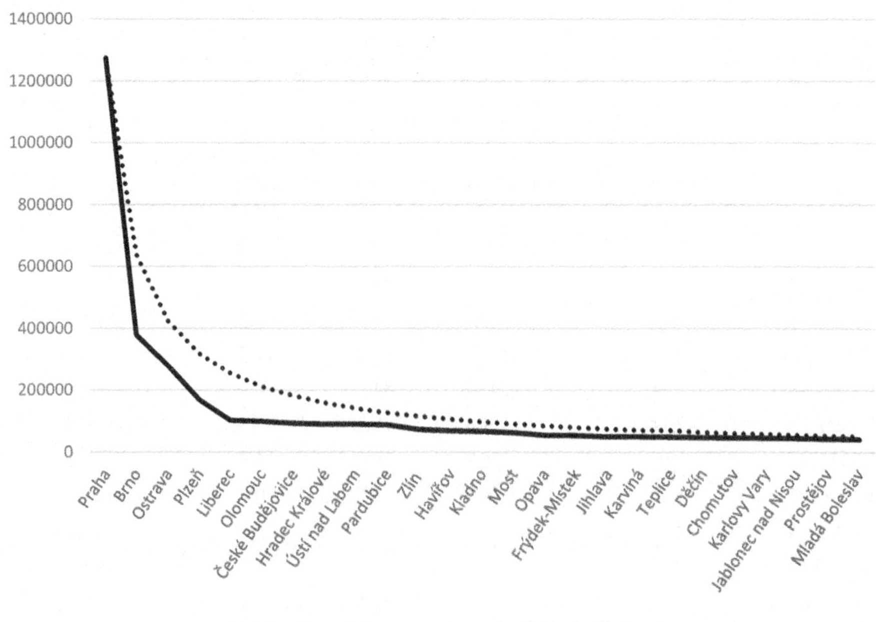

Fig. 1 Real population (2021) and potential population by Zipf´s law of the top 25 Czech cities. *Source* Own figure based on data [5]

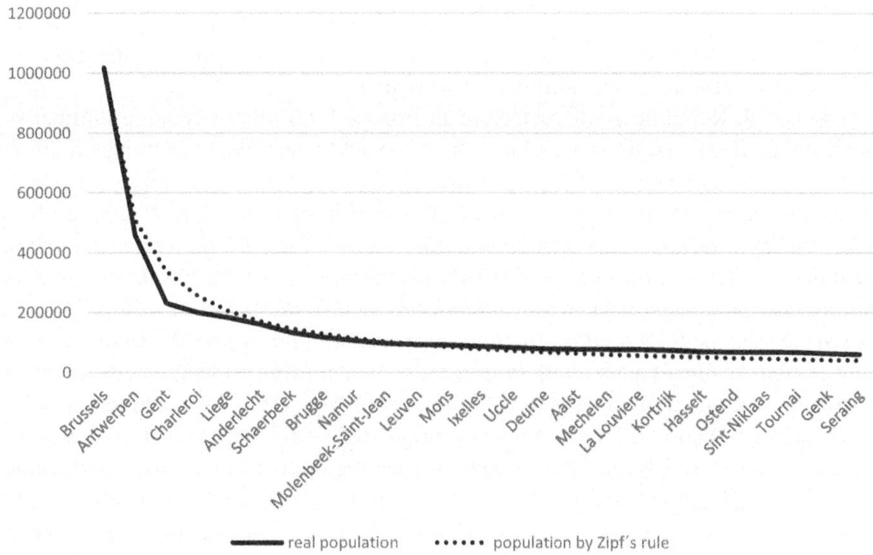

Fig. 2 Real population (2022) and potential population by Zipf´s law of the top 25 Belgian cities. *Source* Own figure based on data world population review

Definition and hierarchy of urban areas in Czech strategic documents

According to the OECD [27], the most important sources of economic growth are the metropolitan areas of the three largest Czech cities, i.e. Prague, Brno, and Ostrava. According to the Czech Regional Development Strategy 2021+ [19], the significantly dominant position of Prague is due to the supra-regional importance of the capital, the political, economic, social, and cultural centre of the Czech Republic. This metropolitan area is characterised by having a many times higher population, compared to the other two metropolitan areas, it is the centre of a significant commute to work and schools, and it is also the largest transport hub in the Czech Republic. The metropolitan area of Brno, the second largest city of the Czech Republic, forms the central part of the South Moravian Region. Brno is an important socio-economic centre, and the seat of some administrative bodies with nationwide jurisdiction. The metropolitan area is characterised, similarly to Prague, by a strong commute to work and to schools in Brno, and shows positive socio-economic development over the last 10 years. The Ostrava metropolitan area was included in the category of metropolitan areas with regard to its population size, extensive urbanised background and cross-border ties. It is located in the Moravian-Silesian Region, which is among the structurally affected regions (coal mining).

Other regional cities create agglomerations, the socio-economic importance of which is lower, even with regard to the number of inhabitants. But, even among these agglomerations, they are significant, and influenced by the size of the population of their centres, which ranges from approx. 45,000 (Karlovy Vary) to 168,000 (Pilsen) [5]. On a European scale, these are smaller cities, but this corresponds to the specifics of the Czech settlement and administrative structure.

The Czech Republic is characterised in Europe by a relatively large number of small and medium-sized towns, which form the backbone of the residential structure, and at the same time enable the polycentric development of the territory [32]. These towns are centres of functional regions which, according to ([34], p. 7), can be characterised "*by internal complexity (containing all or most basic functions needed in the daily life of a population), integrity (these functions are strongly mutually related* via *the concrete daily activities of inhabitants within the territory of a functional region) and centrality/nodality (the functional division has a specific spatiality that relates central places with their hinterlands).*" According to [32], a city or town whose economy is growing and developing, positively affects its catchment area.

The Czech Regional Development Strategy 2021+ for these functional regions at a lower territorial level defines higher-order regional centres (cities with more than 15,000 inhabitants and a catchment area with at least 30,000 inhabitants) and lower-order regional centres (at least 5,000 inhabitants and a catchment area with at least 10 thousand inhabitants) [19].

The Czech Regional Development Strategy 2021+, which emphasises functionally defined regions, is relatively ambitious in its goals for individual types of territory. In this context, it is necessary to point out, that most of the major actions will be implemented through the EU's cohesion policy. This entails the necessity of certain organisational, strategic, financial, and administratively demanding actions. Therefore, the priority of the Regional Development Strategy [19] is the development of strategic planning, precisely on the basis of functional regions, through tools such as ITI in metropolitan areas/agglomerations, integrated territorial development plans in regional growth poles and their natural catchment areas, and the community-led local development (CLLD) for LAG territories. These tools then help the realisation of larger investments in the given territory. However, especially in connection with CLLD strategies, [6] recommends simplifying the process of their approval, so that there is no delay or slowing down of CLLD project implementation. This could jeopardise the fulfilment of the set goals.

Programming of territorial and global challenges in urban areas in the period 2021–2027

From Sect 1., it follows that the Czech Republic, like other countries of the European Union, faces significant challenges in relation to the European Green Deal, which is particularly reflected in the programming of cohesion policy, rural development policy, and transition of coal regions. The Czech Republic, like other member countries, has prepared a Partnership Agreement for the implementation of cohesion policy in the period of 2021–2027, which reflects five basic goals: Smarter Europe, Greener, low-carbon Europe, More connected Europe, Social Europe, and Europe closer to citizens. These goals will be fulfilled through operational programmes (hereinafter, termed as OP): OP Technology and applications for competitiveness, Integrated Regional Operational Programme (hereinafter, termed as IROP), OP Jan Ámos

Komenský, OP Environment, OP Transport, OP Employment plus, and OP Technical assistance [20]. The three urbanised coal mining regions (NUTS 3), the Ústecký, Karlovy Vary, and Moravian-Silesian regions, face specific problems, which must cope with the transition to a carbon-free economy. The Just Transition Programme financed by the JTF [20], is intended for these three regions.

For the definition of cohesion policy programs, the government of the Czech Republic chose the strategy of "*evolution, not radical revolution, because despite some problems with implementation in the past, the system functions as such*" ([21], p. 45), i.e. it follows on from the definition of operational programmes from the period of 2014–2020. In this context, it is necessary to point out the problem that is associated with the fact that the individual objectives of the EU cohesion policy are always fulfilled by more than one programme (see [19]). This can make it difficult to evaluate the implementation of cohesion policy in the Czech Republic during the period of 2021–2027, including climate goals.

Urban regions in the Czech Republic, as well as in other countries, are struggling with the effects of climate change, which paralyse life in cities (energy demands of urban space, problems with providing public transport, lack of greenery, etc.). In the Czech Republic, they can draw help in the field of climate change from the programmes of Technology and Applications for Competitiveness, Integrated Regional Programmes, Environment, Transport, and Just Transition. All of these programmes have undergone assessment of climate action in ERDF/CF, with the aim of assessing their setting and the potential of meeting the EU's climate goals. In this assessment, in which the authors of this chapter took part, it became clear that the Czech Republic is not always able to use the relevant financing options for climate actions, and to link the implementation of cohesion policy with the fulfilment of the goals of national strategies, i.e. Climate Change Adaptation Strategy for the Czech Republic 2021–2030, and the National Energy and Climate Plan. Basically, for all programmes, it was recommended to strengthen climate spending. On the other hand, it should be noted that Czechia belongs to the new member countries which were admitted to the EU in 2004, which are still struggling with a deficit of infrastructure investments, caused by the legacy of development in the second half of the twentieth century (see e.g. [12, 18, 37]). It is important for these countries to balance climate actions with these infrastructure needs, in terms of EU cohesion policy settings.

Another problem is the competences of the Czech ministries (the governing bodies of the programmes), when sub-parts of the issue of climate change are divided between several ministries. This is then reflected in the setting of the programmes, and climate actions are thus "chopped up" into individual programmes, so that they correspond to the competences of the ministries. For example, support for climate actions for municipalities is included, both in the OP Environment (reducing the energy demand of public buildings and public infrastructure, water quality and water supply, municipal waste, care of protected areas), and in IROP (green infrastructure in public spaces of municipalities and cities for the improvement biodiversity, pollution reduction, and stormwater management). Even if activities do not duplicate and overlap, this is a complication for applicants, as well as for identifying and evaluating climate action synergies between programmes. From the logic of the matter, it would be most appropriate, and he easiest method, to solve the issue of combatting climate change within one programme, but given the "puzzle" system of competences of the ministries of the Czech Republic, this is not realistic.

Another issue that affects the current environment, and not only in the Czech Republic, is the effort made to cope with the effects of the Covid-19 pandemic. This should be helped by the implementation of the National Recovery Plan, whose measures should contribute to the resilience of the Czech economy, and to the digital and green transformation [22]. The National Renewal Plan summarises the challenges that are faced by urban areas in the Czech Republic, in accordance with their definition in the Czech Regional Development Strategy [19]. In connection with the fight against climate change, in urban areas it is mainly a *green transition*, associated with the support of efficient and ecological transport, and the relevant infrastructure, improvement of air quality, support of renewable sources, and a reduction of the energy demand of buildings [22]. The National Recovery Plan thus follows on from the Regional Development Strategy 2021+, and addresses similar topics as EU cohesion policy programmes. In this context, the readiness of stakeholders to use all of these possibilities is very important. It is a big challenge for urban areas (due to certain reprogramming), also considering their role in the territory. Successful urban areas will strengthen their role and, in accordance with ([32], p. 154), "*deepen city-rural ties reflecting social and technological developments as well as challenges to environmental protection.*"

When it comes to the challenges faced by Czech urban areas, in connection with climate change, and the possibilities regarding how to solve them, it can be summarised that a relatively wide range of support activities is offered. However, due to the fragmentation of climate actions into different programmes (EU cohesion policy programs, JTF for coal mining regions, implementation of the RRF -Recovery and Resilience Facility, EU framework programmes, etc.), it will be complicated to look for synergies between implemented actions. The described *overprogramming* and *dismemberment* of climate actions in various programmes, can thus become a risk, not only for the actual implementation of climate actions and the fulfilment of EU climate goals, but also for the evaluation of their fulfilment in the Czech Republic.

4 Urban and Regional Futures According to the Real Ability of EU Cohesion Policy to Change Anything: Context and Critical Reflection

The contextual challenges of urban regions linked to megatrends are many, and the EU is clearly aiming to adapt to this at both programme and project levels, especially in relation to environmental challenges. The reality of potentially expected changes is identified in evaluations, the importance and scope of which has increased dramatically, especially in the second decade of the new millennium [33]. This has been pursued precisely through the wider application of not only qualitative evaluation methods, but also through the increased use of quantitative approaches (e.g. counterfactual evaluation), with a focus on the use of a mix of evaluation methods in recent years. This shift in policy-making is highly commendable and moves in the right direction for the substantive focus of relevant policies, and increases the pressure on policy-makers, in terms of accountability for public financial management.

However, there have been long-term critical discussions by some prominent authors, regarding the reality and potential of regional policy to have a real and long-term impact on the development of regions and urban centres. The most significant in this direction are the critical reflections of Markussen [15, 16, 10, 17]. Their criticism of the current approaches that are applied in regional politics, was aimed at the theoretical foundations. The first is Professor Ann [15, 16], who quite seriously questioned the importance of existing regional studies approaches, for understanding regional problems or for influencing policy making. Her criticism was that the regional studies discipline is overwhelmed by a series of vague concepts that lack empirical depth and, in particular, suffer from what she called *policy distance*, in the sense of economic policy.

Hadjimichalis and Hudson [10] pointed out other questionable elements of policy-making, and the context of theoretical approaches, and they relate to the context of the causes of interregional differences, after the economic crisis, from 2008 to 2010. According to their analysis, it emerged that the dramatic consequences of the economic crisis, especially in some EU countries, were caused by deeper systemic problems associated with the asymmetric functioning of the Eurozone (i.e. centrally controlled monetary policy, and still nationally decentralised fiscal policies), and the context of geographies of uneven development. This observation was followed up by Prof. Ron [17], who also pointed to possible deeper causes of uneven development and inter-regional differences. In this context, according to Martin, regional policy does not try to act against the "market", but against systemic features that are related to how the economy in the EU is managed. The fundamental idea in this direction, is a critique of the functioning of overarching policies and related problems, which are subsequently linked to the context of interregional differences.

Martin ([17], p. 265) also believes that, not only the new spatial economics dominantly focused on the importance of urban agglomerations, but also regional studies with an extreme number of concepts and paradigms that lack integration and synthesis, do not provide a particularly convincing basis for the creation of economic policies, which would be able to correct the spatial imbalance. The art of politics is, therefore, what August Losch (1939/1954, The Economics of location) called for in the first half of the twentieth century, in order to "*let space-conditioned particularity grow, without letting the whole run wild –that is 'political art*'". However, this political art requires a convincing and relevant conceptual and empirical foundation.

The views mentioned above may sound rather pessimistic. They clearly ignore the possibilities of forming unique local, regional, or national ties of key actors who can coordinate very well under certain circumstances, and at the same time shape the hitherto untapped potential in the given areas. Nevertheless, it is necessary to perceive them very carefully, because they point to a long-term problem of regional studies, which are characterised by a large number of different theoretical approaches in the basic distinction, between divergent and convergent groups of theories. The attention of individual theories is usually focused only on partial aspects of spatial or territorial development, and they often have a microeconomic basis (i.e. focus on the formation of industrial clusters, related variety concept, regional innovation systems, etc.). For a long time, there has been a lack of a systemic approach and assessment of uneven regional development, as a result of broader systemic problems that are associated with the functioning of key macroeconomic policies at the national and especially transnational level.

At the same time, it is evident from the first section of this chapter that the importance of the environmental dimension of regional development is growing in the long term, which is reflected in the formation of the relevant instruments of cohesion policy. The result is mandatory and normatively determined percentage shares of the budgets of public spending programmes, that are dedicated to the mitigation or adaptation of global megatrends. In the cohesion policy which is conceived in this way, and which in the long term presents itself as focusing more on urban regions, there is an evident retreat from the foundations of economic theories, which is also caused by the fragmentation of the main directions of economically-oriented theories of regional development. Urban regions are, and will undoubtedly be, under greater pressure from the effects of climate change, and it will be necessary to spend more funds, at least on projects that are aimed at adapting to these trends. Opportunities will also increase as a result of technological progress and the development of digitisation. Long-term socio-economic problems of urban regions, and the causes of inter-regional disparities, in general, will have an increasingly systemic background.

5 Concluding Remarks

Shaping sustainable urban futures will become increasingly more complex, as evidenced by the analysis of the context of the meaning and role of internal and external factors of local and regional development. The paradigmatic content shift in the formation of originally regional and currently more complexly focused cohesion policy, leads more towards normatively conceived policy-making, without deeper economic foundations. Regional development and territorial cohesion programming are, and will be, under great pressure from the thematic focus of key environmental megatrends.

Project management in cohesion policy, especially in the case of infrastructure investments, has moved further towards the need to respect the greening process and, in particular, to reflect climate-neutral approaches. Urban centres, in particular, will have to carefully evaluate their future investment plans in relation to climate goals in projects that are focused on environmental and transport infrastructure, relevant equipment, building renovations, and also in the heating industry.

The complexity of programming, and rather reprogramming, of a large number of different financial flows can be demonstrated on the example of the Czech Republic, which stands out by European standards for its highly fragmented residential structure, and high number of small and medium-sized cities. Although, in principle, individual operational programmes must not overlap in terms of content, certain similar types of climate actions are handled competently between different departments, and thus also ministries. This also complicates the system for implementing specific projects. It is also necessary to emphasise that even if Czech cities and smaller towns face the same challenges of climate change as Western European urban centres, there is still a certain infrastructure deficit from the second half of the twentieth century. As a rule, these investment needs are not fully in line with normatively established climate goals. Generally, this is a problem in all Central and Eastern European countries of the EU.

The whole context of a sustainable urban future is also complicated by the fragmentation in partial approaches of regional development theories, which have a rather microeconomic focus. A more comprehensive view of the systemic causes and factors that are connected to the socio-economic problems of urban centres and also to the configuration of inter-regional disparities has also been lacking for a long time.

References

1. Agarwal S, Rahman S, Errington A (2009) Measuring the determinants of relative economic performance of rural areas. J Rural Stud 25:309–321
2. Bičík I, Jančák V (2005) Transformační procesy v českém zemědělství po roce 1990, Praha: Katedra sociální geografie a regionálního rozvoje, PřF UK. *(Transformation processes in Czech agriculture after 1990, Prague: Department of Social Geography and Regional Development, Faculty of Science, Charles University.)* ISBN: 80–86561–19–4
3. Carayannis EG, Barth TD, Campbell DK (2012) The Quintuple Helix innovation model: global warming as a challenge and driver for innovation. J Innov Entrep 1:2: 1–12
4. CZSO (2021) Statistická ročenka Karlovarského kraje 2020. CZSO, Praha. Available online at https://www.czso.cz/csu/czso/statisticka-rocenka-karlovarskeho-kraje-2021
5. CZSO (2022) The 50 largest cities by population. Available online at https://vdb.czso.cz/vdbvo2/faces/en/index.jsf?page=vystup-objekt&pvo=RSO18&pvokc=&katalog=30829&z=T
6. EACE (2020) Ex-ante evaluace Integrovaného regionálního operačního programu pro období 2021–2027. ZÁVĚREČNÁ ZPRÁVA. Available online at https://www.dotaceeu.cz/cs/evropske-fondy-v-cr/narodni-organ-pro-koordinaci/evaluace/knihovna-evaluaci/ex-ante-evaluace-irop-pro-programove-obdobi-2021-2
7. EEA (2015) The European environment—state and outlook 2015: synthesis report, European Environment Agency, Copenhagen. Luxembourg: Publications Office of the European Union, 2015. ISBN 978–92–9213–515–7. https://doi.org/10.2800/944899
8. European Commission (2014) Integrated Territorial Investment. Available on-line: https://ec.europa.eu/regional_policy/sources/docgener/informat/2014/iti_en.pdf
9. European Commission (2021) Technical guidance on the climate proofing of infrastructure in the period 2021–2027. European Commission, 16. 9. 2021 (2021/C 373/01)
10. Hadjimichalis C, Hudson R (2014) Contemporary crisis across Europe and the crisis of regional development theory. Reg Stud 48:208–218
11. Hill B (2012) Understanding the common agricultural policy, Routledge, Earthscan. ISBN 978-1-84407-777-9

12. Kopp J, Kureková L, Hejduková P, Vogt D, Hejduk T (2021) Relationships between Insufficient Drinking Water Supply and the Socio-Economic Development of Small Municipalities: Mayors' Opinions from the Czech Republic. Water 13:2098. https:// doi.org/https://doi.org/10.3390/w13152098
13. Long A, Ploeg JD (1994) Endogenous development: Practices and Perspectives. In: Ploeg JD, Long A, Born from within: Practice and perspectives of endogenous rural development, Uitgeverij Van Gorcum, 1994, ISBN 90–232–2893–6.
14. Lösch A (1939/1954) The economics of location. English Translation, 1954. Yale University Press, New Haven, CT
15. Markusen A (1999) Fuzzy concepts, scanty evidence, policy distance: the case for rigour and policy relevance in regional studies. Reg Stud 33:869–884
16. Markusen A (2003) Fuzzy concepts, scanty evidence, policy distance: the case for rigour and policy relevance in critical regional studies. Regional Studies 37(2003): 6–7
17. Martin R (2015) Rebalancing the spatial economy: the challenge for regional theory. Territory, Politics, Governance 3(3):235–272. https://doi.org/10.1080/21622671.2015.1064825
18. Mittwallyová H, Janković V (2020) Serbia and the Czech Republic: comparison of railway structure. Int J Bus Man 8(1):107–121. https://doi.org/10.20472/BM.2020.8.1.006
19. MMR (2020) Strategie regionálního rozvoje České republiky 2021+. Available online at https://mmr.cz/cs/microsites/uzemni-dimenze/regionalni-rozvoj/strategie-regionalniho-rozvoje-cr-2021
20. MMR (2021) Dohoda o partnerství. Available online at https://www.dotaceeu.cz/cs/evropske-fondy-v-cr/kohezni-politika-po-roce-2020/priprava-obdobi-2021-2027/s/verejna-konzultace-k-prvnimu-navrhu-dohody-o-partn
21. MMR-NOK (2018) Koncepce politiky soudržnosti po roce 2020. Available online at https://www.dotaceeu.cz/cs/evropske-fondy-v-cr/kohezni-politika-po-roce-2020/koncepce-cr-pro-politiku-soudrznosti-2021
22. MPO (2021) Národní plán obnovy. Plán pro oživení a odolnost České republiky. Konsolidované znění po konzultacích s Evropskou komisí. Available online at https://www.planobnovycr.cz/dokumenty
23. Naisbitt J (1988) Megatrends. Ten new directions transforming our lives. Grand Central Publishing. ISBN-13: 978–0446356817
24. Hampl M, Müller J (2019) Vývoj geografického rozmístění obyvatelstva v Česku po r. 2000: post-transformační tendence?, Geografie 124 (4):385–409
25. OECD (2013) Putting green growth at the heart of development, oecd green growth studies. OECD. ISBN 978-92-64-18114-4 (PDF)
26. OECD (2014) Innovation and Modernising the Rural Economy. OECD Publishing. ISBN 978–92–64–20539
27. OECD (2016) OECD Regions at a Glance 2016. OECD Publishing, Paris. ISBN 978-92-64-25679-8
28. OECD (2018a) Key data on Local and Regional Governments in the European Union (brochure). OECD, Paris. Available online at www.oecd.org/regional/regional-policy
29. OECD (2018) OECD Environmental performance reviews: Czech Republic 2018. OECD Environmental performance reviews. OECD Publishing, Paris. https://doi.org/10.1787/9789264300958-en
30. Ourednicek M, Simon M, Kopecna M (2015) The reurbanisation concept and its utility for contemporary research on post-socialist cities: The case of the Czech Republic. Moravian Geographical Reports 23(4):26–35
31. Ouředníček M, Špačková P, Novák J (eds.) (2013) Sub urbs: krajina, sídla a lidé. Academia, Praha ISBN: 978-80-200-2226-4

32. Pělucha M et al (2012) Venkov na prahu 21. století. Alfa Nakladatelství, Praha. ISBN 978–80–87197–49–3
33. Pelucha M, Kveton V (2017a) Evaluation culture within institutional and methodological context: the case of EU structural funds in the Czech Republic. Evaluační teorie a praxe/ Evaluation Theory and Practice [online]. 2017, roč. 5, č. 2, s. 1–26. eISSN 2336–1158. ISSN 2336–114X. Available at https://www.evaltep.cz/inpage/autumn-2017/. Special English 5(2), Autumn 2017
34. Pelucha M, Kveton V (2017) The role of EU rural development policy in the neo-productivist agricultural paradigm. J Reg Stud 51(12):1860–1870. https://doi.org/10.1080/00343404.2017.1282608
35. Pelucha M, Kasabov E (eds) (2020) Rural development in the digital age. Routledge, London and New York. ISBN 978-0-367-35658-3
36. Ray C (2006) Neo-endogenous rural development in the EU, In: Cloke P, Marsden T, Mooney P (2006), Handbook of Rural Studies. Sage Publications. ISBN 0–7619–7332
37. Sýkora L, Mulíček O (2009) The micro-regional nature of functional urban areas (FUAs): Lessons from the analysis of the Czech urban and regional system. Urban Research & Practice 2(3):287–307. https://doi.org/10.1080/17535060903319228
38. Vaishar A, Zapletalová J, Nováková E (2016) Between urban and rural: sustainability of small towns in the Czech Republic. European Countryside 8(4):351–372. https://doi.org/10.1515/euco-2016-0025
39. Viturka M (2011) Integrační teorie regionálního rozvoje—představení a aplikace. Politická ekonomie 59(6): 794–809. https://doi.org/10.18267/j.polek.822
40. Witz P, Leviäkangas P, Łukasiewicz A, Szekeres K (2015) Implementation of transport infrastructure PPPs in the Czech Republic, Finland, Poland and Slovakia—a comparative analysis on national contexts. Int J Manage Net Econ 3(3): 220–237.World Bank. Urban population (% of total population). Available online at https://data.worldbank.org/indicator/SP.URB.TOTL.IN.ZS [cit 2022–08–03].
41. Woods M (2011) Rural. Routledge, Taylor & Francis Book. ISBN 978–0–415–44240–4
43. Wokoun R, Mates P, Kadeřábková J, Grospič J, Khendriche-Trhlínová Z, Kouřilová J, Macháček J, Peková J, Pělucha M, Postránecký J, Toth P (2011) Základy regionálních věd a veřejné správy [Basics of regional sciences and public administration]. Plzeň: Aleš Čeněk, 2011, 474. ISBN 978–80–7380–304–9
43. World Population Review (2022) Population of cities in Begium. Available online at https://worldpopulationreview.com/countries/cities/belgium. cit 2022–08–05

Transitioning to Circular Plastic Economies in Cities: Conceptual and Policy Implications of a Practice Perspective

Olamide Shittu and Christian Nygaard

Abstract Changing the nature of how plastic is used and circulated through economic systems is central to transitioning to a circular plastic economy (CPE) in cities. The theoretical and analytical underpinnings of the CPE are embedded in the scholastic and policy discussions on the need for cities to transition from a linear to a circular economy. However, current frameworks for understanding CPEs are economic-centric, overly focused on physical features and conceptualised mainly through circular models. Moreover, the extant analyses of such a transition in socio-technical systems have focused on waste infrastructure, recycling technology and behavioural change. However, studies show that these approaches may also reinforce linear models of production and consumption and are inadequate in galvanising a shift to the CPE. This chapter explores an alternative conceptual analysis of the CPE through the practice perspective. Specifically, the chapter adopts the 'zooming out and in' methodology to conceptualise the CPE as a teleoaffective formation or the constellation of practice complexes oriented towards sustainable plastic production and consumption. Although focusing on plastic within the circular economy framework could be considered simplistic, such an analytical approach is important to dissect the complex lifecycle of each material resource and design policies that make them sustainable. Therefore, this chapter is positioned within the circular economy discourse by exploring practice formations at various levels related to plastic use and their implications for sustainability transitions. As these concepts are further clarified in the chapter, they are applied to explain how plastic interconnects integrative practices bound by the general understanding of sustainability discourse across the social sphere. The need to make plastic use sustainable and circular is additionally enhanced by the discursive and non-discursive formations of resource reuse, repurposing, remanufacture and recycling. On the other hand, zooming into the household domain in the CPE shows how plastic contributes to the performance of household practices and how they, in turn, enhance the spatial–temporal mobility of plastic circularity. The influence of practice complexes outside the household

O. Shittu (✉) · C. Nygaard
Centre for Urban Transitions, Swinburne University of Technology, PO Box 218, Hawthorn, VIC 3122, Australia
e-mail: oshittu@swin.edu.au

Z. Allam (ed.), *Sustainable Urban Transitions*, Urban Sustainability,
https://doi.org/10.1007/978-981-99-2695-4_14

domain on household plastic use is also expanded upon. The chapter then closes with a theoretical explanation of how cities could transition from the current linear model of plastic use to the CPE. In this regard, there is a need to transition the existing compound or disintegrated set of sustainability practices into an integrative or well-defined and formalised web of sustainability practices that enable the circularity of plastic materials. The implications of such transitions for research and policy are then discussed.

Keywords Plastic consumption · Circular economy · Household practice · Practice theory · Cities

1 Introduction

Climate change, ocean plastic pollution and overexploitation of natural resources have alerted urban policymakers and scholars to the environmental impact of unsustainable plastic production, use and disposal [1, 21, 28]. The volume and intensity of consumption activities and socio-material processes in cities result in the high leakage of materials (including plastics) from disposal mechanisms and into the environment [14, 33]. It is widely recognised that solutions are required to transition current plastic-related resource consumption in cities to low-impact models, specifically the circular plastic economy (CPE) [43]. A CPE is a system where plastic consumption is minimised, the lifecycle of plastic materials is maximised, and plastic waste is eliminated. Extant low-impact models in cities include reforms along the linear value chain with better waste management and recycling and an emphasis on attitude-behaviour-choice (ABC) approaches [38], but with less focus on reduction and reuse, which are integral aspects of the CPE [29].

Moreover, reformed linear models often face contamination, North–South dumping and low product demand issues [7], and ABC models often encounter a disconnect between household attitudes and their practices [20, 24]. A critical insight from the literature is, therefore, that existing implementation frameworks are inadequate to deliver CPE transitions in cities, particularly concerning the day-to-day activities of households and firms [23, 33]. However, the circular plastic economy (CPE) concept in cities is also not without some analytical challenges in theory and practice.

First, the framework to understand the CPE as a societywide system is currently inadequate. As the name implies, the CPE can be regarded as an economic strategy. There is a growing consensus in circular economy literature for transforming business processes such as product design, manufacturing, energy use, resource consumption and waste management through technological innovations, including alternatives to problematic materials like plastic [22]. While technological innovations provide an avenue to transform production systems towards circularity with businesses at the forefront [27], they are not enough for circular economy transitions, which require restructuring activities across all domains, including households.

Second, there is a tendency for extant CPE frameworks to be overly focused on the physical aspect of plastic, thereby ignoring other aspects of its materiality, such as mediating affective attributes [34, 36]. Plastic's ubiquitous and tenacious nature in production and consumption activities in cities proves that the material might not be easily eliminated or replaced by alternatives [9]. Consequently, there is a need to provide a holistic conceptualisation of plastic that delves into the intricacies of the material's embeddedness in day-to-day activities, especially in the social (or household) domain.

Third, the transition of cities to CPEs is, to a large extent, currently conceptualised in terms of implementing circular models such as reduce, reuse, repair, remanufacture and recycle [5]. However, as several studies have shown, embedding circular solutions into day-to-day activities presents more complications in practice [5, 47]. Besides, as later discussed in this chapter, expanding existing circular models is only an aspect of the transition of cities to CPEs. Therefore, in addition to developing and implementing alternative and novel circular solutions, it is also important for existing linear models and processes that cannot be replaced to be transformed.

This chapter adopts practice theory to address some of the above-outlined CPE analytical issues for theory and practice. We aim to provide a practice-based conceptual framework for configuring the CPE ecology. Using the 'zooming in and out' method, we conceive of the CPE through a practice perspective [25]. Although this significantly simplifies the complexity that a circular economy would exhibit, it nevertheless allows us to focus on the specifics of daily routines while at the same time engaging with how each routine relates to, conditions and impacts other routines. Also, utilising [46] concept of teleoaffective formation, we conceptualise the CPE as the constellations of general understandings, practice complexes and normative orders that achieve both circularity and sustainability. According to Welch and Warde (2016, pp. 1–2), general understandings could be understood as:

> … ideational elements… [that are] common to multiple practices, condition the manner in which practices are carried out and are expressed in their performance… General understandings are formulated in *both* doings and sayings… [and] include such things as concepts, values, and categories. Specific candidates might include: collective concepts, such as nation, state, economy or organisation; membership categories, such as ethnicity or gender; fundamental, culturally structuring concepts, such as animal\human or private\public; and diffuse but culturally significant understandings, such as notions of convenience, cosmopolitanism or authenticity.

Furthermore, a shift from behaviours, technological approaches and market mechanisms to practices provides a means of analytically disaggregating plastic consumption to focus on routinised activities of households and firms. Two practice theory concepts underlying the CPE analysis presented in this chapter are 'integrative practices' and 'compound practices' [31, 44]. Integrative practices are the bundle of routines with defined skills and performance processes that jointly deliver a specific outcome. For instance, the morning routines that result in someone getting to work involve formalised practices such as official wear and driving. Conversely, compound practices may also result in a specific outcome but do not involve a formally defined set of know-hows or explicit socially agreed meanings [44].

Schatzki [31] and Warde [44] consider the difference between compound and integrative practices to be their extent of formalisation and cohesiveness (e.g., eating and driving). In this chapter, we draw on the degree of formalisation and cohesiveness to extend these concepts into household activities for which common parlance is not readily available. For instance, cooking or driving (integrative practices) readily bring to mind a set of activities and their goals and commonly understood meanings. Sustainability practices do, however, not similarly bring to the mind a set of individual or multiple activities that, societally, are associated with a common goal or widely understood meaning. By extending both concepts to such activities, the study views existing plastic-related sustainability practices to be compound in the sense that there are no current widely adopted processes around sustainable plastic use in cities (e.g., sorting and recycling).

This, however, raises the question of how the routinised activities of households and firms assemble to form a constellation of practices that individually and jointly constitute the CPE ecology. We contend that the challenge of achieving the CPE as a teleoaffective formation is transforming compound sustainability practices into integrative sustainability practices as potential intervention points for delivering plastic-related sustainable and circular economy outcomes [31, 44, 46]. Extending this discussion, [35] examines grassroots strategies that can be implemented to achieve broader CPE transitions.

The remainder of the chapter is structured as follows. Section 2 discusses the key conceptual terms and foundations of a practice-based perspective. Section 3 zooms out and analyses the CPE as a teleoaffective formation [46]. Section 4 focuses on day-to-day plastic-related household routines. We here zoom in on the actualisation of the CPE as a set of practices performed in the daily lives of the practice carriers (that is, people) to achieve circularity and sustainability. Section 5 discusses the implications of conceptualising CPE from a practice perspective for policy and practice. Section 6 concludes the chapter.

2 Practice Theory, Sustainability Transitions and Circular Economy

Scholars have employed the practice theory to explain diverse social phenomena ranging from daily activities to global consumption patterns. This endeavour is partly motivated by the need to provide alternative theoretical explanations of social configurations and changes different from traditional theories of social formations, such as institutionalism and interactionism [38]. Both theories have been collectively critiqued for their over-emphasis on an aspect of the social world while ignoring practices [38]. Practices are the unit of analysis and serve as the building blocks of social domains and interact to create social phenomena [42].

Ontologically, practice configurations exist as a flat constellation interweaving multiple elements together in a dynamic and organised order [32]. In other words, the bundles of activities that constitute our everyday life are set up as in a web of practices with varying degrees of complexity. In the societal system, practices exist as spatial–temporal entities that arose from an enduring process of aggregating daily activities [30]. These activities are achieved by combining practice elements, including materials, competences, and meanings. Thus, a *practice* can be defined as a bundle of activities that are affectively orientated towards achieving some set of goals. The normativised ordering of goals of a practice and the associated affectivities constitute what Schatzki refers to as 'teleoaffectivity' [46]. Moreover, practices possess 'teleoaffective structures' or "a range of normativized and hierarchically ordered ends, projects, and tasks, to varying degrees allied with normativized emotions" [31], p. 80).

Meanwhile, sustainability transitions have been a preoccupation of scholars in the practice theory domain [6, 37]. Perhaps the most empirically studied phenomenon in practice theory literature is unsustainable consumption which manifests in modern-day societal patterns and its effect on climate change [48, 49, 41]. Theoretically, practice theory scholars such as [17, 39] and [6] have continued to interrogate how practices' historical construction and reconstruction can lead to social change, particularly sustainability transitions. Of note here is also [13] conception of sustainability as a practice composed of routinised activities which enable socio-ecological processes. Sustainability transitions to circular economies, in this sense, would be the reconfiguration of unsustainable practices and the normativisation of sustainable practices (see Fig. 2). Understanding CPEs from a practice theory lens presents a new paradigm that provides a nexus between the systemic and agency, to address not only the materiality of plastic but also the consumer practices that drive and are associated with this.

The CPE is in response to a renewed call for the transition from a linear economy to a cyclical mode of production and consumption locally and globally [29]. Shifting to the CPE is a promising alternative model that could ensure that waste and pollution are marginalised from the system by the continual loop of plastic materials through socio-technical and ecological processes [19]. The predominant literature on circular economy has focused on the systemic analysis of such transitions in cities, countries and a global scale. Such analysis explores the socio-technical elements required to achieve a circular economy through frameworks such as the multi-level perspective with little consideration for the elements that connect those structures to the day-to-day activities, thereby reinforcing and regenerating the system.

3 Zooming Out Beyond Technological Analysis: Conceptualising the CPE as a Teleoaffective Formation

Welch's reconceptualisation of teleoaffective structure guides the CPE framework presented in this section. [46] argues that Schatzki's concept of teleoaffective structure does not "capture cultural configurations that cross multiple practices" (p. 5).

Although Schatzki's concept of a 'teleoaffective regime' represents the bundle of teleoaffectivities that exist across multiple practices, [46] contends that it creates issues for the application of practice theory in consumption studies. Of particular concern is the difficulty in separating 'teleoaffective regime' from 'general understandings', two concepts Schatzki identified as embedded in practices but which also exist outside of and regulate practices. Therefore, [46] proposes 'teleoaffective formation', a related concept employed in consumption studies to understand practice complexes in multiple social domains. According to [46]:7), teleoaffective formation is:

> … a configuration across multiple practices, conditioned by a relational nexus of general understandings, that enjoins those practices to common ends and normatively orders the orientations and affective engagements of those practices.

The above definition serves as a basis for understanding the circular (plastic) economy. We define a *circular economy* as a configuration composed of sustainability practice complexes across all social domains aimed at eradicating waste, optimising resource use and prolonging material lifecycle.

Taking [46] definition of teleoaffective formation, the CPE is thus the arrangement of multiple sustainability practices oriented towards sustainable plastic production and consumption. For any practice to be part of the CPE, plastic as a material (in any form) must compose part of the elements that make up its performance. In the societal system, the general understandings that condition the CPE include the notions of ecological sustainability, sustainability transitions and ongoing public discourse about the environmental pollution resulting from indiscriminate plastic disposal. In Fig. 1, a conceptual representation of the circular (plastic) economy, these general understandings are categorised as sustainable lifestyles, sustainable socio-economic development, and sustainable socio-ecological integration. As depicted in Fig. 1, these general understandings integrate domains to form the CPE and are found across the circular economy spectrum. It should be noted that the general understandings of sustainability would take different forms from expert to lay understandings across each domain.

Furthermore, the general understanding of sustainable plastic use intersects other discursive and non-discursive formations, for instance, around the well-known three 'Rs' of reduce, reuse and recycle. However, the CPE underscores the importance of sharing, maintaining, reducing, reusing, remanufacturing, and repairing plastic materials above recycling and disposal [10]. It should be noted that the CPE is not only meant to eliminate plastic waste, but it is also a system championed to protect the environment, minimise resource use, deliver more employment opportunities, promote sustainable manufacturing, design, and consumption, and grow the economy in general [12].

As [46] argues, teleoaffective formations are configurations that exist across multiple scales of analysis. Depending on the framework adopted, teleoaffective formations can be analysed based on the practice complexes existing in different domains (household practices are further explored below). Thus, primary social institutions such as the economy, governance and households involve constellations of

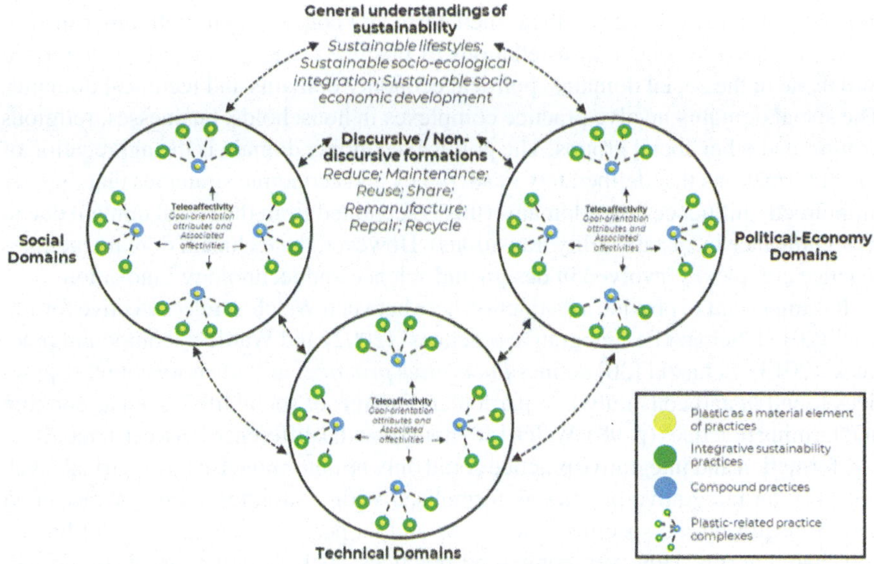

Fig. 1 Circular plastic economy framework from a practice perspective

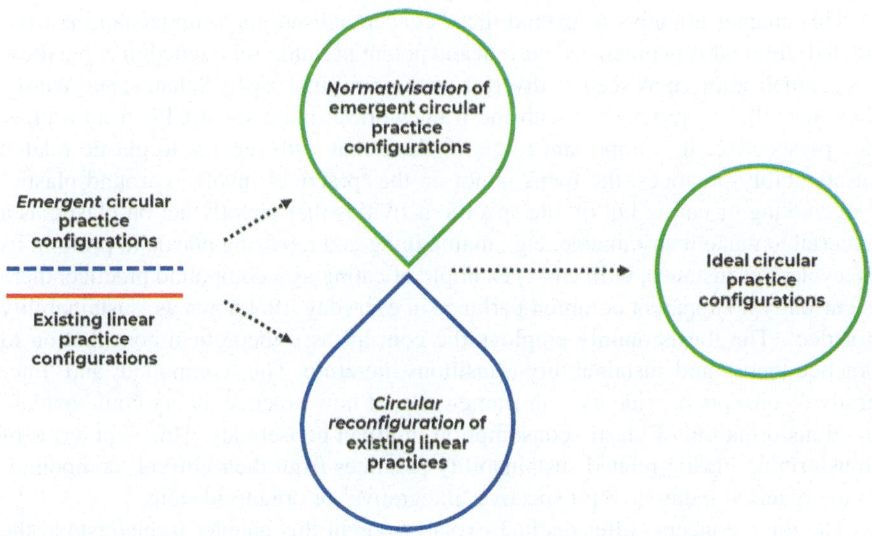

Fig. 2 Transitioning from linear to circular plastic economy from a practice perspective

practice complexes that grant them their form and connect them with other institutions. As shown in Fig. 1, constellations of practice complexes can be categorised into those in the social domains, political economy domains and technical domains. The social domains involve practice complexes in households, businesses, religious groups, and other social groups. The political economy domain is the aggregation of social formations that deliberately make policy and economic strategies that directly or indirectly influence other domains (this is separated from the social domain due to its significance in sustainability transitions). However, the technical domain includes practice complexes involved in design and science and technology innovation.

It is important to provide a distinction here between Welch's 'teleoaffective formation' (2017), Schatzki's 'integrative practices' (2002) and Warde's 'compound practices' (2013). Schatzki [30] defines *integrative practices* as "the more complex practices found in and constitutive of particular domains of social life", such as cooking and farming practices (p. 98). Welch [46] notes that the difference between teleoaffective formation and integrative practice could only be determined at an empirical level. For [44], an integrative practice is formalised with a straightforward or measured way of acquiring the specified know-how and its performance carries coordinated and shared norms. However, compound practices, such as eating, are disorganised, but connect several integrative practices with different spatial–temporal evolution together [44].

This chapter attempts to extend these conceptualisations to understand plastic-related sustainability practices' current and potential state. Acknowledging that these conceptualisations may seem to divert from their original use by Schatzki and Warde, they nevertheless provide us with the framework to understand CPE from a practice perspective. It is important to clarify here that with regards to plastic-related sustainability practices, the focus is not on the 'practices involved around plastic' like cooking or eating but on the specific activities that directly act on plastic as a material to make it sustainable, e.g., maintaining and repairing plastic to prolong its lifecycle. For instance, unlike [44] example of eating as a compound practice, there is currently no apparent common parlance in everyday life known as 'sustainability practice'. The thesis mainly employs the concept as a theoretical contribution to practice theory and sustainability transitions literature. The 'compound' and 'integrative' concepts provide us with frameworks of how practice theory could explain the transformation of plastic consumption in urban households. This is in terms of transforming plastic-related sustainability practices from their current 'compound' or unorganised nature to a prospective 'integrative' or organised state.

The three concepts offer distinct explanations in this chapter to understand the CPE and avoid conceptual ambiguities. In Fig. 1, teleoaffective formations exist in the CPE's overarching societal system (not to be confused with the social domain) and those bounding constellations of practice complexes. They are composed of general understandings that link complex practices and discourses together to achieve some set of goals. As explained by [46], promoting sustainable consumption is a teleoaffective formation integrating business practices, policy practices, consumer culture and civil society organisations. The CPE could also be a spatially bound teleoaffective formation such as local, city, regional, national, or international geographical boundaries. As depicted in Fig. 1, integrative practices are those complexes that are relatively structured, have standard conventions and can be learnt in an organised way. As [31] notes, these practices are found in specific domains and combine noncomplex practices. For example, cooking, bathing, decorating, recycling, parenting, and homemaking are some integrative practices that can be found in the household domain.

In Fig. 1, compound practices are disorganised in structure but generally recognised as binding multiple integrative practices together to achieve a project [44]. Plastic-related sustainability practices, defined as practices that directly ensure the sustainable use of plastic materials, can be categorised as compound in their current form. Aside from recycling, the other aspects of plastic-related sustainability practices, such as reuse and reduction, still lack systematic processes, including how, when, where or what to reuse and reduce. There is also no generally accepted standard or formalised means to learn what sustainable plastic use is.

Given the current level of unsustainable plastic consumption and lack of adequate regulation on plastic use (Blue [4], plastic-related sustainability practices are compound practices that "may be considered as failed, or not yet worked up, integrative practices" [44], p. 26). However, in the CPE, plastic-related sustainability practices would manifest as integrative practices within domains or teleoaffective formations at societal system scales with clearly defined goals, organised with supportive sociomaterial arrangements, and connected by general understandings of sustainability. Transforming the current 'compound' plastic-related sustainability practices to 'integrative' plastic-related sustainability practices is one of the tasks of stakeholders involved in sustainability transitions to the CPE.

Therefore, the transition of cities to the CPE would require understanding how practices constitute the socio-technical domains and the role of plastic material in interlinking practices to achieve projects or goals. The following section employs this approach by zooming in on plastic's role in forming and changing the practices around plastic consumption in households.

4 Zooming in Beyond Physical Features: Examining the Complexity of Plastic-Related Practices in Households

Analysing the role of plastic in the performance of household practices requires centralising plastic materials from other practice elements, as depicted in Fig. 1. The form the CPE takes in a domain (such as households) would be somewhat similar to the framework in Fig. 1, albeit at a smaller scale. A significant influence in how plastic-related household practices are formed and performed (as with other domains) is the teleoaffectivities that enjoin the practices together in achieving household projects. Teleoaffectivities can be understood as the manifestation of general understandings and (non-)discursive formations articulated in the performance of practices [46]. Figure 1 shows that simple practices are (re)combined to form more complex practices that interact to create the household teleoaffective formation.

Furthermore, plastic-related practices composing the CPE are connected to discursive and non-discursive formations such as sharing, maintaining, reusing, repurposing, and recycling around general understandings of sustainability, as shown in Fig. 1. This means that plastic-related household practices in the CPE would be integrative practices with established guidelines and conventions on sustainable plastic use [31].

In [40] categorisation of practice elements, plastic represents a material element that embodies and conveys the meanings (or affective aspects) and competences (or skills) of the carrier through practice performance. By combining with other material and non-material elements, plastic serves as a means through which carriers express affective engagements oriented towards achieving a set of goals. For instance, [34, 36] found that plastic assists low-income urban households in conveying meanings such as mobility, convenience, and accessibility in their daily performance.

Furthermore, plastic mobility within and outside of the household spatiotemporal boundary can also be influenced by the meanings embedded in its physical form. These embedded meanings reflect the value placed on plastic as a material addressing multiple conceived needs. However, some plastic materials, such as single-use and disposable plastics, carry less post-consumption value or meaning in their physical form. Thus, they are highly likely to move out of the household domain. While they may be transitory, these types of plastic may not have an enduring or predominant role in the material arrangements of the CPE, especially at the household level, thereby requiring their elimination or recycling.

In addition to mediating meanings, the current indispensability of plastic to household practices is derived from its potential to manifest in different forms physically and provide high spatiotemporal mobility. Presently, plastic materials could be employed in performing household practices such as cooking (plastic kitchenware), eating (plastic tableware), bathing (bathroom beauty products), refrigerating (plastic packs), cleaning (washing machines), decorating (plastic arts), gardening (plastic flower vases) household entertainment (plastic appliances), shopping (plastic bags), parenting (plastic toys) and so on.

Another point to consider is that a household's plastic use in the CPE is also influenced by the complex interaction of household practices, the general level of know-how in performing those practices and the socio-economic and cultural background within which the household is located. For instance, households with young people working long hours and possessing little sustainability education would more likely eat out often and purchase packaged foods but may not be able to recycle plastic materials appropriately.

This implies that the (non)performance of practices not directly related to plastic by a household would influence their (non)performance of practices related to plastic use. For example, households living near fresh food supplies in city centres or supermarkets might find it easier to purchase fresh foods frequently to reduce plastic use. Nevertheless, a household's purchase of fresh instead of pre-packaged foods could be influenced by the importance attached to the emotional expression and goals of sustainable lifestyles. In comparison, other households may value the convenience offered by pre-packaged foods above the general environmental costs, particularly within the context of time pressures and low economic resources. The preceding examples show that due to limited spatial–temporal and (non)material resources, households' performances of practice instead of another depends on the importance carriers attach to the practice's teleoaffectivity.

Concerning the role of socio-economic status in influencing the importance placed upon sustainable practices, Australian data suggests that environmental concern is most evident among those within the highest personal income quintile [2]. On the other hand, [33, 35] reveal that low-income households experience several environmental justice issues that prevent them from adopting plastic-related sustainability practices. First, unlike higher-income households that may afford existing sustainability solutions, which often are costly, low-income households lack access to adequate socio-economic resources and socio-material arrangements that enable the adoption of sustainability practices. Second, low-income households are often excluded from policy initiatives and social groups promoting sustainability education and upskilling, consequently inhibiting them from adopting sustainable lifestyles.

Moreover, income class considered more broadly, as well as gender expectations, appear to play a role in avoiding pre-packaged foods. For instance, middle-class motherhood in the United Kingdom appears to be associated with expectations of providing home-cooked, fresh, and healthy food, in sharp contrast with take-away and junk food viewed with disgust by this group [26]. Similarly, age may play an interesting role in convenience food. To illustrate, age appears to play a more significant role than gender, education level or socio-economic status among Danish youth, the majority of whom consume convenience food regularly [16].

Lastly, and as the above exemplifies, teleoaffective formations [46] and broader socio-economic practices outside the household domain, such as business practices and policies, influence plastic consumption in the CPE. Specifically, the sociomaterial practices that a business adopts will to an extent, shape the performance of household practices such as shopping. These business practices include offering plastic bags to consumers, providing reusable bags, serving as deposit points for plastic recyclables, supporting organic and fresh food supply, and engaging in green marketing activities, among others.

Moreover, there is a need to transform supply chain processes in transitioning cities to CPEs holistically. For instance, many fresh products that are presented to the consumer in a market style (i.e., not individually wrapped/packaged) often arrive at the retail stage wrapped up in many single-use plastics that are removed before final sale. Consumers may choose agricultural produce that is pre-packaged or packed at the point of sale, while others may decide to avoid plastic packaging altogether. In this sense, 'sustainable' shopping practices of consumers may not significantly affect their plastic footprint compared to 'unsustainable' shopping practices.

The following section focuses on the policy implications of the transitions to the CPE.

5 Transitioning to a Circular Plastic Economy: Implications for Policy and Practice

Transitioning to the CPE requires understanding the interconnections among existing elements, practices, teleoaffectivities, nexuses and large practice formations. This necessitates the holistic consideration of teleoaffective formations [46] and the trajectory of material and non-material arrangements within the system. Uncoordinated or little changes in practices or the constituted elements, such as in isolated local initiatives, may not have the significant effect required to create a shift from a linear to a circular economy.

Given the heightened intricacies of the globalised world economy, any isolated effort by an individual, business or city administrator to implement the CPE policies may be overwhelmed by the entrenched and unsustainable practice complexes that perpetuate the current linear economy. In this regard, a transition to the CPE involves the purposeful and active involvement of all stakeholders who could influence practices at varying societal scales. From a practice perspective and deriving from the above discussions, the CPE can be achieved from two aspects: transforming existing plastic-related practices to become sustainable and engendering integrative plastic-related sustainability practices.

The first aspect refers to the redesign and reconfiguration of elements of existing plastic-related practices (material, meaning and competence) circularly and sustainably. As a material element in the performance of a myriad of practices across social, political-economy, and technical domains, plastic use may not be circular or sustainable if the performance of those related practices continues to be linear and unsustainable. This also implies that focusing on an aspect of the 'R's of sustainability may not deliver the circular economy objectives.

Therefore, there is also a need to rethink and redesign production, consumption, packaging, and agricultural processes, among others. A significant means that this could be done would be the implementation of initiatives that promote service models and producer responsibility for plastic-related goods. Government bodies could facilitate such transition by collaborating with non-governmental organisations, research organisations, businesses, and community associations on scaling up the CPE initiatives. For large manufacturing organisations, other sustainable alternatives to plastic could also be identified through research and development.

Establishing plastic-related sustainability practices is the most important aspect of shifting to the CPE. As argued in this chapter, plastic-related sustainability practices are those integrated practices mainly performed to make plastic use sustainable and circular without leaking into the environment. Therefore, sustainability stakeholders must collaborate to establish infrastructures, rules, processes, skills, emotional affectivities, and other elements that promote and advance the sharing, maintaining, reduction, reuse, repurposing, remanufacturing, repairing, and recycling of plastic materials across all social domains. Aside from this, more attention should be paid to socio-economic and environmental inequalities and injustices arising from the conflicts and contradictions of existing teleoaffective formations that may prevent the transition to the CPE.

Transitioning from the linear plastic economy to the CPE would be as depicted in Fig. 2. The three panels in Fig. 2 are connected by arrows symbolising the transition process. Panel A illustrates emergent circular practices and existing linear practices. Panel B illustrates the normativisation of emergent circular practice configurations and the circular reconfiguration of existing linear practices. Finally, panel C illustrates the final stage of circular transitions. In the context of this chapter, the arrows connecting the panels (the transition process) in practice represent changes in general understandings, (non-)discursive formations and teleoaffectivities.

The framework suggests that the linear trajectory of current practice complexes would require reconfigurations to fit into the CPE. Reconfiguration might mean different things. For instance, some practices and their constituting elements could disappear or lose carriers. Other practices or practice elements would be rearranged to either form new practices that combine in new ways or recruit new carriers to conform to the CPE. On the other hand, there would be a few extant practice configurations, most likely on a small scale, that reflect the CPE, such as some innovative small business practices. These emerging circular practice configurations need to be normativised, performed continuously and expanded across all institutional domains to entrench them into the system. The final transition state would encompass practice complexes that emerge or are created and performed to achieve a waste minimization strategy for long product lifecycles and sustainable socio-ecological integrations.

6 Conclusion

This chapter presented a theoretical framework for transitioning cities to the CPE based on a practice theory perspective. This is accomplished through a zoom-in-and-out approach which explores the socio-material arrangements composing the CPE at different societal scales. While the application of practice theory to sustainability transitions is still evolving, this chapter more specifically expounds on the transition to CPEs from a practice theory perspective. In doing so, it provides an extension, and thus a scholarly contribution, to practice theory.

First, the chapter employs the concept of teleoaffective formation to analyse the CPE as the complex interconnection of social, political-economic, and technological domains [46]. This shows that the CPE is to be understood as a system extending beyond technological innovations to include the nexus of practice complexities enjoined by the general understandings of sustainable lifestyles, sustainable socio-economic integration, and sustainable socio-economic development. In turn, general understandings condition the discursive and non-discursive formations guiding circular practices.

In the CPE, sustainability is the general understanding that guides all plastic-related practices. Propagating this general understanding becomes one of education, capabilities, and resources. General understandings are, however, but one component of the CPE. A transition strategy is also about developing infrastructure, design and related processes that enable households (or practitioners) to gradually adopt sustainability practices within a framework that nevertheless delivers the CPE. Capability-enhancing infrastructure, design and processes provide a top-down opportunity to transform practices and enable a shift or realignment in meanings and preferences over time. These meanings are then articulated in emerging plastic-related sustainability practices.

The conceptual framework presented in this chapter provides a pivotal contribution to the circular economy literature. Specifically, applying a practice-based perspective illustrates how the daily activities in an institutional domain (for instance, household consumption) are embedded within the socio-technical and practice complexes of the CPE. Also, the conceptual framework presents a transition pathway from linear practice configurations to circular practice configurations in cities. Conceptualising the transition to the CPE will benefit urban policymakers, organisations and scholars when adopting either top-down or bottom-up approaches to implementing circular economy strategies.

Acknowledgements The authors express gratitude to the reviewer for their constructive and insightful feedback on the early versions of the chapter. This chapter is based on the lead author's doctoral research, which was supported by Swinburne University of Technology, Australia, through the Swinburne University Postgraduate Research Award (SUPRA) and a Travel Grant from the Centre for Urban Transitions, Swinburne University.

References

1. Akanle O, Shittu O (2018) Value chain actors and recycled polymer products in Lagos metropolis: Toward ensuring sustainable development in Africa's megacity. Resources 7(3):55. https://doi.org/10.3390/resources7030055
2. Australian Bureau of Statistics (2012) Environmental views and behaviour, 2011–12, viewed 16 Feb 2020. https://www.abs.gov.au/ausstats/abs@.nsf/Lookup/4626.0.55.001main+features32011-12
3. Bartolotta JF, Hardy SD (2018) Barriers and benefits to desired behaviors for single use plastic items in northeast Ohio's Lake Erie basin. Mar Pollut Bull 127:576–585
4. Blue Environment (2019) Data on exports of Australian wastes. Blue Environment, viewed 26 March 2019. http://www.environment.gov.au/system/files/resources/3e7b446f-0c76-492c-93e6-4ad15b473a30/files/data-exports-australian-wastes-issue-3.pdf
5. Bocken N, Strupeit L, Whalen K, Nußholz J (2019) A review and evaluation of circular business model innovation tools. Sustain 11(8):2210
6. Borch A, Vitterso G, Sto E (2015) Studying sustainable change: From ABC to practice. Gaia 24(2):102–107

7. Brooks AL, Wang S, Jambeck JR (2018) The Chinese import ban and its impact on global plastic waste trade. Sci Adv 4(6):eaat0131d
8. Chib S, Sheikh H (2016) Eco friendly carry bags: Shop owner's perception. Splint Int J Profs 3(6):7
9. de Sousa FDB (2021) The role of plastic concerning the sustainable development goals: The literature point of view. Clean Responsible Consum 3:100020
10. Ellen MacArthur Foundation (n.d) The circular economy in detail. https://www.ellenmacarthurfoundation.org/explore/the-circular-economy-in-detail
11. Ertz M, Huang R, Jo M, Karakas F, Sarigollu E (2017) From single-use to multi-use: Study of consumers' behavior toward consumption of reusable containers. J Environ Manage 193:334–344
12. European Academies Science Advisory Council (2016) Indicators for a circular economy. German National Academy of Sciences, Leepoldina
13. Gäbler K (2015) Green capitalism, sustainability, and everyday practice. Global Sustainability: 63–86
14. Godfrey L (2019) Waste plastic, the challenge facing developing countries—ban it, change it, collect it? Recycling 4(1)
15. Guldmann E, Huulgaard RD (2020) Barriers to circular business model innovation: A multiple-case study. J Clean Prod 243:118160
16. Halkier B (2017) Normalising convenience food? Food, Culture & Society 20(1):133–151. https://doi.org/10.1080/15528014.2017.1265835
17. Hargreaves T (2011) Practice-ing behaviour change: Applying social practice theory to pro-environmental behaviour change. J Consum Cult 11(1):79–99
18. Hasson R, Leiman A, Visser M (2007) The economics of plastic bag legislation in South Africa. S Afr J Econ 75(1):66–83
19. Korhonen J, Honkasalo A, Seppala J (2018) Circular economy: The concept and its limitations. Ecol Econ 143:37–46
20. Lavelle MJ, Rau H, Fahy F (2015) 'Different shades of green? Unpacking habitual and occasional pro-environmental behavior. Glob Environ Chang 35:368–378
21. Martin C (2019) Plastic world. Curr Biol 29(19):R950–R953
22. Merli R, Preziosi M, Acampora A (2018) 'How do scholars approach the circular economy? A systematic literature review. J Clean Prod 178:703–722
23. Millar N, McLaughlin E, Börger T (2019) The circular economy: Swings and roundabouts? Ecol Econ 158:11–19
24. Newton P, Meyer D (2013) Exploring the attitudes-action gap in household resource consumption: Does "environmental lifestyle" segmentation align with consumer behaviour? Sustainability 5:1211–1233
25. Nicolini D (2009) Zooming in and out: Studying practices by switching theoretical lenses and trailing connections. Organ Stud 30(12):1391–1418
26. Parsons JM (2016) When convenience is inconvenient: 'health' family foodways and the persistent intersectionalities of gender and class. J Gend Stud 25(4):382–397. https://doi.org/10.1080/09589236.2014.987656
27. Pieroni MP, McAloone TC, Pigosso DC (2019) Business model innovation for circular economy and sustainability: A review of approaches. J Clean Prod 215:198–216
28. Rochman CM, Browne MA, Halpern BS, Hentschel BT, Hoh E, Karapanagioti HK, Rios-Mendoza LM, Takada H, Teh S, Thompson RC (2013) Policy: Classify plastic waste as hazardous. Nature 494(7436):169
29. Sauvé S, Bernard S, Sloan P (2016) Environmental sciences, sustainable development and circular economy: Alternative concepts for trans-disciplinary research. Environmental Development 17:48–56
30. Schatzki T (1996) Social practices: A Wittgensteinian approach to human activity and the social. Cambridge University Press, Cambridge
31. Schatzki T (2002) The site of the social. Penn State University Press, University Park, PA

32. Schatzki T (2016) Practice theory as flat ontology. In: Practice Theory and Research, pp. 44–58. Routledge
33. Shittu O (2020) Emerging sustainability concerns and policy implications of urban household consumption: A systematic literature review. J Clean Prod 246:119034. https://doi.org/10.1016/j.jclepro.2019.119034
34. Shittu O (2021) 'Almost everything in the house now is plastic': Foregrounding plastic materiality in household routines and practices. Sociological Research Online. https://doi.org/10.1177/13607804211034887
35. Shittu O (2022) Grassroots strategies for environmental governance and circular cities: Lessons from Lagos and Melbourne. In: McCormick K, Evans J, Palgan YV, Frantzeskaki N (eds.), Research agenda on sustainable cities and communities. Edward Elgar Publishing, manuscript submitted for publication
36. Shittu O, Nygaard C, Bailey A (2021) Sustainability practices and materiality: Transforming plastic consumption in urban households, manuscript submitted for publication
37. Shove E, Walker G (2010) Governing transitions in the sustainability of everyday life. Res Policy 39(4):471–476
38. Shove E (2010) Beyond the ABC: Climate change policy and theories of social change. Environ Plan A 42(6):1273–1285
39. Shove E (2019) Connecting practices: accumulation, circulation, interweaving and convergence viewed 10 June 2019. https://practicetheorymethodologies.wordpress.com/2019/06/10/elizabeth-shove-connecting-practices-accumulation-circulation-interweaving-and-convergence/
40. Shove E, Pantzar M, Watson M (2012) Making and breaking links. In: Shove E, Panter M, Watson M (eds) The dynamics of social practice: Everyday life and how it changes. SAGE Publications Ltd., London, pp 21–42
41. Sole T, Wagner C (2018) Understanding domestic fuel use practices in an urban township. Building Research and Information 46(2):220–230
42. Strengers Y, Maller C (2014) Social practices, intervention and sustainability: Beyond behaviour change. Routledge, London, UK
43. Vanapalli KR, Samal B, Dubey BK, Bhattacharya J (2019) Emissions and environmental burdens associated with plastic solid waste management. In: Plastics to Energy, pp 313–342. William Andrew Publishing
44. Warde A (2013) What sort of a practice is eating?', In: Shove E, Spurling N (eds), Sustainable practices: Social theory and Climate Change. Routledge, pp 33–46
45. Welch D, Warde A (2015) Theories of practice and sustainable consumption. In: Handbook of Research on Sustainable Consumption. Edward Elgar Publishing
46. Welch D (2017) Consumption and teleoaffective formations: Consumer culture and commercial communications. J Clean Prod 0(0):1–22
47. Winans K, Kendall A, Deng H (2017) The history and current applications of the circular economy concept. Renew Sustain Energy Rev 68:825–833
48. Yates L, Evans D (2016) Dirtying linen: Re-evaluating the sustainability of domestic laundry. Environ Policy Gov 26(2):101–115
49. Zhang H, Lahr ML (2018) Households' energy consumption change in China: A multi-regional perspective. Sustain 10(7): 2486

The Brunswick Spine: An Example of Green TOD in Melbourne, Australia?

Rajjan M. Chitrakar, Peter Hogg, and Dillan Eshow

Abstract In recent years, Australian governmental policy agenda for urban development has embraced transit-oriented development (TOD) initiatives driven by green concerns. There is a growing consensus that the combination of green urbanism and TOD offers social and environmental benefits and can significantly reduce the carbon footprint associated with urban development. Despite the rhetoric from the government, green TODs have seldom been put into practice. In this chapter, we explore the prospect of delivering green TOD in Melbourne, using the case study of the 'Brunswick Spine', a linear strip of the urban area between Sydney Road and a major suburban railway line in Melbourne's inner north, which represents a classic example of the nineteenth century transport-based urbanism. The Brunswick Spine not only presents exciting possibilities as a linear TOD, with excellent public transport links at a variety of scales, but the emergence of green buildings as evident in the newly built housing complexes such as the Nightingale Village indicates a new impetus in achieving green urbanism in an existing TOD-friendly urban area. We argue that the spine shows some promising signs of a green TOD and sustainable urban (re)development that may set an example to be emulated elsewhere in Australia.

Keywords Transit-oriented development · Green urbanism · Green TOD · Brunswick · Melbourne

1 Introduction

Every city aspires to achieve smart and sustainable urban growth. To this end, many contemporary cities around the world have employed the principles of transit-oriented development (TOD) to deal with the challenges posed by the proliferation of suburban development in the post-war era. The TOD concept, which was first presented in 1993

R. M. Chitrakar (✉) · P. Hogg · D. Eshow
Department of Business and Construction, Melbourne Polytechnic (Epping Campus), Corner of Cooper St. and Dalton Rd., Epping VIC 3076, Australia
e-mail: rajjanchitrakar@melbournepolytechnic.edu.au

Z. Allam (ed.), *Sustainable Urban Transitions*, Urban Sustainability,
https://doi.org/10.1007/978-981-99-2695-4_15

by American Architect and Urban Planner Peter Calthorpe, is a technique to "create more compact, walkable communities" configured around rail transit stations [4], p. 25), with a focus on the growth-friendly components, such as density, mixed-use and public spaces [18]. According to [5], p. 210), over the past three decades, TOD has proved to be a promising tool to address "the vicious cycle of sprawl and car dependence feeding off of each other, replacing it with a virtuous cycle", in which trips are increasingly replaced by mass transit systems and development centred around transit stations helps slow the spread of sprawl. This growth model is now viewed favourably as a sustainable approach to urban development.

Australia is a largely suburban nation, where the relentless growth of urban sprawl has been a key characteristic of the urban development of its major cities. Clearly, the viability of such a pattern of urban development is increasingly problematic, and TOD appears to be ever more relevant to the Australian context. There is no doubt that Australian cities could benefit from the application of TOD to help tackle the problem of over-reliance on private vehicles. But the evidence shows that TOD has only occurred sporadically and is somewhat limited in practice in Australia [9]. Bajracharya et al. [2] argue that TOD has often been adopted as a fad rather than a comprehensive strategy to cope with sprawl. Consequently, Australian cities have been less than successful in reaping the benefits of TOD until now. There is a lack of execution of the integral factors for its success, such as rapid transit, density, and mixed land use, as these cities have largely failed to strategically and statutorily integrate TOD into urban planning and development processes [24].

In recent years, however, Australian governmental policy agenda for urban development has embraced TOD initiatives driven by green concerns [2]. This comes at a time when there is a growing consensus worldwide that the combination of green urbanism and TOD offers social and environmental benefits and can significantly reduce the carbon footprint associated with urban development. Not only does this bring the opportunity to integrate green urbanism into the urban development process in Australian cities, but it can potentially help accomplish green TOD, a concept that has evolved as a more adaptive and responsive approach to urban growth that incorporates green dimensions into the conventional TOD.

Despite the rhetoric from the government, the green dimensions of urbanism are yet to be fully realised in Australian cities, with little actual green TOD appearing within the scope of the current urban development process. While green urbanism is increasingly gaining popularity among urban researchers and practitioners in Australia, it becomes critical to explore how it can aid in achieving sustainable urban development when the application of TOD concepts alone has not been successful in doing so. Our initial observations indicate that there is a potential to promote and execute green TOD in the urban redevelopment process in the inner suburbs of Australian cities, particularly in Melbourne, as these suburbs, with their relatively compact and walkable neighbourhoods, respond better to the TOD principles and have seen the emergence of green buildings along with other green infrastructure in the recent years [6]. There is already evidence of infill development, including the conversion of former industrial sites into residential ones, that has been taking place for several decades, leading to higher urban density [29]. Can we then achieve green

TOD in such inner-city areas of Australian cities? If so, how and what are the barriers to it? In this chapter, we explore the prospect of delivering green TOD in Melbourne's inner suburbs. We present findings from the case study of the 'Brunswick Spine', a linear strip of the urban area between Sydney Road and a major suburban railway line in Melbourne's northern suburb of Brunswick. This represents a classic example of the nineteenth century transport-based urbanism, with a good provision of public transit systems, and the potential for effective TOD.

2 Green Urbanism and the Environmental Benefits of Its Integration with TOD

Green urbanism is a topic of growing importance in contemporary urban development practices. In the simplest terms, green urbanism may mean bringing nature and/or natural lifestyle into the city. But more than that, green urbanism challenges our current understanding of the urban planning discipline and emphasises issues that have a direct impact on resilience and sustainability, such as green buildings, water-sensitive urban design, sustainable transport, walkable cities, high-density and infill development, liveability and healthy communities. Beyond these considerations, green urbanism can take a restorative approach to the natural environment. Lehmann [19] defines green urbanism as a "conceptual model for zero emissions and zero waste" that contributes to a more energy efficient city with a lighter ecological footprint that is more beneficial to the environment and the people. According to [25], p. 149), green urbanism may refer to settlements that are "smart, secure and sustainable". He adds:

> They are smart in that they are able to adapt to the new technologies of the 21st century, secure in that they have built-in systems that enable them to respond to extreme events as well as being built to last, and sustainable in that they are part of the solution to the big questions of sustainability, such as climate change, peak oil, and biodiversity (p. 149).

Green urbanism stresses re-evaluating our current approaches to city building by considering energy consumption and prioritising green areas to achieve environmentally and socially sustainable cities [32]. In addition to its underlying focus on reduced energy use, green urbanism adopts a more sustainable approach to urban design and landscape architecture to help shrink the environmental footprint associated with urban development [17]. On the one hand, it aims to improve access to green spaces by increasing the area of open space and community gardens within the development area. At the same time, it seeks to promote walkability by creating a network of green transit lines [5]. While much of the development area could be taken up by roads, green urbanism ensures that enough space can be provided to build high quality pedestrian pathways, with the addition of urban landscaping and outdoor seating areas to encourage public use and activity [14] that are inviting, visually stimulating and well-lit at night [2]. The inclusion of urban green spaces aids in the economic value and the sociability of the area, and promotes environmental benefits within the same framework that encourages pedestrians and cyclists for a better and more efficient environment.

Consideration of green urbanism within the TOD strategies was relatively undervalued until the recent past. However, the last few years have seen a growing emphasis on the principles of green urbanism in examining the role of TOD and its wider impact on the quality of the built environment. In line with this, green TODs have emerged as a combined effort with additional environmental benefits. According to [5], p. 210), two of its early and strong proponents, green TOD is a more environmentally friendly version of TOD, in which the combination with green urbanism can "create synergies that yield environmental benefits beyond the sum of what TOD and green urbanism offer individually". Likewise, [27], p. 1) suggest that green TOD is "an evolution of the TOD theory, influenced by sustainable development and green urbanism" and "expands the environmental and ecological dimensions of conventional TOD". According to [20], p. 36), green TOD delivers a highly environmentally friendly urban form by embedding ecological planning features into TOD such as "reducing energy consumption, improving the efficiency of energy usage, and enhancing the productivity of renewal energy basis on the high density feature, providing chances to integrate energy resources and foster electric vehicles industry through the mix-use pattern, increasing biophilic open space diversely via ecological planning approach, conserving and reusing energy by green techniques, architecture design, etc.".

The carbon footprints of green TOD can be up to 35% less than those of conventional developments, largely achieved by significantly improving walkability and carbon neutral mobility [5]. This is possible because:

> green TOD emphasises the coordinated development of transportation, society, and the environment, to build a sustainable and liveable transit-oriented community. Conventional TOD injected new thinking at the social and environmental dimensions by integrating green urbanism and ideas associated with an ecological community [27], p. 1).

Cervero and Sullivan [5] make a similar observation on the environmental benefits of green TOD, which offers:

> a form of urbanism and mobility that could confer appreciable environmental benefits. It emphasises pedestrian, cycling and transit infrastructure over automobility. It mixes land uses, which not only brings destinations closer but also creates an active, vibrant street life. And through building designs and resource management systems, it embraces minimal waste, low emissions, and to the degree possible, energy self-sufficiency (p. 216).

Cervero and Sullivan [5], p. 210) suggest that while the benefits of combining TOD with green urbanism can deliver "energy self-sufficiency, zero-waste living and sustainable mobility", these may be realised in terms of (a) higher densities, (b) mixed land uses; (c) reduced surface parking and impervious surfaces; and (d) solar energy production at stations (Table 1).

Table 1 Environmental benefits of green TOD [5]

TOD	Green urbanism
Mobile sources	Stationary sources
Transit design: World-class transit (trunk and distribution); station as hub; transit spine	Energy self-sufficient: Renewably powered—solar, wind, organic waste converted to biogas; energy efficiency; district heating/cooling; combined heat and power
Non-motorised access: Bike paths; pedestrian ways; bike-sharing/carsharing	Zero waste: Recycling and reuse; methane digesters; rainwater collection for irrigation and grey water use; bioswales
Minimal parking: Reduced land consumption; building Massing and impervious surfaces	Community gardens and open space: Composting; tree canopies; water-table recharging
Compact development mix of uses	Buildings: Green roofs; orientation (optimal temperatures); insulation; glazing; air-tight construction; low-impact and recycled materials

3 Green Urbanism in the Australian Context

Australia has some of the highest per capita carbon emissions in the world. In 2018, Australia's greenhouse gas (GHG) emissions were 15.5 metric tons per capita as compared with 5.5 in the UK and 8.6 in Germany [31]. Some reasons for this include large dispersed cities, energy inefficient buildings, reliance on fossil fuels for most energy generation, high volumes of fossil fuel exports and emissions from agriculture and land clearing. Transport makes up 18% of total Australian emissions and half of that or around 9% is estimated to be due to private car usage alone (Fig. 1) [8]. Buildings also contribute to a significant amount of GHG emissions. It is estimated that the construction, maintenance and operation of buildings emit around 25% of all greenhouse gases in Australia [21]. Large house size, poor passive solar design and poor insulation coupled with a coal-based electricity supply, widespread use of gas for heating and the ubiquitous split-system air conditioner all contribute to this.

Buildings and cities are emerging as a significant environmental problem both globally and in Australia [28]. Given the very high level of energy usage in all aspects of urban living, green urbanism is a highly relevant strategy to promote sustainable urban growth in Australian cities. It can assist in improving urban sustainability outcomes in several ways—from reduction in car usage to more energy efficient building design. In practice, however, while a reduction in car usage may be achieved by adapting green urbanism principles, we argue that other sustainability goals have been more elusive. For instance, most multi-residential construction in Australia is privately developed for profit, where there is a tendency for many developers to 'max out' the site and to build as tall as possible in order to maximise returns for investors. This tends to produce high and medium rise apartment buildings with

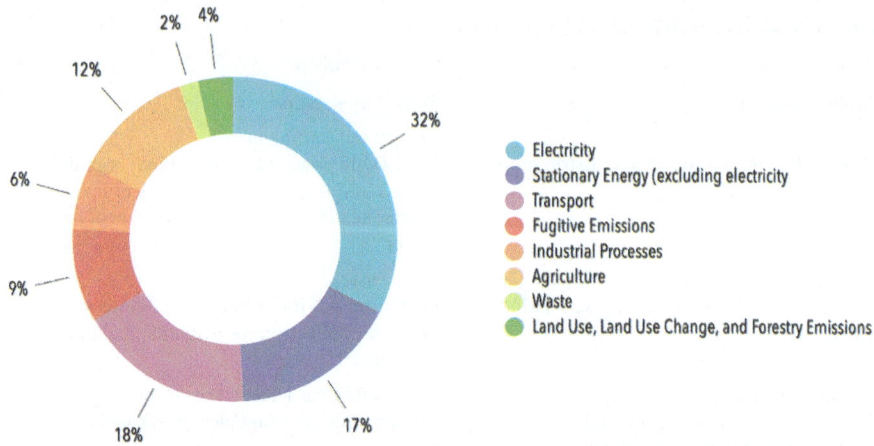

Fig. 1 Contribution to GHG emissions in Australia by sector, year to December 2017 [8]

a high embodied energy (the primary materials being concrete, steel, aluminium and glass), which require energy-hungry elevators, air conditioning and heating to function. The favoured façade type is the (single glazed) curtain wall, which has very poor thermal performance, made worse by the widespread failure to use appropriate shading and passive solar design measures (Fig. 2).

At least in theory, however, multi-residential developments (i.e. flats and apartments) should be more sustainable than stand alone houses as they are generally smaller, and therefore, consume fewer resources to build and service. Furthermore, multi-residential developments often have a lower ratio of external walls in proportion to floor area than a stand alone house, as in most cases each apartment in large part abuts the adjacent unit, and thus becomes to some degree self-insulating. On the other hand, despite the best intentions of local planning schemes, tall buildings overshadow the ground plane in many cases and can create wind tunnels, often making streets uninviting in winter. In addition, the street interface is often not 'activated'

Fig. 2 The effect of high-rise apartment towers at ground level, Melbourne CBD (Authors)

with retail or other activity, leading to barren streetscapes in most high-density residential developments. As is the case in many other cities, most notably London, an investor-led approach to urban development leads to 'land banking' and too many empty apartments, in turn, leading to dead and empty retail precincts [1].

Being profit-driven, private developers are often reluctant to contribute to the greater common good beyond the confines of their development sites. Return to investors is the prime concern and wider community benefit is far down the list of concerns. This leaves the provision of essential public amenities such as schools, child care, parks and 'green infrastructure' to local and state governments, who are constantly playing catch up in the provision of these non-profit making services. In a landmark study of development in one part of the Melbourne Central Business District (CBD), Leanne Hodyl documented how developers in Melbourne were allowed to more intensively develop sites, with much higher plot ratios and far less give back to the community, than in 'high density' global cities such as Hong Kong, New York, Vancouver, Seoul and Tokyo (Hodyl, 2014). Much of the inner city of Melbourne, and indeed some suburban development areas, are now increasingly characterised by very tall residential tower developments (heights of 50 and 60 stories are not uncommon) built very close together, with no additional provision of open spaces, green or otherwise. The environmental credentials of TOD designed along these lines is highly questionable. Density does not, of its own accord, equal sustainability.

Beatley and Newman [3] argue that Australia has both the resources and capabilities to move away from fossil fuels toward sustainable energy self-sufficiency. Installing solar panels can assist in achieving this goal. By 2018, more than 2 million domestic rooftop solar PV systems had been installed on Australian houses [7]. However, the uptake of solar in the multi-residential market has been much slower, as a cursory glance at an aerial photograph of a newly developed area will demonstrate. In the first instance, there is proportionally much less roof as compared to façade in multi-storey buildings as opposed to a single-family home, and so proportionally less space for solar panels. Secondly, the rooftops of many multistorey developments are dominated either by roof terraces or by services such as heating and air-conditioning equipment and elevator plant rooms, again leaving little room for solar panels. With towers packed close together in many locations and developers building ever taller, high-rise towers frequently overshadow one another, meaning photovoltaic electricity generation is ineffective.

4 Urban Development in Melbourne and the Prospect of Green TOD

Melbourne is a sprawling multicultural metropolis of more than five million people, which has been experiencing rapid population growth and development over the past 30 years. In the decade from 2010 to 2022 (until COVID and associated border closures) an average of 180,000 people moved to Melbourne per year, either from regional Victoria, from interstate or, more importantly, from other countries. This rapid growth rate dropped to near zero with the closure of the international borders in response to the Covid 19 pandemic in 2020 and 2021, but is expected to resume as the borders reopen. Projections for Melbourne's population by mid-century vary from 8 to 10 million.

Despite a boom in high-rise apartment towers in the inner city, and some suburban centres, the bulk of new dwellings are to be found in the new, low-density suburbs on the urban fringe, much of which have been built on the fertile farmlands and remnant native grasslands of the western plains or pushing into the bushland of the east and north (Fig. 3). The new outer suburbs are poorly served by public transport, few people walk anywhere and most trips are by car, resulting in high carbon emissions and high fuel bills.

Founded in 1835 on Naarm, part of the the lands of the First Nations Kulin people, the modern city of Melbourne consists of a relatively dense nineteenth century urban core surrounded by successive belts of low-density automobile-dependent suburbs, in places stretching more than 50 km from the city centre. With the exception of the City Loop, an underground rail line consisting of five stations in the CBD, the suburban railway system is entirely radial—all suburban rail lines radiate from the city centre out to distances of up to 40 km; there are no cross connections. The present state government has undertaken an extensive (and expensive) at grade level crossing removal program since it was first elected in 2014, elevating railway lines and opening up the land formerly occupied by the tracks for public use. This project, which continues to be rolled out across the metropolitan areas, has dramatically

Fig. 3 Tract housing, Truganina, 17 km west of the Melbourne CBD (Authors)

improved traffic flows and the urban environment in the affected areas, and largely dispensed with Melbourne's 'big country town' image. An even more ambitious project, the 'suburban rail loop', proposed by the state Premier in the run up to the 2018 state election (without a supporting business case, or in-depth transport planning study) aims to link up and cross-connect the radial rail lines as they pass through the middle ring suburbs and so create an actual network of railway lines. While sound in principle, detail is lacking. The suburban rail loop is not expected to be completed until the 2050s, and, with estimated costs of between A$50 and A$100 billion, the project remains contentious. The suburban rail loop may not eventuate in full given the often-fractious nature of state politics.

In addition to the rail 'network', Melbourne has an extensive network of trams and buses. Based on the nineteenth century cable car network, the tram system is largely confined to the inner ring of suburbs built before World War II. While there have been many additional suburbs added since then, the tram network extends no more than 15 km from the city centre, covering around one-tenth of the metropolitan area. Trams compete with cars for space on roads, and are, therefore, often slow and subject to traffic conditions and best suited for short-distance trips. In the nineteenth century, in the absence of other motorised road transport, the tram network led to the development of linear shopping strips in the inner suburbs based around the tram lines.

Alongside walking, cycling is the most healthful and environmentally benign mode of transport. Bike culture is strong in Melbourne, and commuter cycling is particularly common in the inner suburbs. The bicycle path network has improved greatly in recent years, but still leaves much to be desired. Often, bike paths are merely markings on the roads, and cyclists have to mix in with car traffic. Many bike paths are discontinuous, so the network is often fragmented, and is largely non-existent in the outer suburbs. Where possible, bicycle traffic should be separated from vehicular traffic, and to address this concern, increasingly 'Copenhagen' style bike lanes are being introduced, greatly improving rider safety. In addition, there is a growing network of bike paths that run alongside creeks, tram and train lines, and through parks and along beaches. While there is still some way to go in improving the cycling network, much progress has been made in recent years. Train, tram, bus, bike and foot travel operate at different urban scales (metropolitan, suburban and local), and all play a part in developing TOD. Encouragingly, most or all are present in Melbourne's inner suburbs.

TODs have been a central feature of Melbourne metropolitan planning since the release of the Melbourne 2030 report in 2002 [12]. In the report, there is a plan for metropolitan Melbourne along with its various revisions and updates (e.g. Melbourne @ 5 Million, 2008) that have been issued since then and are generally referred to as 'Plan Melbourne'. Six Central Activity Districts (CADs) were identified as sites for TOD in the report—the aim being to transform Melbourne into a polycentric metropolis. The 2017 version of Plan Melbourne refers to Metropolitan Activity Centres (MACs), and identifies several more 'MACs' than in the 2008 version, as well as some future MACs (Fig. 4) [11]. All MACs lie on major rail lines and connect to various other modes of transport such as buses, trams, cycle paths, pedestrian networks, and private cars. All the CADs identified for development and intensification utilise existing suburban nodes as their 'seed stock'. Disappointingly, only two of the six CADs will be connected to the proposed suburban rail loop. The nine MACs are augmented by a hierarchy of numerous smaller activity centres, which also have the potential to incorporate some elements of TOD and green urbanism.

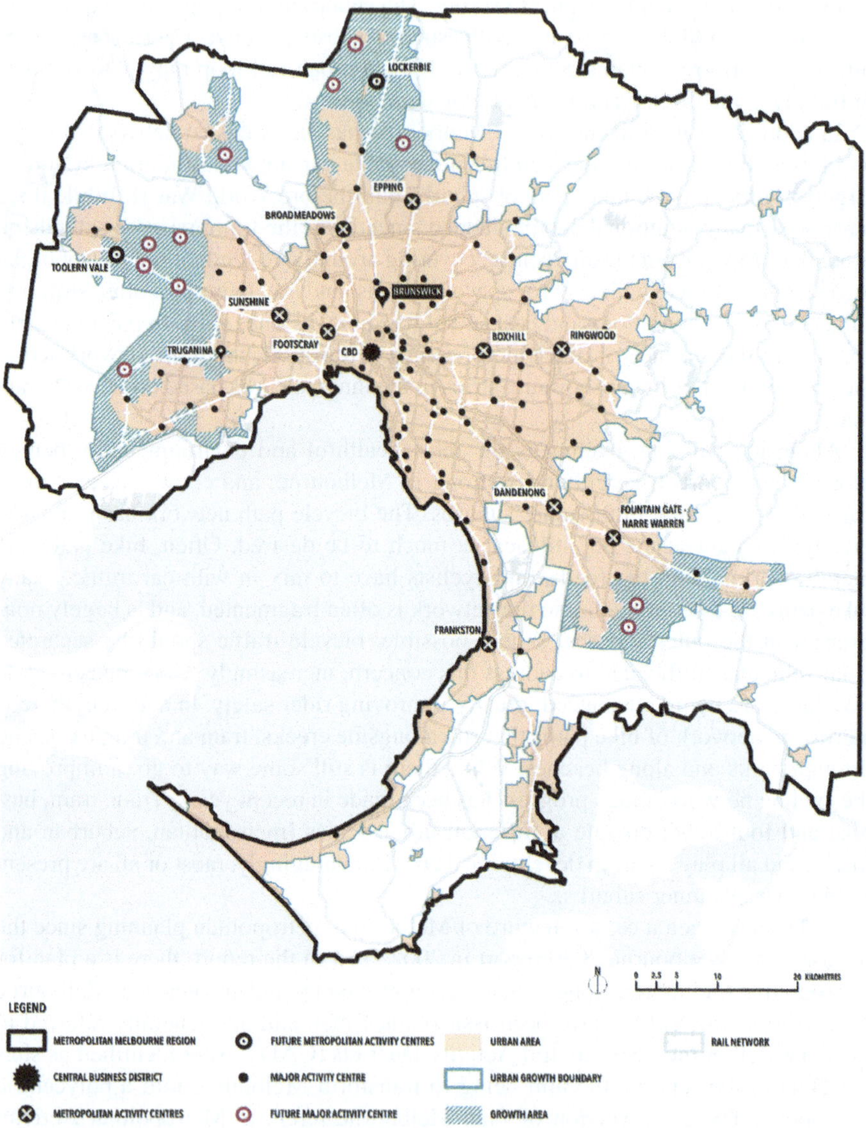

Fig. 4 Metropolitan activity centres (Adapted from Plan Melbourne, 2017)

It is, therefore, fair to say that Plan Melbourne has formalised a process already well underway as most MACs were pre-existing suburban centres and several of which had histories as independent towns before they were swallowed up by the expansion of the suburbs. By identifying the MACs as sites for high-density development, the state government may simply have given the 'green light' to yet more speculative developers through infill developments of the inner-city areas, without necessarily maximising community and environmental outcomes. Nonetheless, the push for infill development and increased density is a positive step in achieving sustainable urban development. Although infill development in Melbourne commenced in the post-war era, several urban consolidation movements have occurred since the 1980s [29] as regulations encouraged such developments allowing subdivision of single-family detached houses and encouraging medium to higher density developments near activity centres [13]. According to [26], approximately 35% of new residential development can be attributed to infill development in the existing suburbs of Australian cities.

4.1 The Brunswick Spine

Brunswick is a rapidly gentrifying 'hip' inner suburb located 4 km north of the Melbourne CBD within the City of Moreland Local Government Area (LGA) (Fig. 4). From the early 20th Century, Brunswick was a mostly light industrial suburb that employed a largely working-class population. In the post-war era, Brunswick became a largely migrant area with the population largely drawn from southern Europe and the middle-east. While much of the old ethnic diversity remains, giving the suburb much of its vibrancy and colour, the post-war working-class population of Brunswick is being increasingly displaced as gentrification and 'hipsterfication' proceeds apace. Former factories and small family homes are being displaced by apartment developments, while 'groovy' cafes and bars proliferate.

The linear shopping strips developed along tram lines in the nineteenth century are characteristic of much of inner Melbourne. The Sydney Road Brunswick shopping strip is one such tram-based shopping precinct. Running north–south, and more than 4 km in length, it is the longest in Melbourne and is a classic example of nineteenth century transport-based urbanism. West of Sydney Road, and roughly parallel to it, is the Upfield Railway line. The distance between the two varies from 200 to 250 m. The strip of land between Sydney Road and the railway line forms a 'Spine' of mostly former industrial land, mixed in with other land uses, that runs from Brunswick Road in the south to Moreland Road in the north at a distance of around 2.4 km (Fig. 5). There are three railway stations, Jewell, Brunswick and Anstey, in the Spine. Moreland Station, just north of the study area, has recently been redeveloped and elevated as part of the level-crossing removal project and the track raised. Alongside the railway line is an Upfield bicycle path, connecting with the University of Melbourne and medical precinct and the Melbourne CBD to the south and Coburg to the north. Six bus lines cut across the Spine in an east–west direction.

Nowhere in the Spine is much more than 100 m from either a railway station, a bike path, bus, or tram.

Over the past several decades, the industrial sites in the Brunswick Spine have been increasingly redeveloped for residential use as the state government encouraged urban renewal and infill development projects as part of Plan Melbourne's compact city policies. The last 10 years have seen a steady increase in new housing supply in and around the Brunswick Spine [10]. Housing development data for the Moreland City Council LGA show that there is an average annual increase in dwelling stock of 1,080 dwellings per annum between 2006–15, with the highest increase occurring in Brunswick [10]. For the same period, 47% of new dwellings were developed in this

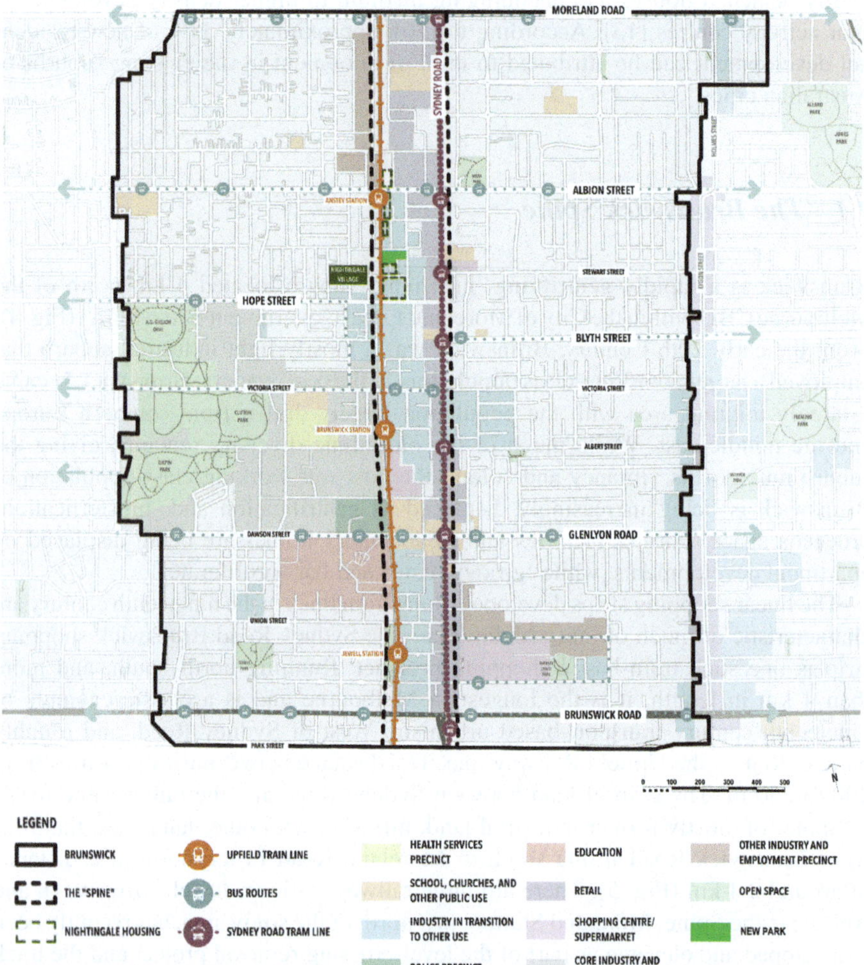

Fig. 5 The brunswick spine (Modified from Google Maps, 2022)

LGA as a result of urban infill development projects, with a single lot accommodating up to nine dwelling units [10]. Moreover, a large number of new dwellings are being developed near activity centres or transit lines in Brunswick and the surrounding suburbs. For instance, 50% of net new dwellings added between 2005—16 within the Moreland City Council LGA are located in and within 400 m of an activity centre [10]. This has become possible despite many residents' resistance to densification in the inner suburbs, including Brunswick [30]. Observations show that the dominant mode of new development has been medium-rise buildings of 5–10 storeys in height, with architectural and sustainability outcomes of mixed quality.

There is little green open space in the Spine however. While there is a group of parks about half a kilometre to the west of the rail lines, and expansive parkland a similar distance to the south, there are no wildlife corridors. Habitat for native animals and birds in the Spine is limited. Effective green urbanism needs to address all of these issues.

Mitigating the environmental impacts of buildings is a key focus of green TOD. In Australia, buildings are evaluated using green building rating systems through third-party verification [16]. While all new residential developments must meet minimum environmental standards, most do no more, and enforcement of even the minimum standards is often inadequate and patchy. Sustainability often appears to be an afterthought, the use of high embodied energy materials such as concrete and aluminium is widespread, and double glazing is uncommon. While some apartment blocks have solar PV panels on the roof, often the orientation of windows is determined more by site constraints and maximising yield than by basic passive solar design principles. Insulation and appliance efficiency are generally the minimum required by building codes and no more. Most new apartment blocks have balconies, but many do not provide any additional open space either at ground level or on the roof. The poor environmental performance of many buildings is thus a major barrier to sustainable urban outcomes, but developers have little incentive to improve sustainability. The predominant profit-based approach to development in Australia encourages developers to cut construction costs by providing only the minimum of environmentally sustainable design (ESD) techniques required by law, while the high energy bills that result for the lifetime of the development are problems left to the occupants.

There are, however, signs of change. Starting in 2014 a series of residential apartment blocks have been built in the Brunswick Spine which challenge the prevailing development model. The Commons, an architect-initiated development based on the German Baugruppen model, took advantage of a 'problem site' adjacent to the railway line, which was cheap to buy because of perceived noise issues, and where a lack of neighbours allowed a greater volume than might be allowed elsewhere in the Spine (Fig. 6). This type of development model is new to Australia and has yet to be widely taken up.

The Commons broke new ground in regards to sustainable architectural development in Australia. Arguing proximity to multiple forms of public transport and the bike path, the Commons was exempted from standard car parking requirements. The building achieves a 7.5-star rating for energy efficiency (2.5 stars higher than

Fig. 6 The commons, Florence street, Brunswick, Breathe architects, 2014 (Authors)

required at the time), uses low toxic materials (for example, chrome was forbidden), and boasts a rooftop communal vegetable garden, BBQ and laundry as well as a bank of solar PV panels. Car ownership is forbidden, and instead of costly underground car parks, there is ample on-site bicycle parking and two shared car spots in the street. Because of its 'triple bottom line' approach to development, a refusal to engage with the more parasitic aspects of property development (estate agents, publicists and marketing), and the elimination of underground car parking (A$40,000 per space according to Breathe founder Jeremy McLeod), the Commons was able to deliver highly sustainable, compact and community-oriented housing at market competitive prices.

A second development based on the Commons, the Nightingale, was built immediately across the road a few years later. The success of the Commons and the Nightingale was used to launch the Nightingale model of housing development. Thirteen projects have now been completed, more are in construction, and more are in preparation. Many of these have been built near the Commons and the original Nightingale in the Brunswick Spine at the new Nightingale Village. With each project designed by a different architect, many of the leading lights of the Melbourne architectural community are now involved in Nightingale projects. As with earlier Nightingale projects, the buildings in the new Village include high levels of environmentally sustainable design—all are certified carbon neutral and powered by Green Power and use passive solar design and cross ventilation (Fig. 7).

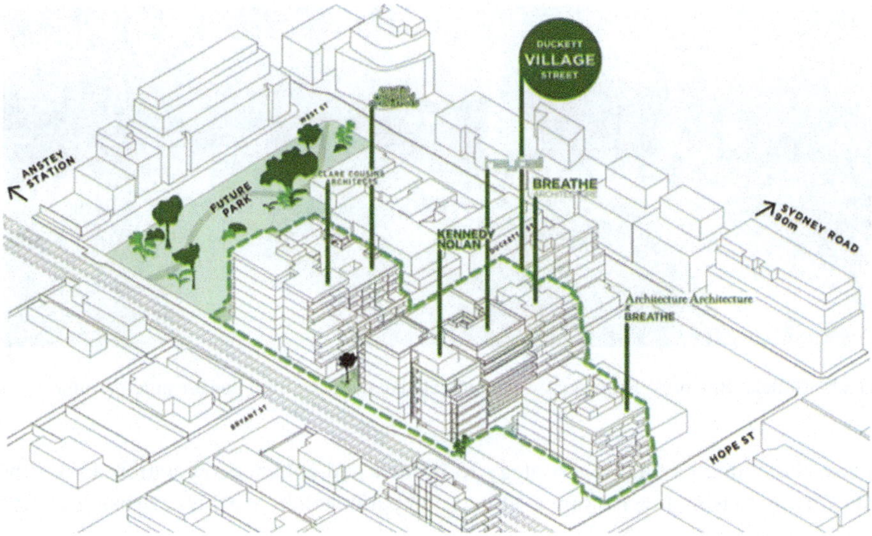

Fig. 7 The nightingale village in the brunswick spine (Nightingale Housing, 2021)

Although there is a general lack of green open space in the Spine, the development of new parks is proceeding through the initiatives of the local council. The construction of Bulleke-Bek Park in the Nightingale village demonstrates, in miniature, a vision for future sustainable development, combining the benefits of TOD with sustainable, high density development and green urbanism (Fig. 8). Developed by Moreland City Council, following several years of community consultation, Bulleke-Bek park provides much needed green open space in what, in Australian terms, is a high-density urban environment. Further, in adopting an indigenous planting scheme, Bulleke-Bek helps to provide habitat for native animals and plants. If this model were to be adopted more widely, both in the Spine and beyond, it would help improve the urban environment, provide high quality sustainable housing and establish a new benchmark for sustainable urbanism in Australian cities. The Moreland City Council is planning for a further five new parks, including three in or near the Spine [22].

The local council's response in providing more green open spaces comes with the realisation that the existing urban open spaces (both public and private) can be impacted by infill development and densification. Witheridge [29] argues that on the one hand, the infill development of inner suburbs has modified both the quantity and quality of private open spaces, whereas, on the other hand, there is an uneven distribution and a shortage of public open space per capita across most inner suburbs. There is, therefore, significant pressure on local governments to provide greater amenities and ecosystem services in public open spaces. This requires updating of the current open space policy mechanism to regulate infill development to allow for better provision of, and equal access to, green open space to cater for the needs of the growing population [29]. On the other hand, while urban areas are prone to the urban heat island effect as they become denser, together with an increase in built-up

Fig. 8 Bulleke-Bek park, Brunswick—the upfield bike path and train line are at left (Authors)

areas, the council's ongoing strategy to protect and enhance vegetation across the LGA by increasing tree canopy in streets and parks will help achieve urban greening to overcome such effects [23].

4.2 Towards a Green TOD?

The Brunswick Spine presents exciting possibilities as a linear TOD. It has excellent public transport links at a variety of scales. The Spine is highly walkable, the topography is gentle, with a slight uphill slope to the north, and there is no, or very little, need to use vehicular-oriented transport to access most services. Nowhere is far from the activity of Sydney Road. This compact and pedestrian-friendly urban district has seen a significant increase in infill development along with the construction of several new parks in the last decades, contributing to mixed land uses and higher residential density. In addition to the excellent urban design outcomes, the emergence of green buildings in the Spine as evident in the newly built housing developments such as the Nightingale Village indicates a new impetus in achieving green urbanism in an existing TOD friendly urban area. The architectural design response of these new developments, despite the existing site constraints, has been highly favourable towards achieving energy-efficient and sustainable building outcomes.

This leads to an important question: is the Brunswick Spine a genuine example of green TOD? We argue that the Spine shows promising signs of being a green TOD and a sustainable urban redevelopment precinct that may set an example to be emulated elsewhere in Australia. However, there are several barriers to this uptake in the wider context of Australian cities. As already stated, the prevailing residential development model is profit driven by private developers, with little new public or social housing being provided and little giving back to the community. Housing cooperatives are unknown. Unless it can be somehow leveraged to increase returns, private development tends to view sustainability as an add-on luxury and concepts such as community as irrelevant. Green TODs, such as that potentially developing in the Brunswick Spine, will only be successful if state and local governments act to ensure the wider public benefit by providing the green infrastructure, public services and amenities and transport needed, or by insisting that private developers do so.

Green urbanism is multi-dimensional. While energy efficiency in buildings and the transport system is important, it is not the only consideration. Green TOD is prone to 'greenwashing' from developers and the government. A whole-of-life cycle approach to the sustainability of buildings and urban infrastructure is important, otherwise the net ecological impacts of high embodied energy and end-of-use waste products can undo the positives. Appropriate urban design is critical; walkable neighbourhoods must be safe and welcoming and provide a sense of community, in addition to being compact.

Appropriate landscape design is also critical to restorative green urbanism and green TOD. Tree cover provides shade and a pleasant urban environment, while permeable ground surfaces reduce loads on the stormwater system and help replenish groundwater. Australia has a unique flora and fauna, but traditional street planting and suburban gardens have generally adopted 'exotic' plants from Europe and elsewhere. While often attractive and providing environmental services such as shading, water transpiration, imported plants and trees are ecologically useless to Australian native birds and animals, and as a consequence, many Australian cities are largely ecologically barren. Green TOD should therefore aim to provide habitat for native species at the same time as providing a pleasant and functional urban environment.

5 Conclusion

The Brunswick Spine incorporates many of the aspects we would hope to see in a TOD precinct. The Spine is well provided with a variety of transport options, operating at all scales, all within easy walking distance of where people live or work. The provision of parks and the planting of street trees, and especially in the establishment of native plants and trees to provide habitat for urban wildlife, is a further step towards green urbanism. With the Nightingale Village, and other low-carbon sustainable developments in the area, the Brunswick Spine demonstrates that green TOD is possible in Australia, given a bit of vision. While all of the key features of green TOD are present in the Brunswick Spine, all are capable of being taken much further. However, not all developments in the Spine are as sustainable as the Nightingale, and the Spine is still short of green open space and wildlife habitat and transport links can always be improved.

Despite some shortcomings, the Spine shows promising signs of green urbanism and sustainable development and sets an example to be emulated elsewhere in Australia. If it can happen in one part of Brunswick, why not elsewhere? What needs to happen to ensure that TOD and green urbanism are widely adopted? To achieve green TOD in Australia, state and local governments need to deliver policy measures and planning guidelines that ensure the wider environmental benefit by providing green infrastructure, public services and amenities and energy responsive building design outcomes. More importantly, the models by which urban development takes place in Australia need to change. Alternative development models need to be pursued and encouraged. Where development is purely profit-driven, developers need to recognise the value of good, sustainable design and of vibrant, walkable and sustainable neighbourhoods. It is to be hoped that the Brunswick Spine does succeed and becomes a sustainable and thriving community to provide a living example of the benefits of green TOD and sustainable design and urbanism.

References

1. Atkinson R (2018) London's extraordinary surplus of empty luxury apartments revealed. The Conversation. https://theconversation.com/londons-extraordinary-surplus-of-empty-luxury-apartments-revealed-97947
2. Bajracharya B, O'Hare D, Byrne J (2010) Greening transit-oriented development and subtropical design. The Bellwether Zone: Planning South-East Queensland
3. Beatley, T., & Newman, P. (2012). *Green urbanism down under: Learning from sustainable communities in Australia.* Island Press.
4. Calthorpe P (1993) The next American metropolis: Ecology, community, and the American dream. Princeton Architectural Press
5. Cervero R, Sullivan C (2011) Green TODs: marrying transit-oriented development and green urbanism. Int J Sust Dev World 18(3):210–218
6. Chitrakar RM, Hogg P, Eshow D (2022) Integrating green urbanism into the transit-oriented development in Australia. Paper presented at the *Ecocity* World Summit 2022. Rotterdam, The Netherlands and online
7. Clean Energy Council (2019) Clear energy council report 2019. Clean energy council
8. Climate Council (2017) Transport emissions: Driving down car pollution in cities. The climate council
9. Curtis C (2012) Transitioning to transit-oriented development: The case of Perth, Western Australia. Urban policy and research 30(3):275–292
10. Delwp (2016) Housing Development data. Department of environment land water and planning. https://www.planning.vic.gov.au/land-use-and-population-research/housing-development-data
11. Del WP (2017) Plan Melbourne 2017–2050: Metropolitan planning strategy. Department of environment land water and planning
12. Department of Infrastructure (2002) Melbourne 2030: Planning for sustainable growth. Department of infrastructure
13. Freestone R, Hamnett S (2000) The Australian Metropolis. Allen and Unwin
14. Hendrigan C, Newman P (2017) Dense, mixed-use, walkable urban precinct to support sustainable transport or vice versa? A model for consideration from Perth, Western Australia. Int J Sustain Transp 11(1):11–19
15. Hodyl L (2015) Report by Leanne Hodyl–2014 Churchill Fellow to investigate planning policies that deliver positive social outcomes in hyper-dense, high rise residential environments. Melbourne: The Winston Churchill Memorial Trust of Australia
16. Holz J, Sigler T (2015) Green urbanism in Australia: an evaluation of green building rating schemes. In: Paper presented at 7th State of Australian Cities Conference. Gold Coast, Australia
17. Huang W, Wey WM (2019) Green urbanism embedded in tod for urban built environment planning and design. Sustain 11(19):5293
18. Jamme HT, Rodriguez J, Bahl D, Banerjee T (2019) A twenty-five-year biography of the TOD concept: from design to policy, planning, and implementation. J Plan Edu Res 39(4):409–428

19. Lehmann S (2012) Green urbanism: formulating a series of holistic principles. In: Haas T (Ed.) Sustainable Urbanism and Beyond: Rethinking Cities for the Future. Rizzoli
20. Liu JH, Te Pai J, Lin JJ (2018) Planning strategy for green transit-oriented development using a multi-objective planning model. Int Rev Spat Plan Sustain Dev 6(1):35–52
21. Martek I, Hosseini MR (2019) Buildings produce 25% of Australia's emissions. What will it take to make them 'green'—and who'll pay? The Conversation. https://theconversation.com/buildings-produce-25-of-australias-emissions-what-will-it-take-to-make-them-green-and-wholl-pay-105652
22. Moreland City Council (2012). *Moreland Open Space Strategy 2012–2022.* Moreland City Council.
23. Moreland City Council (2017) Urban Forest Strategy 2017—2027. Moreland City Council
24. Newman P (2005) Transit oriented development: An Australian overview. Transit oriented development: making it happen, 2005. Perth, Western Australia, Australia
25. Newman P (2010) Green urbanism and its application to Singapore. Environ Urban ASIA 1(2):149–170
26. Newton P, Glackin S (2014) Understanding infill: Towards new policy and practice for urban regeneration in the established suburbs of Australia's cities. Urban policy and research 32(2):121–143
27. Niu S, Hu A, Shen Z, Huang Y, Mou Y (2021) Measuring the built environment of green transit-oriented development: A factor-cluster analysis of rail station areas in Singapore. Front Arch Res 10(3):652–668
28. Van der Heijden J (2017) Innovations in urban climate governance: Voluntary programs for low carbon buildings and cities. Cambridge University Press
29. Witheridge J (2015) Infill development and the distribution of open space in Melbourne. In: Paper presented at the 7th State of Australian Cities Conference. Gold Coast, Australia
30. Woodcock I, Dovey K, Wollan S, Robertson I (2011) Speculation and resistance: Constraints on compact city policy implementation in Melbourne. Urban Policy Res 29(4):343–362
31. World Bank (2021) CO_2 emissions. World Bank. https://data.worldbank.org/indicator/EN.ATM.CO2E.PC
32. Zaręba A, Krzemińska A, Widawski K (2016) Green urbanism for the greener future of metropolitan areas. In: IOP conference series: earth and environmental science, vol 5. IOP Publishing, p 052062

Nightlife and Urban Livability—The Case of Mashhad

Kasra Talebian and Muge Riza

Keywords Livability · Night-time economy · Urban night · Urban lighting · Persian culture · Mashhad

1 Introduction

With more than half of the world's population living in urban areas in 2011 from about 30% in 1950, it is projected that the urban areas will host more than 70% of the world's population by 2050. As the United Nations [1] notes, such increasing urban concentration in very large cities is a relatively new phenomenon that the world is experiencing which also highlights the growing need to investigate the quality of living in such cities. Meanwhile, improvements in lighting technologies are changing the night face of cities rapidly by transforming the nightscape from static to dynamic and from dark and monochrome to bright and colourful. The night phase of cities is dramatically changing into a market-oriented festive-like scene while the design and management of urban spaces are mainly focused on the daytime life of cities. Proper planning and management at decision-making levels can and should use this extra source of time and space in favour of increasing the livability of cities. This study aims to shed ~~a~~ light on the area of influence of urban lighting on the livability of cities.

K. Talebian (✉) · M. Riza
Eastern Mediterranean University, Famagusta, Cyprus
e-mail: kasra.talebian@emu.edu.tr

Z. Allam (ed.), *Sustainable Urban Transitions*, Urban Sustainability,
https://doi.org/10.1007/978-981-99-2695-4_16

2 Urban Livability

Urban livability discourse suggests that "there is an ideal relationship between the urban environment and the social life it sustains" [2]. One of the early definitions of livability can be found in Newman's [3] research on sustainability and cities. He has considered the city as a biological system that can be observed through sources input, metabolism and the outputs in an 'Extended Metabolism Model of the City'. While livability is the positive outcome of all resource inputs, there is a negative output alongside, which is the cost of providing livability, the waste output. So we should know that livability is basically a very valuable and costly product (Fig. 1).

Balsas [4] points out different meanings livability may have to different people and the fact that it is a concept that people recognise easily, but it is hard to define in a way everybody understands. Hankin and Powers [2, p. 848] suggest that "there is a complicated social dimension to livable urban space, a social dimension that geographers need to problematize." They believe, "If we examine city spaces that are lauded as highly livable, we can get a sense of who is able to live and how." Answering who are the users of spaces and what they do, plus what they do not do, are also mandatory in reading the livability of urban spaces.

Pacione [5: 1] emphasizes the role of context in addressing livability, "the precise meaning depends on the place, time, and purpose of the assessment and on the value system of the assessor." Hankin and Powers [2] and Ley [6] perceive livability as an essentially ideological mindset in which the definition of it for poor and middle-class is quite different, also for those who live in the village compared to urban dwellers. Ley also points out the competition over defining livability which discloses the eagerness toward defining quality of life for specific (contrasting) benefits of determinants.

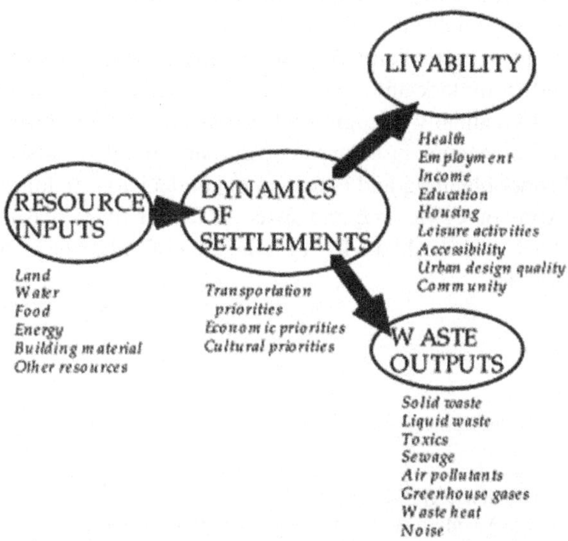

Fig. 1 Livability by Newman's definition has inevitable waste output [3]

So the question of evaluating livability according to which measures is a critical discourse. A city can be livable (obtains high grade from specific indicators), or be livable for a defined range of users (city obtains high grade considering target group's needs). This brings up the necessity of addressing socio-cultural past experience of the target group, which needs extensive research and would be expected to clarify which city/community is highly likely livable for those who are coming from 'X' or 'Y' cities or regions. While some of X's needs are common for all human beings and can be measured independently like safety, income and physical health, some are dependent on the context and personal experience of users like, feeling of safety, cultural adaptation and psychological health. Therefore, it is essential to consider the socio-cultural backgrounds in livability discussions. This hidden layer contains meaning-giving aspects of life which comes right after the primary survival needs, forms our identity and sometimes brings up contradictory values in different cultures. The highest degrees of contrasting cultures appear in extreme cases like displacement of war refugees or forced gentrifications, which makes it hard for a group to adapt in a short period of time and consequently keeps them under constant social and psychological pressure.

Considering the scale of investigating urban livability, two main streams can be identified in the literature. First in the scale of a neighborhood or a district and mostly focused on societal dimensions and walkability. One of the key figures in the first group focused on social aspects of livability is Jacobs [7]. Her point of view on learning livability from socially sustainable everyday life of communities, neighborhoods and districts extends toward extracting frameworks and guidelines (based on new moral, social and cultural norms or recognizing existing ignored norms) in organically formed multi-use districts. She broke the hierarchy of top-down analysis, acknowledged investigator's recognition of vibrant and livable spaces and brought up the necessity of taking a deeper look at local happenings, gatherings and public lives and brought up the necessity to respect diversity and co-existence in urban planning. Specifically her focus on the effect of natural surveillances on safety and perception of safety and 'defensible space theory' with Newman [8] is of great importance when it comes to investigating the night phase of cities.

The first group has a more ideological approach, considering the subject as a framework applicable to any context to improve living experience indicators also referred to as 'quality of life'. While indicators may vary, the core theory is looking at livability as a holistic concept which involves socio-cultural issues, morality and all levels (mostly upper levels) of human needs in Maslow's (1987) hierarchical pyramid. This approach is fractal in nature as Salingaros [9] suggests urban spaces should be, as the core concept targets living in a community at any scale, making it applicable to different scales such as a neighborhood, an urban district or the whole city. This approach addresses both individuals and societies as parcels that are making a city rather than looking at city from the bigger scale in a competition within a broader context. In this approach, the economy is equally important and other factors such as safety, income, and health. Fischer [10] is also a key figure in the first group. She emphasizes the government's role in building 'livable communities' and points out the role of safe streets, good schools, and public and private spaces in breeding a spirit of community and opportunities for urban dwellers to participate in the community. Dumbaugh and Gattis [11] consider walkability, scale, and safety of urban spaces and Wheeler [12] adds healthy environment, decent housing, uncongested roads, parks and recreational opportunities and vibrant social interaction as essential requirements for making urban living an enjoyable experience.

The second group of studies mostly consider cities as products which are subject of branding and economic growth and absorb of resources (financial, social and cultural) from any context above city borders, in national scale or globally. These studies are focused on performances of cities and comparison is the key element in which mostly involves ranking of cities and a framework to evaluate and label cities as more or less livable by combination of indicators. The result of these comparisons primarily targets employers who need a basis to pay their globally mobile human resources (e.g. Mercer's ranking of livable cities) or authorities and governors who are involved with branding their cities in a larger context mostly for financial reasons.

Urban centers are a vital part of investigating livability in both scales; district and city. Characterized mainly by diversity of uses, and a mix of civic, administrative and professional services, very often, these centers experience boom and bust cycles, which negatively affect the livability of adjacent and older urban areas. These cycles are directly related to the economy as well as to the region's social, cultural, historical and political trends. Some examples include the wave of peripheral shopping centers, added personal mobility, suburbanization, discount centers, malls, strip commercial areas, catalogue sales and online shopping. Balsas [4, p. 102] points out how "peripheral centers attract the most mobile and affluent customers, which leaves the old centers dependent on the local, poorer market" and would start a slow downward decline process. Local authorities have made efforts, especially in American cities where decentralization of urban centers has become a critical issue, to bring people back to urban centers in a recentralization process. In most cases main potential has been found to be reactivating evening and night-time economies which has also been the focus of authorities in UK cities [13, 14].

Balsas [4] brings up the question, What makes a city center a livable place? His answer includes five dimensions of Lynch [15, p. 1] to the question, What makes a good city? Vitality, sense, fit, access and control plus Balsas' dimension: viability. In other words, he claims that "if a city center is a vital place, with sense of place and time, where the urban environment fits the human body and its activities as well, it is accessible and can be controlled but does not have the ability to attract continuing investment, it is not necessarily a livable place." From Balsas point of view retailing is a basic, integral component of urban life, perhaps the most critical one.

Zec et al. [16, p. 125] define quality of life as "people's opportunity to access infrastructures, communication and transportation modes, food, clean air, housing, networks of paths, open spaces, facilities, greenery and parks." They refer to walkability as a principle that is getting more and more essential for a livable environment as "livable cities define an urban formation that inspires walkability by linking street patterns and facilities for living, working and recreation within an optimum closeness to encourage security, sustainability and pedestrian-friendly environment." Smith [17, p. 69] discusses dimensions of employment including "hours worked (full-time, part-time, etc.), wage rate, health insurance and retirement benefits, proximity to affordable transit and childcare options, and work safety provisions." Although mentioned items mostly fall under employment subject, especially in the United States, some of them like insurance can be provided at the national level as social welfare services in other contexts. Also, proximity to affordable transit and childcare are two important items but not necessarily related to employment as they can be affected even more in some cultures by social and financial support of families.

Walkability and accessibility are inseparable pillars of livability that relate with people from all ages in the scale of street and district. Walking provides social, environmental and economic benefits [18], because it is a green and simplest mode of transportation [19] and it is healthy and enjoyable [16]. The subject of walkability is highly interlinked with safety, another pillar of livability mentioned by most researchers.

In sum, livability is a complex multifaceted concept. The definition of National Research Council [20] seems to be inclusive and precise,

> [Livability] refers to the extent to which the attributes of a particular place can, as they interact with one another and with activities in other places, satisfy residents by meeting their economic, social, and cultural needs, promoting their health and well-being, and protecting natural resources and ecosystem functions... and helps capture some of the externalities ignored or inaccurately valued by market mechanisms (p. 24).

The missing point in the above definition is the context. The fact that livability is a highly relative term as what would be considered a livable community in one part of the world might be deemed highly unsatisfactory in another. This might be due to cultural differences or to different standards of living that alter expectations for urban design, transportation, other infrastructure, and service provision.

Efforts have been made to collect, in a compact form, key factors and indicators related to livability as defined by prominent scholars in table below (Table 1). These indicators are categorized and grouped into four main dimensions; Health, Socio-culture, Economy and Infrastructure.

Table 1 Livability dimensions, sub-dimensions and indicators in literature [21]

Livability dimensions	Sub-dimension	Indicators
Health 1, 2, 5, 8, 9, 10, 13	Physiological wellbeing 13	Life expectancy *1, 2, 3, 4, 10—infant survival rate *4, 5, 6—hospital beds per person 1, 2, 10—low carbon living environment 2, 10, 12—availability and quality of of private healthcare 9—availability and quality of public healthcare 9 sport facilities, 9, 10 time/week for sleep 10, obesity rate 10—air and water quality 10, 12 food availability 10—health insurance 14—clean environment 15
	Safety 6, 11, 15	Safe public spaces 5, 6, 11—crime rate 9, 10—car accidents—military conflict 9—terror 9—transportation fatalities 10, number of police officers 10—number of firefighters 10, response time for emergency calls 10 work safety 14
Socio-culture 3, 4, 5, 6, 7, 8, 10, 11, 13	Recreational and attention restorative opportunities 1	Parks and natural environments 8, 10—recreational opportunities 5, 8—cultural activities (including festivals and holidays) 15—time/week for leisure 10, beautiful places 15 scales fitting human body 15
	Social and mental wellbeing 2, 13	Education 1, 6, 8, 9, 10—sense of community 1, 5, 6, 11, 15—perception of safety (person, neighborhood) 10, 11—sense of belonging 10, 11—sense of place and time 15—social vibrancy (+vitality) 10, 11, 15—having social support members (relatives, friends) 10—regularity of face to face conversation 10—child care services 10, 14—elderly services 10—suicide rate 10
	Diversity 11	Activities 4, 11—ethnographic (age, gender) 10—multi-cultural environment 4, 9, 11—food and drinks 9, 12—consumer goods and services 8, 9, 10
	Political environment 8, 10	Passport power, political parties
Economical 8, 10, 13, 15	Income 10, 14	Income over expenses rate (affordability) 1, 10, 14, 15—employment rate 1, 10, 14—diverse employment opportunities 15—grow of number of businesses 10, 15—income of households, 10—availability of online work/work from home 10—retirement benefits 14—ability to absorb investment 15
	Housing 1, 5, 8, 10, 12	Living area per person 1, 2, 3, 10—housing quality 9, 10—housing affordability 10

(continued)

Table 1 (continued)

Livability dimensions	Sub-dimension	Indicators
	Council or authorities spending 10	Spendings on services sector, cultural facilities, planning services, education services, highway and public transport services, social care services, 10
Infrastructure 4, 10, 11, 12	Infrastructures 9, 10, 12, 15	Walkability 1, 5, 10, 11, 12, 15—accessibility to public or affordable transport 1, 5, 10, 12, 14, 15—accessibility to green and blue spaces 10, 12, 15 open spaces 12—time spent on commuting 10—road quality and network 1, 5, 8, 9, 10, 12—availability of resources and quality of energy, electricity and water 9, 10—traffic
	IT and telecommunication provision 9, 10	Mobile access 10—Internet access 10—public/ social media censorship
Numbered references	1. Newman, 2. Leach, 3. Oktay, 4. Jane Jacobs, 5. Wheeler, 6. Fischer, 7. Hankin and Powers, 8. Mercer, 9. Economist Intelligence Unit, 10. Leach (CAM method), 11. Jane Jacob, 12. Zec, Erem and Colakoglu, 13. National Research Council, 14. Smith, 15. Balsas	

3 Urban Night and the 24/7 City

Within a century after the invention of electric street lighting, the magical and mystical effects of having lit streets at night decayed. As the third and fourth generations have been born to see illuminated streets at night, like other similar infrastructures such as highways and bridges, street lighting is now considered as an essential element of urban life without its glorious effect. Urban lighting, however, is now becoming more than just streetlights. Colourful and dynamic lights in greeneries and on urban facades together with a wide range of night-time activities seem to attract the curiosity of new generations.

The term, 24-h city, has been used to label cities known to be active late into the night hours [22]. Early signs of 24-h city urban spaces were observed in many European cities in the Baroque period. Alewyn [23] was the first to point out a shift in the history of the night: Between the fifteenth and the eighteenth centuries, princely celebrations show a slow shift from the street to the court, and from day to night. This was the sharpest break in the history of celebrations in the West, "….marking a new era in the history of the night" [23, pp. 37–39]. Port cities like Amsterdam have traditionally been active on a 24/7 basis due to the arrival of ships at various times. However, new means of transportation and technology provided the chance for a city like Las Vegas to become a prominent 24/7 city in the middle of a desert. Other cities like Lyon-France and Eindhoven-Netherlands have benefited from this term (24/7 city) as part of their city branding attempts while they are not actually twenty-four hours active during the whole year. It seems that in larger cities, the urban nightlife is occurring with or without planned decisions specially wherever population or wealth increases, the number of night-time activities and length of their extension after dark grows relatively unless there is a rule or force to control or limit it. These regulations and restrictions are mainly implied by authorities due to lack of budget for providing safety or lack of knowledge from the hidden potentials of the night phase because simply it was not achievable before. However, these limitations do not eliminate the nightlife concept but push it to take place as an underground activity and transfer nightlife from urban spaces toward interior spaces (Fig. 2).

The formation of outdoor 24/7 zones within cities also occurs in response to having free time at night. Especially the young generations expect 'something' more exciting than having streetlights which were once considered luxurious and still are the primary function of urban lighting. In some cases, mainly in cases of smaller cities, activities may extend only a few hours after dusk, or maximum, until midnight. Sharpe ([25], p. 14) calls the night phase a "second city—with its own geography and its own set of citizens" and McQuire ([26], p. 114) states, "there are relatively few accounts to theorize the impact of electric lighting on the experience of urban space."

One of the well-documented and investigated cases in the literature on the subject of night is the fall and rise of 24/7 districts in Britain in the works of Bianchini [27], Heath [22] and Roberts and Eldrige (2009), where limited working hours of 9 a.m. to 5 p.m. applied to central business districts of British cities and later on the 24-h city concept was implied to bring back safety and diversity with the help of night-time economies to these areas. The problem in British cities started by allocating business-only central districts in cities that became empty, undesired, and unsafe after 5–6 p.m. Licensing initiatives in the 24-h city concept included restaurants, bars, and clubs, mostly alcohol-related functions. These initiatives improved safety in these areas (Heath 1997), however, led to colonization of night by youngsters

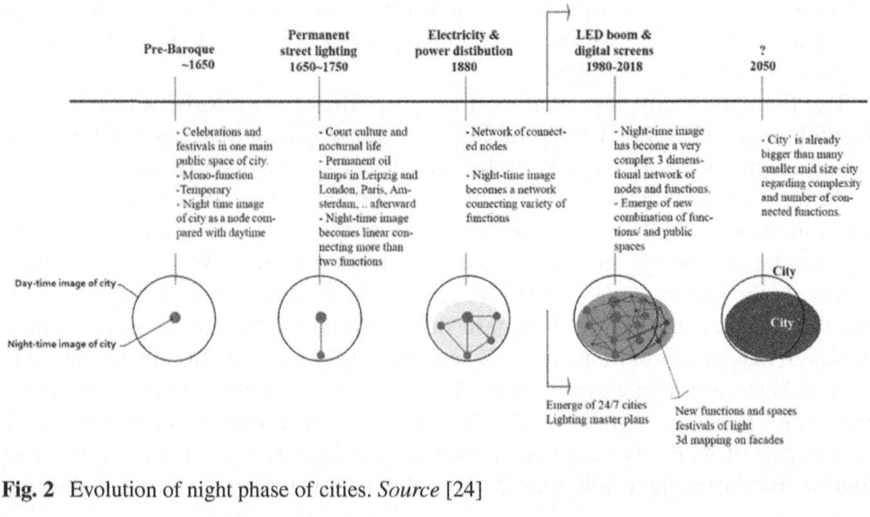

Fig. 2 Evolution of night phase of cities. *Source* [24]

and to some extent male users, leaving no space for elders and families or those who are not in favour of drinking alcohol such as Muslims [28]. Residential blocks within 24-h areas bring conflict of benefits in almost every context, especially when alcohol-related functions are involved. While residents close to active nodes at night have complains about the unrest made by noises at night and urinations in public spaces, existence of residents in such areas provides natural surveillances, and boosts the night-time economy [29] (Heath 1997).

Campo and Ryan [30] investigated 29 American mid-size cities searching for a new urban nightlife which is "neither the sophisticated entertainment of theater, symphony or ballet, nor high-end entertainment serving the needs of corporate clients." They define 'Entertainment Zones' as 'concentrated nightlife districts' that are mostly active in former industrial and commercial buildings that "have outlives their original uses and have survived largescale developments and planning programs." These buildings are located on edges of downtowns around former industrial or commercial districts, which "makes them easily accessible to downtown workers and nearby residents while placing them in a relatively familiar territory for suburbanites" (pp. 292–293).

While for a person living in Berlin, London, or in other European capitals, nightlife is associated with bars, clubs, and alcohol-related activities. On the contrary, for a person living in the Middle Eastern cities such as Mashhad, Mecca or Baghdad, in which Islamic rules apply, nightlife is more associated with public spaces around religious buildings, and night-long traditions and rituals as well as indoor private gatherings and parties [24].

Safety is the most vital function and primary motive behind the development of urban lighting. Darkness is associated with insecure feelings since it lowers visibility and consequently the ability to recognize faces and visibility [31]. A street could be seen as a source of threat by the absence of other people indicating that no one would reach out for help in case of a violent event and the fact that a person walking alone is an attractive target for offenders who have company. Perkins et al. [32] points out three issues associated with fear of crime at night; social and physical disorder, territorial functioning, and defensible space.

The fact that urban lighting affects the livability of man-made environments is not a new topic. It can easily be imagined that without streetlights, the essential spirit of the city would be turned back to the spirit of villages at night. However, the current role of urban lighting is way more than illuminating streets, and the night phase has shown to be more than a complementary part of day activities of cities. While nightlife may exist at some levels in all kinds of cities, for people who have lived in cities that are highly active at night, nightlife and night activities are being seen as an essential component of their routine life not necessarily a luxurious time–space. They depend and rely on night-time activities (Table 2).

4 Mashhad's Urban Night

Mashhad, the second largest city in Iran, is a 24/7 religious Metropol known as the 'city of light', in which ideology has played a crucial role in its formation. Previous studies have shown the potential of urban lighting to highly impact cognitive meanings, cultural identity and public confidence, which can play a crucial role in creating a sense of belonging for the dwellers [31]. In the case of Mashhad, the 24/7 district around the holly shrine of Imam Reza is a physical and visual expression of 'keeping a light on' in Persian culture, which creates a strong cultural identity for citizens [24]. The area around the shrine is highly active at night by male and female users who are shopping, participating in rituals, shopping or just visiting and praying in public spaces of shrine complex together with their children after midnight (Figs. 3 and 4). The shrine is not the only illuminated element, but perhaps the most influential one. The city's official buildings, banks, residential apartments, transportation infrastructures and urban sculptures are illuminated. From an economic point of view, nightlife plays a crucial role for absorbing tourists since they are less involved with working hours and day/night boundaries and are more likely to be users of night services while putting less load on the routine day function of the city (Fig. 5).

The livability concept partially depends on users' backgrounds and living experiences. However, there are common grounds that can enhance livability in any context. A pedestrian-friendly environment is one of the key factors mentioned by many scholars which can be enhanced and improvised with the help of urban lighting. A road that provides vehicle access during the day can play the role of a pedestrian path and might turn into an urban open space by the help of a network of night-time activities (Fig. 6). In other words, safe and vibrant nightlife acts like a layer that is applied into existing infrastructures and increases livability factors by getting the most out of twenty-four hours of day/night limit. The temporal dimension of nightlife brings

Table 2 Area of influence of urban lighting on livability [21]

Livability dimensions	Sub-dimensions	Indicators	Descriptions/instances
Health	Physiological wellbeing	• Availability of medical facilities at night	• 24/7 hospital/health services
		• Availability of sport facilities at night	• Basic outdoor gym facilities in public spaces • Illuminated and functional jugging and bicycle paths • 24-h indoor sport facilities such as gyms, swimming pools, sport salons

(continued)

Table 2 (continued)

Livability dimensions	Sub-dimensions	Indicators	Descriptions/instances
		• Loss of darkness (sensorial) • Excessive light intake surpasses melatonin production in brain, weakening body's immunity system • Darkness limits vision and provokes other senses, excessive illumination without preserving urban dark, semi-dark areas surpasses other senses rather than vision	• Dark and semi-dark areas in urban nightscape • Light trespass trough residential windows caused by streetlights and façade lightings
	Safety	• Crime rate • Safe public spaces	• Illuminated streets and public spaces reduces crime rate by decreasing hiding spots for offenders while increases possibility of facial recognition
Socio-culture	Recreational and attention restorative opportunities	• Being in close proximity to natural environment • Possibility of outdoor activity in less polluted air condition • Leisure activities • Restorative environment and recreational activities; possibility of distancing person's attention from daily routines (being away) or get their effortless attention (fascination), and supports people's intended activities (compatibility)	• Illumination of urban greeneries • Changing the focus of light to greeneries instead of hard surfaces wherever applicable (parking lots) • With the help of lighting existing urban spaces can host leisure activities at night • Illumination of publicly available natural resources such as seas, lakes, gardens, and/or adjacent public areas • Illuminated restorative natural and built environment • Illuminated public/private landmarks act as ornaments of nightscape, hence as a soft entertainment tool for both pedestrian and vehicle users

(continued)

Table 2 (continued)

Livability dimensions	Sub-dimensions	Indicators	Descriptions/instances
	Social and mental wellbeing	• Active indoor and/or outdoor nightlife • Psychological relief of pressures • Reduction in fear, stress and sadness • Safety, security, fear of crime • Number of crimes and incidents at night fall during festivals and public events • Sense of place is increased by festivals and urban celebrations • Incrementing safety of public spaces by increase in number of natural collective surveillances/ observers • Illuminated pathways increases perception of safety • Visibility of paths and accesses reduces stress • Cultural identity and sense of belonging • Urban lighting creates a strong city image • Sense of belonging to a larger group • Public and private space that help foster a spirit of community	• Illumination of public landmarks instead of private buildings • Lit streets and pedestrian accesses • Surfing the city at night as a leisure activity • Having a place to go at night • Perception of safety increases in illuminated urban spaces • Increased number of users reduces fear of crime due to natural surveillances/observers • Number of public festivals/events and active nights per year • Availability of educational opportunities • Vibrant social interaction at night and being in a crowd of people (group ecstasy effect) • Vibrant nightlife can boost cultural identity • Common public experience that forms a sense of belonging to a group and/or boosts cultural identity through collective (or cumulative) public events such as concerts, festivals, religious events/rituals • Visual or cognitive channels such as national/religious landmarks

(continued)

Table 2 (continued)

Livability dimensions	Sub-dimensions	Indicators	Descriptions/instances
	Diversity	• Multi-cultural environment • Ethnographic diversity (age, gender, culture) • Diversity of activities and functions	• Diversity of illuminated landmarks to relate to various backgrounds (educational, national, historical, financial, cultural, religious and monumental landmarks) • Possibility of being in contact with people from different age, gender and cultural background • Possibility of attending or having the option to choose between variety of activities
Economy	Income	• Absorbing tourists and foreign investments • Income opportunities	• Direct investment from other cities/countries in forms of buying lands, starting businesses and any other clues that reveals investments have been made by non-locals especially in 24 h economies • Illumination of landmarks boosts positive city image and absorption of tourists and foreign investments • Since tourists are free of working hour routines, any form of 24/7 active areas are popular touristic attractions
	Employment	• Flexible working hours • Providing job opportunities for mid and low-income group	• Authorities' approach toward vendors and night markets (relaxation of regulation regarding use of public spaces for vendors and night markets) • Extended open hours regulations • Availability of night-markets

(continued)

Table 2 (continued)

Livability dimensions	Sub-dimensions	Indicators	Descriptions/instances
Infrastructure and legibility	Accessibility, legibility, protection of natural resource	• Accessibility of 24/7 activities • Nightmap: diverse and unique lighting on building or urban structure helps orientation and recognition by differentiating districts (specific color, visually unique approach toward buildings, specific lighting fixtures)	• Availability and accessibility of active nodes at night by foot and vehicle • Visually rich cityscape as a potential for cruising the night with vehicles as a recreational activity • Illuminated landmarks makes the city more readable • Cityscape as a sight worthy of respect, even admiration
		• Reduction in transportation/commuting by accessibility of facilities within walkable distances • Flexible use of empty lands and parking lots without extending urban footprint	• Active nodes at night in the scale of districts increases accessibility by walk and reduces in-city travels • Flexible use of empty lands as night markets or food streets can be achieved by providing light in potential lots and urban public spaces boosts land value
	Light pollution and energy consumption	• Lowering consumption of energy • Sky glow and disruption of circadian cycle	• Use of LEDs for street-lightings which holds majority of electricity consumption of city at night • Downward cutoff streetlights • Effective regulations on façade and urban structure lightings (downward illumination sources, controlled or no upward lights) • Light trespass on residential windows

Fig. 3 Holy Shrine of Imam Reza in Mashhad—Iran provides a solid cognitive meaning for the whole city—photo by https://unsplash.com/collections/1828476/imam-reza

Fig. 4 Presence of female users and children in urban spaces around the shrine at 2:00 a.m. The illuminated building, lighting sculpture and strings of light bulbs above the street are common features of nightlife in Mashhad. *Source* Author

Fig. 5 Remaining ruins of old houses are bought by the municipality and, with the help of lighting and proposals of local artists, these lost spaces are revived as 'Patogh', traditional term for neighbourhood's gathering spaces. *Source* Author

Fig. 6 Informal temporary use of urban open space in front of a mosque extends to sidewalks and boulevard medians during Ramazan month—Mashhad, Iran. *Source* [24]

the flexibility to transform and enhance spaces that are not considered as contributors to livability by their daytime characteristics, such as parking lots, vehicle paths and leftover spaces around urban structures. These lost spaces can turn into safe, walkable, and economically active areas without expanding cities' footprints.

Brätt et al. [33] investigated the placemaking efforts cities have undertaken using lighting, pointing out how "unsightly or ill-functioning lights can be a detriment to placemaking". The example below (Fig. 7) is the lighting of one of the main transportation bridges in Mashhad city becoming a chaotic, randomly coloured lighting scene due to lack of proper maintenance. This affects the perception of people from proper lighting and normalizes the lack of maintenance in private projects. A commercial building that followed the same scenario was observed in the case of Mashhad, a hint that big projects can establish a lighting culture that might be followed by the private sector and smaller businesses.

Norouzian-Maleki et al. [34] concluded that among factors affecting urban livability, urban lighting has been perceived the most important factor by Iranian interviewees even before the presence of people in public spaces, while Estonian interviewees ranked it the least important factor. This comparison again reminds the fact that livability is a relative concept that is highly dependent on the context. Also, a confirmation that lighting bears a special meaning in Persian culture [24].

Fig. 7 Randomly colored lighting as a result of lack of maintenance in Sayyad bridge (top) and Kian Shopping Center and Hadis restaurant (bottom), Mashhad, Iran. *Source* [24]

5 Conclusion

Urban public spaces are products of cities. Considering the city as a living phenomenon creates misconceptions related to the potentials and requirements of urban open spaces. Like a machine that does not necessarily need to rest the way organic creatures do, if no harm occurs to the environment, urban public spaces do not need to have resting hours. Urban lighting and various functions can boost a sense of belonging and ignite higher usage of urban spaces during night hours. This may create a foundation for forming a 24-h area depending on the local culture and its happenings, events, and memories that have shaped the place's cultural identity. Organically formed 24-h urban spaces have already been oriented to their surroundings, both functionally and financially. However, light alone cannot fulfill great expectations of newly formed urban areas. Urban spaces need to also adapt to the extensive use of portable, self-illuminated digital devices and the internet of things (IoT) to enhance and create new forms of 24-h areas where distinctions between physical and cyber space and day and night are blurred.

There seems to be similarities in some aspects of nightlife regardless of geographical and cultural differences. For instance, different forms of gatherings are common. A possible reason is that humankind still depends on naturally developed sense of safety being in a crowd provides. Another common activity is the early morning eateries and breakfast mini-restaurants that seems to be formed for those who get to work very early and don't eat breakfast at home. A prominent example of permanent customers of early breakfast eateries include the working class in general and construction and municipality workers in street cleaning and garbage collecting sections specifically.

Beside natural resources such as parks, canals, and seafronts act as a ground for further functions and gatherings to urge, in man-made environments, public places such as cinemas, theaters, libraries, universities campuses, cafes and many similar functions have the potential to be active after midnight by considerably low electricity costs and public desire shows hints of demanding such spaces (Fig. 8). However, in reality, number of such functions active at night is very limited in all contexts.

There have been growing concerns regarding the consequences of over-illuminated urban spaces and loss of darkness. Edensor [36] acknowledges the positive effects of darkness as "the potential for conviviality and intimacy to be fostered in the dark, the aesthetics and atmospherics of darkness and shadow, the possibilities for apprehending the world through other senses and the dismissal of the star-saturated sky".

One of the characteristics of 24/7 areas is their temporal nature as they are almost invisible during the day and do not have a physically dominant presence and unplanned and self-organizing nature. The vital role of governmental bodies

Fig. 8 An example of public desire for diverse night-time activities shared on Twitter. *Source*: [35]

and authorities in conducting the wave of nightlife in urban spaces needs to be addressed. The formation of nightlife may occur without the interpretation of authorities. However, when the potential is revealed, it can be accelerated, organized, and managed to bring up the most for the benefit of the city and its dwellers by public authorities.

Alcohol consumption and related anti-social behaviour have become a critical concern in the Western context's night-time management of public spaces. Robert and Turner [37] called attention to the depression of desire of any social group, other than youth, to go into city centers at night. Eldridge [13] pointed out the contrasts between a desirable, safe, and comfortable city center at night and alcohol-related problems such as violence, fear of crime, and public urination. The study by Jayne et al. [28] pointed out the exclusion of Muslim youth from the night-time economies of some city centers due to their culture's abstinence from alcohol. Finally, Van Liempt et al. [38] concluded that "Discourses of disorder, anti-social behavior and the 'alcoholization' of urban nightlife constitute a danger for any city that wishes to appear as an innovative, exciting, creative and safe place in which to live, visit, play and consume." An alcohol-free zone within the public urban spaces of European countries may help attract a range of users that feel excluded from public spaces during the night due to their beliefs, personal experiences, gender, age, or fear of crime.

Achieving a comprehensive understanding of 24-h areas needs precise observations of existing examples around the world. The story of Mashhad may be the story of many similar cities that have collected and developed solutions for using urban spaces during the night. Mashhad is among the lucky cases that have received enough attention to be partially documented in Persian literature. It is one among many interesting stories of local cultures, some of which are not being told beyond their borders.

References

1. United Nations Department of Economic and Social Affairs/Population Division, World urbanization prospects: the 2011 revision
2. Hankins KB, Powers EM (2009) The disappearance of the state from "livable" urban spaces. Antipode 41(5):845–866
3. Newman P (1999) Sustainability and cities: extending the metabolism model. Landsc Urban Plan 44(4):219–226. https://doi.org/10.1016/s0169-2046(99)00009-2
4. Balsas C (2004) Measuring the livability of an urban centre: an exploratory study of key performance indicators. Plan Pract Res 19(1):101–110. https://doi.org/10.1080/0269745042000246603
5. Pacione M (1990) Urban liveability: a review. Urban Geogr 11(1):1–30
6. Ley D (1990) Urban Livability in context. Urban Geogr 11(1):31–35. https://doi.org/10.2747/0272-3638.11.1.31
7. Jacobs J (1961) The death and life of great American cities. Random House, New York
8. Newman O (1976) Design guidelines for creating defensible
9. Salingaros NA (2003) Connecting the fractal city. In: Keynote speech, 5th Biennial of town planners in Europe, Barcelona

10. Fischer EE (2000) Building livable communities for the 21st century. Public Roads 63(6):30–34
11. Dumbaugh E, Gattis J (2005) Safe streets, livable streets. J Am Plann Assoc 71(3):283–300. https://doi.org/10.1080/01944360508976699
12. Wheeler S (2001) Livable communities: creating safe and livable neighborhoods, towns, and regions in California
13. Eldridge A (2010) Public panics: problematic bodies in social space. Emot Space Soc 3(1):40–44
14. Roberts M (2006) From 'creative city' to 'no-go areas'–The expansion of the night-time economy in British town and city centres. Cities 23(5):331–338
15. Lynch K (2013) "Dimensions of performance": from Good City Form (1981). In: The urban design reader. Routledge, pp 229–234
16. Zec G, Erem O, Colakoglu B (2018) Livable city: a student design project on Ayvalik, Turkey. Türkiye Bilim Akad Kültür Envant Derg 124–145. https://doi.org/10.22520/tubaked.2018.18.008
17. Smith DM (1973) The geography of social well-being in the United States: an introduction to territorial social indicators. McGraw-Hill, New York
18. Forsyth A, Southworth M (2008) Cities afoot—Pedestrians, walkability and urban design
19. Oktay D (2012) Editorial: quality of life in cities. Proc Inst Civ Eng - Urban Des Plan 165(3):125–126. https://doi.org/10.1680/udap.2012.165.3.125
20. National Research Council (2002) Community and quality of life: data needs for informed decision making. National Academies Press
21. Talebian K (2021) Developing a method for analyzing and documenting impact of lighting on livability of cities, PhD thesis published by Eastern Mediterranean University, Famagusta-Northern Cyprus
22. Heath T (1997) The twenty-four hour city concept—a review of initiatives in British cities. J Urban Des 2(2):193–204
23. Alewyn R (1989) In: Alewyn R, Sälzle K (eds) Das grosse Welttheater. Die Epoche der höfischen Feste in Dokument und Deutung, 2nd edn. C.H.Beck, Hamburg, pp 37–39 [with plates]
24. Talebian K, Riza M (2020) Mashhad, city of light. Cities 101:102674
25. Sharpe WC (2008) New York Nocturne: the city after dark in literature, painting and photography, 1850–1950. Princeton, NJ: Princeton University Press
26. McQuire S (2008) The Media City: Media, Architecture and Urban Spaces. London: Sage
27. Bianchini F (1995) Night cultures, night economies. Planning Pract Res 10(2):121–126
28. Jayne M, Valentine G, Holloway SL (2010) Emotional, embodied and affective geographies of alcohol, drinking and drunkenness. Trans Inst Br Geogr 35(4):540–554
29. Bianchini F (1995) Night cultures, night economies. Plan Pract Res 10(2):121–126
30. Campo D, Ryan B (2008) The entertainment zone: unplanned nightlife and the revitalization of the American downtown. J Urban Des 13(3):291–315. https://doi.org/10.1080/13574800802319543
31. Painter K (1996) The influence of street lighting improvements on crime, fear and pedestrian street use, after dark. Landscape Urban Planning 35(2–3):193–201
32. Perkins DD, Wandersman A, Rich RC, Taylor RB (1993) The physical environment of street crime: defensible space, territoriality and incivilities. J Environ Psychol 13(1):29–49
33. Brätt J, James G, Price R, Sewell J (2010) Best practices in place making through illumination. Virginia Tech Urban Affairs and Planning Program Report
34. Norouzian-Maleki S, Bell S, Hosseini S, Faizi M (2015) Developing and testing a framework for the assessment of neighbourhood liveability in two contrasting countries: Iran and Estonia. Ecol Ind 48:263–271. https://doi.org/10.1016/j.ecolind.2014.07.033
35. @erinroseglass on twitter!: Library, relatable, Tumblr posts (2020) Pinterest. Available at: https://www.pinterest.com/pin/erinroseglass-on-twitter--337770040804851370/(Accessed: 30 May 2023)
36. Edensor T (2015) The gloomy city: rethinking the relationship between light and dark. Urban Stud 52(3):422–438

37. Roberts M, Turner C (2005) Conflicts of liveability in the 24-hour city: learning from 48 hours in the life of London's Soho. J Urban Des 10(2):171–193
38. Van Liempt I, Van Aalst I, Schwanen T (2015) Introduction: geographies of the urban night. Urban Stud 52(3):407–421
39. Maslow A, Lewis KJ (1987) Maslow's hierarchy of needs. Salenger Inc 14(17):987–990
40. Roberts M, Eldridge A (2009) Planning the night-time city. Routledge

Forecasting Urban Solutions

Actualising 'Autonomous Cities' Via Artificial Intelligence for Achieving Netzero Urban Futures

Zaheer Allam and Aditya Dixit

Abstract Rapid Urbanisation coupled with climate change is causing a number of issues relating to urban governance and financing, which ultimately impacts on the liveability levels of urban areas. As the world turns towards smart cities to better increase economic performance of cities, there seems to be a lack of urban governance policy regarding climate change. Technology can aid on this front but must not be designed as a product solely for private economic gains, and instead designed to aid the public good. Smart Cities, being a concept already making use of technology can be redesigned to better make use of its technological background to ensure a better response to contemporary urban challenges, including that of climate change.

Keywords Autonomous cities · Artificial intelligence · Smart cities · Sustainability · Climate change · Technology · Resilience · Urban policy · Future cities

1 Introduction

The global population is gradually witnessing a growth rate of 1.06% yearly according to the latest report by UN Habitat [58]. As a result, it is projected that the global population will reach a high of about 8.5 billion people by 2030, and by 2050, it will have risen to over 9 billion people. It has also been reported that currently, approximately 56.2% of the global population is now residing in urban areas, and the number will increase to 60.4% by 2030, and to approximately 68% of the global

Z. Allam (✉)
Chaire Entrepreneuriat Territoire Innovation (ETI), IAE Paris—Sorbonne Business School, Université Paris 1 Panthéon-Sorbonne, 75013 Paris, France
e-mail: zaheerallam@gmail.com

Live+Smart Research Lab, School of Architecture and Built Environment, Deakin University, Geelong, VIC 3220, Australia

A. Dixit
Department of Mechanical Engineering, Friedrich Alexander University Erlangen-Nuremberg, Schloßplatz 4, 91054 Erlangen, Germany

Z. Allam (ed.), *Sustainable Urban Transitions*, Urban Sustainability,
https://doi.org/10.1007/978-981-99-2695-4_17

population by 2050 [67]. Such trends have been observed to be resultant impacts of the rapid urbanisation that is being witnessed across the globe, prompted by factors like increasing urban sprawl and promises of better liveability status in urban areas compared to rural setups. As more people migrate to cities, the consumption rate of resources has also increased, with urban areas now reported to be consuming more than 60% of the resources, and subsequently, producing over 75% of the global greenhouse gases (GHG) emissions. As such, while the urban areas have been attractive to a majority of the global population, especially the youthful population, as they pursue to partake in the diverse activities and opportunities, both economically and socially, they are also becoming victim of the negatives confronting cities.

The challenges that urban residents face include increasing cost of living following rise in demand for housing, energy, basic infrastructures, security and others. They are also affected by deteriorating air quality, scarcity of resources such as clean water, enough food supply, sufficient recreation and green spaces among others [49]. Urban areas are now being associated with high traffic crisis, with increasing automobiles (over 1.42 billion cars reported in urban cities in 2021 [28] and many other issues that have potential to compromise the liveability status, and resilience of cities. Such urban challenges were anticipated and well covered in the New Urban Agenda (NUA) document and also in the Sustainable Development Goal (SDG) 11, where in principle these policies encourage a paradigm shift that would allow cities to become more socially inclusive, sustainable, adaptable and resilient for existing and anticipated future residents.

One particular reason why there is an urgent need for a paradigm shift in urban areas on areas like energy production, use of automobiles, consumption of resources in construction industry, manufacturing sectors and other is the worsening case of climate change prompted by negative impacts of those sectors and others. On this, according to the latest IPCC report on global warming [37], it is feared that there is a high probability that the 2 °C rise in temperature that the Paris Agreement sought to prevent would not be tenable, and instead, by mid-century, temperatures will have risen to over 2.7 °C and by approximately 3.5 °C by 2100. With such, situations in urban areas will worsen, rendering numerous global agreements, accords and policies such as the SDGs [30], the NUA [65], the Paris Agreement [66], the Addis Ababa Action Agenda of the Third International Conference on Financing for Development [64], the African 2063 Agenda [2] and many others of less influence. However, despite the eminent rise in global temperature, the cited global policies and agreements still have the opportunities to be achieved, if concerted efforts and proactive approaches are urgently adopted, as was discussed in the latest climate change conference COP 26, where there was an increased emphasis on deep decarbonization [62].

In addition to the above global agendas, at the local and national levels, there has been notable efforts in adopting strategies that would help in addressing the menace of climate change. One notable potential solution, that has gained in popularity, is the Smart City planning model, that many cities across the globe have been reported to have embraced [12]. As documented in a wide range of literature [3, 4, 6, 15, 23], this planning model has the potential to not only address issues of climate change, but is helping urban managers find solutions for other urban challenges, such as energy, transport, security, waste management and others. However, like many other models that have been in existence previously, the smart city concept has been facing unprecedented challenges especially in regard to financing. These challenges, as expressed by Tompson [56] are multi-pronged. On one end, being a new concept that has not been implemented fully in global cities, it faces financial challenges as it is a capital-intensive endeavor making it difficult to be finance fully from public budgets. On the other end, this concept is reported to be capital intensive and many urban cities may not have capacities to raise enough financial resources to successfully it. On these two issues, it is only the private sector that has been observed to be willing to partner with local governments. But, one downside of this partnership is that it has been viewed as a profit oriented agenda by the private sector [5]. However, this challenge is surmountable by ensuring the partnership is crafted as a social partnership rather than a private enterprise, hence, guarantee maximum benefits to both residents and local governments [9, 10, 14].

Following the realities brought about by the emergence of the COVID-19 pandemic, and also the increasing concerns of climate change, implementing the Smart City concept and others like the 15-minute city planning model [47], closely influenced by the Smart City concept, will be perceived as being critical. However, these will need to be fashioned such that the aspect of automation is emphasized, banking on availability of advanced technologies such as the Artificial Intelligence (AI), Internet of Things (IoT), Cloud Computing and others. Such will not only allow for efficiency and sustainability agendas to be pursued, but will allow residents to build more social interactions, enjoy higher liveability levels and others as they also continue being accustomed to the new reality of social distancing, working from home and other new normal trends prompted by the pandemic. In right to this background, this chapter will explore how new technologies can allow the actualisation of 'autonomous' cities where deep decarbonisation, resilience and adaptation agendas are pursued.

2 Climate Change and Urban Infrastructural Losses

In the last few decades, the impacts of climate change have increased both in frequency and intensity, reaching even countries that were traditionally not severely impacted. For instance, in the recent past, it has become apparent that even the developed economies that have invested in advanced infrastructures, and have financial capacities are also experiencing diverse impacts of climate change; claiming both lives and damaging properties and infrastructures in wide scale [16, 26]. Events such as pronounced and severe floods, extreme temperatures, erratic precipitation, and outbreak of tropical diseases is no longer unusual in regions like Europe and North America. For instance, in Mid-2021, different parts of Western Europe are reported to have experienced unprecedented flooding, killing over 200 people [24], and causing damages worth over \$11.8 billion, with approximately \$4.7 billion to \$5.9 billion incurred in Germany alone [34]. Such happenings have prompted organisations such as UNFCCC, IPCCC and other bodies to raise alarms and calls for urgent interventions to mitigate the situation. As noted above, if the status quo is to be maintained, the global temperatures would reach a high of 2.7 °C by mid-century, and the climate impacts would only intensify [63]. Interventions in the urban areas in particular as prescribed in different prominent global documents such as SDGs, NUA, Sendai Framework [60] and Paris Agreement [66] need to be given priority, especially in relation to fostering sustainability dimensions.

While there is sufficient evidence that different countries have managed to achieve considerable outcomes in the implementation of those global policies [53], efforts, as emphasized in COP26 need to be redoubled to make up for the shortcomings experienced in the course of implementing different global policies. Increasing the commitment efforts will have significant impacts in reducing the consequences of the different natural disasters that have been increasing in the recent past in different parts of the globe. For instance, in the African regions, which could be argued to have been relatively from pronounced climate change events, are increasing both in frequency and intensity. A case in point are the 2019 disasters instigated by cyclone Idai and Kenneth, both affecting the Southern part of Africa, leaving over 1,000 people dead, and other 2.2 displaced from their residences [59]. In addition, the two prompted massive damages of properties and infrastructure—over $773 million worth of damages by cyclone Idai, and over $100 million worth of destruction by cyclone Kenneth [1]. In Europe, as noted above, over 200 people lost their lives from excessive floods with properties and infrastructures worth over $11 billion destroyed in 2021 alone [31]. In Australia, impacts of climate change exacerbated the 2020 wild fires, leading to deaths of billions of wild animals [51], and loss of vegetation in over 24.5 million acres of land [18]. These examples are just a small portion of real impacts that different countries has been experiencing, and with the projected temperature increase in the future the consequences may be extreme, especially on cities and urban areas located within the low-laying lands and coastal regions. For instance, the SIDS, which have been identified and acknowledged to be extremely vulnerable to the impacts of climate change may experience even more extreme consequences such as the submersion of habitable areas and in some cases, some islands states may be fully submerged [61].

The anticipated extreme climate events call for urgent and concrete interventions that would ensure urban communities are adaptable and resilient to climate change. One of the interventions being championed is investment in infrastructure projects with capacities to guarantee urban areas withstand the diverse climate change scenarios. On this, there will be need for financial support, especially on developing, LDCs and SIDS countries which have been experiencing unprecedented high debt levels; that have been argued to derail their infrastructure investment agendas [19, 50].

3 Climate Change and Urban Technologies

While the impacts of climate change are causing negative consequences across the globe, large ICT corporations involved in finding solutions for climate change events on the other hand have been observed to be profiting from climate change mitigation activities [44]. As a result, these corporations have been investing substantial financial resources in Research and Development (R&D) to ensure their products and services are of higher demand than their competitors, hence, gain more market control. Such competition has continually prompted the growth of ICT firms market such that before the emergence of COVID-19 pandemic, investment in the sector were $239 billion (2019), an increase of $10.7 billion from the 2018 figure ($228.3 billion) [25]. This impressive growth is cited to be prompted by the vast rate at which the smart city concept and the urban digital solutions have been gaining popularity. For instance, in 2019, it is reported that the market value for Smart Cities had reached a high of $83.10 billion, and is expected to grow at a steady compounded annual growth rate (CAGR) of 20% by 2026 to reach approximately $297.7 billion [29]. A similar report conducted by IMARC Group (2020) showcased that the market value for Smart Cities was $312.4 billion in 2018, and with a CAGR of approximately 17.6%, the market value for global Smart Cities would reach a high of $826.3 billion by 2024 and $3.48 trillion by 2026 [55]. These two reports, though giving different figures confirms the popularity of the Smart City concept, especially due to the promise of giving solution to a wide range of urban challenges.

With the Smart Cities planning model concept gaining traction in different parts of the world, the number of technologies and urban digital solutions will keep on increasing. On this, there are diverse 'smart' technologies targeting to give the best solutions for urban challenges, especially in regard to automation. For instance, in the transport sector, that has been identified to require an overhaul to guarantee urban resilience and adaptability, technologies such as electric vehicles, smart biking technologies, car sharing and others are on the rise, providing road users with alternatives to fossil fuel powered vehicles, and also options for not using private vehicles [20]. Such are also seen to have positive effects in helping reduce emission of which, it is reported that cars are responsible for approximately 14% of the total GHG emissions [69]. Further, alternative technologies in the transport sector have been hailed as alternatives to help reduce urban road accidents, which, as reported by Wiggins [69] have been on the rise, but with availability of alternative technologies like biking, and also availability of data, such as expected to reduce. On this, the target, as captured in a UN General Assembly [57], is for road fatalities to be reduced by 50% by 2030, and the use of urban digital solutions is one option of achieving this [20]. In other sectors such as waste management, agriculture, health, education and others, use of modern technologies is seen to have positive impacts. In particular, such are being influenced by technologies such as AI and Big Data that not only enhances real-time data collection and analysis, but also in helping real-time computation for quick decision making [45]. For instance, Libelium [42] highlights how technology is being deployed in water sector in the city of Barcelona to optimize access to quality water by all the residents.

Despite the advantages derived from the diverse smart technologies associated with the Smart City concept, there are some challenges that require to be overcome to ensure that total benefits are accrued. These appertains to data storage and computations, which, in most cases are usually handled by contracted large ICT Corporations. In other cases, the data analysis is handled by urban administrators in the public sector, whom, in most cases have no sufficient capacities or are poorly equipped to generate deep insights; hence making optimal use of datasets [8]. In cases where ICT corporations are mandated to manage urban data, there are reports of concerns and issues, especially in regard to security and privacy of data and access of the same by other stakeholders like small, local startups involved in digital urban solution innovations [32]. Hollands [33] further argues that due to the profit orientation that is associated with most large ICT firms, there has been criticism of commercialization of data, aimed at increasing profits, and such has led to privacy concerns by residents. Further, small startups have raised concerns of difficulties in obtaining access to data. On this, Ismagilova et al. [38] contend that urban databases need to be structured such that they allow for open-data platforms ensuring that all stakeholders have unmetered access. Besides data access, there are concerns of lack of standardization of protocols in relation to different smart devices and sensors that are in use, hence, compounding the challenge experienced in designated urban areas. As a result of those challenges, it is probable that a number of projects may not be fully embraced and accepted by locals, especially in cases where there are fears or suspicions of privacy infringement [33]. Further, lack of acceptability may also be prompted by lack of participation of local companies or startups, hence, residents and other players feel as though they are being exploited by large private sector groups, especially by their private information being use for commercial purposes. In right to all these negative aspects associated with Smart Cities, it would be imperative for urban planners to come up with strategies that would render more efficient and resilient programs without losing the trust of the citizenly. Such strategy could involve embarking on total automation of processes, hence, no, or limited requirements, for human interventions [12, 13]. Further, democratization of technologies deployed in smart urban projects, such that issues like data access could be made open, hence, promote a more inclusive urban landscape.

4 On the Ethics of Climate Action and Urban Policies

The responsibility of coming up with solid climate change mitigation policies, and the solutions thereof is a global issue, and as of now, the global community has made some significant efforts, though much need to be done. This is affirmed by the diverse policies, including groundbreaking ones like the Kyoto Protocol, the Paris Agreement, the SDGs, NUA and the Agenda 2030 for Sustainable Development amongst others. That said, it is the responsibility of the local and national governments to actualize and implement the policies at their level, and in relation to their specific local challenges. This is critical as different regions, and local areas experience different climate change challenges and vulnerabilities whose solution cannot be duplicated from elsewhere. Further, the financial and resource availabilities in different regions differs significantly; with urban areas in regions like developing, LDCs and SIDS experiencing notable financial constraints compared to their counterparts in the developed regions [50]. Those factors notwithstanding, it is paramount to ensure that environmental sustainable gains already made in different regions, especially in urban landscape are not eroded by climate change events, for such would prompt a reversal or slowdown in economic frontiers, hence, negatively impacting on urban resilience, adaptability and liveability [26]. Moretti and Loprencipe [48] expressed further that the impacts of climate change, more so on the critical urban infrastructure, have the potential to attract further costs incurred in the course of maintenance, repair and replacements, hence, attracting unwarranted extra financial demands. Such has in turn been argued to plunge different cities and urban areas into debt crises, which when accumulated, have led economies across the globe into state of indebtedness, thus, risking critical public infrastructures that often fronted as collateral for the loans [52, 68]. With debt challenges and insufficient infrastructural development, cities then would be unable to attract enough Foreign Direct Investments (FDIs), local private investors or tourism activities. Urban areas would also continue to experience unprecedented climate migration, which has become a common phenomenon in the recent decade [36].

On the environment, climate change has prompted numerous challenges like sea level rise, unpredictable and extreme weather conditions, excessive flooding, desertification, acidification of water sources, emergence of invasive species and many others. Such in turn have increased human vulnerability in terms of reduced food security, availability of clean water, health, damages of residential areas, properties and infrastructure and others. On the political sphere, climate change has contributed to escalation of tension between countries, as they pursue decarbonisation agendas, come up with climate change mitigation policies and source for financial resources to support and fund climate change mitigation programs. The tension is also observed in the process of apportioning climate responsibility in respect to how much GHG each country emits. In this case, even low emitters like SIDS and LDCs that cumulatively contribute to emission of less than 1% of the total global emissions, are seen to bear huge burdens due to their geographical locations that are in the gateway to most climate change events [46].

The diverse consequences of climate change on the global community and the subsequent efforts being made to overcome the challenges, are crowded by some ethical concerns of climate change and the need for urgent mitigation. The attention has been particularly on urban areas with diverse objectives, including increasing the safety of residents, allowing them pursue different economic activities without disruptions, and ensuring that they build capacities for resilience and have opportunities to enjoy higher liveability status and others. But, the achievement of global agendas would need total commitment from different players and also regular review of the same to assess the progress made at different time spans. A case in point on this is the COP 26 which did not only emphasize on the deep decarbonisation agenda, but also gave a platform for countries to update their Nationally Determined Contributions (NDCs)—reviewed every 5 years since 2015 when it was enacted [63]. Achievement of those policies need to be actualised by embracing modern strategies like adoption of emerging technologies such as AI [21], IoT, Big Data [39], Cloud Computing [35, 40, 41] and other. Such have been harnessed especially in actualizing diverse smart city objectives [7, 8, 11], especially automation of different processes and activities. Additionally, such have been seen to contribute significantly in quicker decision making and in elimination or minimization of human interventions, which have been seen to derail the achievement of different globally agreed policies [43].

Automation, as emphasized in this chapter needs not be seen as a utopian idea in its promise to make cities more liveable, resilient, adaptable, inclusive and sustainable. Already, there has been some progress made, especially in fields like robotics as was observed in the height of COVID-19 where robots were being deployed for different purposes like delivery of medical supplies in remote areas [54]. Such only portray a glimpse of how much automation can be achieved in different urban arenas, including in areas like early warning for events like climate change instigated ones [17, 22], help in prompt and efficient disaster risk response, especially due to ubiquitousness of technologies such as 3D printing [27]. Automation is also expected to enhance security in urban areas with increase in use of drones and other unmanned aerial vehicle. But such would, as expressed by Macrorie et al. [43], be informed by availability of data, which has been observed to be hampered by apprehensiveness of local residents to share their private information due to privacy concerns. On this, Macrorie et al. [43] argued that use of robots and machines in urban areas has the potential to encourage more data sharing, and this would have significant impacts in enhancing operations in different urban frontiers. But even with automation, it will be paramount to ensure the privacy and security of personal information is piously guarded to void any chances of compromising the trust the people have on the automation technologies.

References

1. AFDB (2019) Programme post cyclone Idai and Kenneth emergency recovery and resilience programme for Mozambique, Malawi and Zimbabwe (PCIREP). African Development Bank Group
2. Africa Union (2021) Agenda 2063: the Africa we want. African Union. https://au.int/en/agenda2063/overview. Accessed 13 Oct 2021
3. Allam MZ (2018) Redefining the smart city: culture, metabolism and governance. Case study of Port Louis, Mauritius. PhD, Curtin University
4. Allam Z (2018) Contextualising the smart city for sustainability and inclusivity. New Des Ideas 2:124–127
5. Allam Z (2019) The emergence of anti-privacy and control at the nexus between the concepts of safe city and smart city. Smart Cities 2:96–105
6. Allam Z (2019) Identified priorities for smart urban regeneration: focus group findings from the city of Port Louis, Mauritius. J Urban Regen Renew 12:376–389
7. Allam Z (2020) Cities and the digital revolution: aligning technology and humanity. Springer International Publishing

8. Allam Z (2020) Data as the new driving gears of urbanization. In: Allam Z (ed) Cities and the digital revolution: aligning technology and humanity. Springer International Publishing, Cham
9. Allam Z (2020) On culture, technology and global cities. In: Allam Z (ed) Cities and the digital revolution: aligning technology and humanity. Springer International Publishing, Cham
10. Allam Z (2020) Privatization and privacy in the digital city. In: Allam Z (ed) Cities and the digital revolution: aligning technology and humanity. Springer International Publishing, Cham
11. Allam Z (2020) Urban chaos and the AI Messiah. In: Allam Z (ed) Cities and the digital revolution: aligning technology and humanity. Springer International Publishing, Cham
12. Allam Z (2021) Introducing the concept of autonomous city. In: The rise of autonomous smart cities: technology, economic performance and climate resilience. Springer International Publishing, Cham
13. Allam Z (2021) Big data, artificial intelligence and the rise of autonomous smart cities. In: The rise of autonomous smart cities: technology, economic performance and climate resilience. Springer International Publishing, Cham
14. Allam Z, Dhunny ZA (2019) On big data, artificial intelligence and smart cities. Cities 89:80–91
15. Allam Z, Newman P (2018) Economically incentivising smart urban regeneration. Case study of Port Louis, Mauritius. Smart Cities 1:53–74
16. Asariotis R, Kruckova L, Nay VM (2020) Climate change impacts and adaptation for coastal transport infrastructure: a compilation of policies and practices. UNCTAD, New York
17. Balbi S, Villa F, Mojtahed V, Giupponi C (2014) Estimating the benefits of early warning systems in reducing urban flood risk to people: a spatially explicit bayesian model. In: 2014 proceedings of the 7th international congress on environmental modelling and software, San Diego, CA, USA. SSRN
18. BBC (2020) Australi fires: a visual guide to the bushfire crisis. BBC News. https://www.bbc.com/news/world-australia-50951043. Accessed 18 Jan 2020
19. Bulow J, Reinhart C, Rogoff K, Trebesch C (2020) The debt pandemic: new steps are need to improve sovereign debt workouts. The International Monetary Fund. https://www.imf.org/external/pubs/ft/fandd/2020/09/debt-pandemic-reinhart-rogoff-bulow-trebesch.htm?utm_medium=email&utm_source=govdelivery. Accessed 15 Apr 2020
20. Butler L, Yigitcanlar T, Paz A (2020) How can smart mobility innovations alleviate transportation disadvantage? Assembling a conceptual framework through a systematic review. Appl Sci 10
21. Calo R (2017) Artificial intelligence policy: a roadmap. SSRN Electron J 1–28
22. Ciffey K, Haile M, Halperin M, Wamukoya G, Hansen J, Kinyangi J, Tesfaye K (2015) Expanding the contribution of early warning to climate-resilient agricultural development in Africa. Research Program on Climate Change, Agriculture and Food Security (CCAFS), Copenhagen, Denmark
23. Dabeedooal JY, Dindoyal V, Allam Z, Jones SD (2019) Smart tourism as a pillar for sustainable urban development: an alternate smart city strategy from Mauritius. Smart Cities 2
24. Davies R (2021) Western Europe—storms case floods in Switzerland, Germany and UK. Floodlist. https://floodlist.com/europe/storms-floods-switzerland-germany-uk-july-2021. Accessed 10 Aug 2021
25. Duffin E (2019) ICT research and development expenditure in U.S. and worldwide 2015–2018. Statista. https://www.statista.com/statistics/732308/worldwide-research-and-development-information-communication-technology/. Accessed 18 Jan 2020
26. Forzieri G, Bianchi A, Silva FBE, Marin Herrera MA, Leblois A, Lavalle C, Aerts JCJH, Feyen L (2018) Escalating impacts of climate extremes on critical infrastructures in Europe. Glob Environ Chang 48:97–107
27. Gahren I (2018) How 3D printing is revolutionizing disaster relief. TBD. https://www.tbd.community/en/a/how-3d-printing-revolutionizing-disaster-relief. Accessed 4 July 2021
28. Global Cars Brands (2021) How many cars are there in the world?. Global Cars Brands. https://www.globalcarsbrands.com/how-many-cars-are-there-in-the-world/. Accessed 5 Nov 2021

29. Globe News Wire (2021) At 20% CAGR, smart cities market size & trends to soar globally, expected to touch a value USD 297.7 billion by 2026. Facts & Factors. https://www.globenewswire.com/en/news-release/2021/04/29/2219867/0/en/At-20-CAGR-Smart-Cities-Market-Size-Trends-to-Soar-Globally-Expected-to-Touch-a-Value-USD-297-7-Billion-By-2026-Says-Facts-Factors.html. Accessed 6 Nov 2021
30. Habitat UN (2021) Sustainable development goals cities. UN Habitat, Nairobi. https://unhabitat.org/programme/sustainable-development-goals-cities. Accessed 15 Apr 2021
31. Henson B, Masters J (2021) Central Europe staggers toward recovery from catastrophic flooding: more than 200 killed. Yale Climate Connections. https://yaleclimateconnections.org/2021/07/central-europe-staggers-toward-recovery-from-catastrophic-flooding-more-than-200-killed/. Accessed 6 Nov 2021
32. Hollands RG (2014) Critical interventions into the corporate smart city. Camb J Reg, Econ Soc 8:61–77
33. Hollands RG (2015) Critical interventions into the corporate smart city. Camb J Reg, Econ Soc 8:61–77
34. Howard LS (2021) Germany's floods will see 'sizeable' protection gap. Could insurance demand increase?. Insurance Journal. https://www.insurancejournal.com/news/international/2021/07/22/623877.htm. Accessed 5 Nov 2021
35. Huang T-J (2017) Imitating the brain with neurocomputer: a "new" way towards artificial general intelligence. Int J Autom Comput 14:520–531
36. International Organization for Migration (2019) Climate change and migration in vulnerable countries: a snapshot of least developed countries, landlocked developing countries and small island developing states. IOM, Geneva, Switzerland
37. IPCC (2021) The physical science basis. Contribution of working group I to the sixth assessment report of the intergovernmental panel on climate change. In: Masson-Delmotte V, Zhai P, Pirani A, Connors SL, Péan C, Berger S, Caud N, Chen Y, Goldfarb L, Gomis MI, Huang M, Leitzell K, Lonnoy E, Matthews JBR, Maycock TK, Waterfield T, Yelekçi O, Yu R, Zhou B (eds). Cambridge University Press (in press)
38. Ismagilova E, Hughes L, Rana NP, Dwivedi YK (2020) Security, privacy and risks within smart cities: literature review and development of a smart city interaction framework. Inf Syst Front
39. Ivanov N, Gnevanov M (2018) Big data: perspectives of using in urban planning and management. MATEC Web Conf 170:1–5
40. Kotenko I, Saenko I, Skorik F, Bushuev S (2015) Neural network approach to forecast the state of the internet of things elements. In: XVIII international conference on soft computing and measurements (SCM). IEEE, pp 133–135
41. Lee KY, Chung N, Hwang S (2016) Application of an artificial neural network (ANN) model for predicting mosquito abundances in urban areas. Eco Inform 36:172–180
42. Libelium (2016) Saving water with smart irrigation system in Barcelona. Libelium World. http://www.libelium.com/saving-water-with-smart-irrigation-system-in-barcelona/. Accessed 19 Jan 2020
43. Macrorie R, Marvin S, While A (2019) Robotics and automation in the city: a research agenda. Urban Geogr 1–21
44. Mark R, Anya G (2019) Ethics of using smart city AI and big data: the case of four large European cities. ORBIT J 2:1–36
45. Mcgovern A, Elmore KL, Ii DJG, Haupt SE, Karstens CD, Lagerquist R, Smith T, Williams JK (2017) Using artificial intelligence to improve real-time decision-making for high-impact weather. 98:2073–2090
46. Mead L (2021) Small islands, large oceans: voices on the frontlines of climate change. International Institute for Sustainable Development (IISD)
47. Moreno C, Allam Z, Chabaud D, Gall C, Pratlong F (2021) Introducing the "15-minute city": sustainability, resilience and place identity in future post-pandemic cities. Smart Cities 4:93–111
48. Moretti L, Loprencipe G (2018) Climate change and transport infrastructures: state of the art. 10:4098

49. Nijman J, Wei YD (2020) Urban inequalities in the 21st century economy. Appl Geogr 117:102188
50. Nishio A (2021) Facing substantial investment needs, developing countries must sustainably manage debt. 16 May 2019. https://blogs.worldbank.org/voices/facing-substantial-investment-needs-developing-countries-must-sustainably-manage-debt. Accessed 10 Aug 2021
51. Samuel S (2020) A staggering 1 billion animals are now estimated dead in Australia's fires. Vox. https://www.bbc.com/news/world-australia-50951043. Accessed 20 Jan 2020
52. Sauer N (2019) Mozambique 'faces climate debt trap' as Cyclone Kenneth follows Idai. Climate Home News. https://www.climatechangenews.com/2019/04/26/mozambique-faces-climate-debt-trap-cyclone-kenneth-follows-idai/. Accessed 10 Sept 2019
53. Sinha A, Sengupta T, Alvarado R (2020) Interplay between technological innovation and environmental quality: formulating the SDG policies for next 11 economies. J Clean Prod 242:118549
54. Sison M (2020) Robots, smart helmets deployed in coronavirus fight. Sci Dev Net. https://www.scidev.net/global/health/news/robots-smart-helmets-deployed-in-coronavirus-fight.html?__cf_chl_jschl_tk__=4f53bc7836c350d03e8c8f99eb933ea00c512e23-1595678055-0-AXWtGC_89SN4ZNeDM8_xlZToUaqJJ4ZzUdY6tVwGbQPZ_Dvk-k3rqfW_hhf9pWVSy9h2-irpbLD_PKmba8aKA8_1SsqRMo9x-wUwK6FrbLXOGFpWjJrZZIomyNk-MeZl7d68XOuLQA-S-uRRE3PR-KSZ36CQylExZcSV39DbgEvpzACf_Aq38pVzTMzleNS6HpD-I7fbEvt9bp8Ont_k_pF5FuMQuWzQJt9IZZ28fYadlmR7YIgWGj20DuvY9UKns7FVYGimFDwEN5VRtBmYDUTbwzc2NSEThMGNVQ9XeYzNPw_Tn2fnk6oOWPsWupiE3olVap2FEwl7bo5DmkN2B6-llS2cNDInxZvPXrMksYCoMKuK_wPnmCUKjcudED2qOg. Accessed 2 July 2021
55. Smart Cities Association (2020) Global smart cities market to reach a whopping $3.5 trillion by 2026. Smart Cities Association. https://www.smartcitiesassociation.org/index.php/media-corner/news/1-global-smart-cities-market-to-reach-a-whopping-3-5-trillion-by-2026. Accessed 18 Jan 2020
56. Tompson T (2017) Understanding the contextual development of smart city initiatives: a pragmatist methodology. She Ji: J Des, Econ, Innov 3:210–228
57. UN General Assembly (2019) Road safety. Department of Safety and Security, United Nations. https://www.un.org/en/safety-and-security/road-safety. Accessed 6 Nov 2021
58. UN Habitat (2020) Global state of metropolis 2020. In: Population data booklet. UN Habitat, Nairobi, Kenya
59. UNCHA (2019) Cyclones Idai and Kenneth. United Nations Office for the Coordination of Humanitarian Affairs. https://www.unocha.org/southern-and-eastern-africa-rosea/cyclones-idai-and-kenneth. Accessed 18 Jan 2020
60. UNDRR (2015) Sendai framework for disaster risk reduction, Sendai City, Miyagi Prefecture, Japan, United Nations
61. UNEP (2014) Emerging issues for small island developing states. UNEP, Nairobi, Kenya
62. UNFCCC (2021) Glasgow climate change conference. UNFCCC. https://unfccc.int/conference/glasgow-climate-change-conference-october-november-2021. Accessed 5 Nov 2021
63. UNFCCC (2021) Nationally determined contributions under the Paris Agreement: synthesis report by the secretariat. In: Conference of the Parties serving as the meeting of the Parties to the Paris Agreement. United Nations Framework Convention on Climate Change, Glasgow
64. United Nations (2015) Addis Ababa Action Agenda of the third international conference on financing for development (Addis Ababa Action Agenda). In: Third international conference on financing for development. United Nations, Addis Ababa, Ethiopia
65. United Nations (2017) New urban agenda. Habitat III
66. United Nations Framework Convention on Climate Change (2015). Paris Agreement. UNFCCC. https://unfccc.int/sites/default/files/english_paris_agreement.pdf. Accessed 8 Aug 2021
67. United Nations Human Settlements Programme (UN-Habitat) (2020) World cities report 2020: the value of sustainable urbanization. UN Habitat, Nairobi, Kenya

68. Were A (2018) Debt trap? Chinese loans and Africa's development options. S Afr Inst Int Aff Policy Insights 66:1–13
69. Wiggins B (2020) Cars are a major source of greenhouse gas emissions—some cities are finally taking action. Global Citizen. https://www.globalcitizen.org/en/content/cities-car-bans-greenhouse-gas-emissions/. Accessed 1 Nov 2021

Sustainable Urban Mobility Governance: Rethinking the Links Through Movement, Representation and Practice for a Just Transport System

Eda Beyazit and Fatih Canitez

Abstract Urban mobility systems are the centrepiece of cities. They are often seen as the engines of economic growth and enablers of high quality of life. They shape cities, local economies, social relationships and, in some cases, national competitiveness. Yet, they are also political objects as much as they are spatial. From body to the globe, urban mobility systems are entangled in the continuous reproduction of dynamics which may create uneven geographical development and burden the economy, societies and the environment. In this sense, the sustainable governance of urban mobility systems is critical in shaping sustainable and resilient cities that address today's challenges. Although several governance frameworks have been suggested in the past years regarding urban mobility, a holistic structure that covers the existing challenges of urban mobility, mainly brought by the recent pandemic and the risks of austerity measures in the post-pandemic period, and the new advancements in transport infrastructure are often missing in existing governance networks. This study aims to, first of all, introduce the key concepts that shape the current climate in transport and mobility research and practice under three themes, movement, representation and practice, based on a review of the urban mobility literature; and secondly, develop a framework for the sustainable governance of mobility systems inspired by Sheller's (Mobility justice: the politics of movement in an age of extremes. Verso Books, 2018) nested approaches to just decision-making in mobility and transport. The chapter also discusses the evolving and central role of local governments and city authorities in meeting the growing urban challenges of our century.

Keywords Sustainable urban mobility · Governance framework · Inequalities · Movement · Representation · Practice · COVID-19 pandemic

---

E. Beyazit (✉)
Department of Urban and Regional Planning, Faculty of Architecture, Istanbul Technical University, Istanbul, Turkey
e-mail: beyazite@itu.edu.tr

F. Canitez
Department of Civil and Environmental Engineering, Faculty of Engineering, Imperial College London, London, UK
e-mail: f.canitez@imperial.ac.uk

Z. Allam (ed.), *Sustainable Urban Transitions*, Urban Sustainability,
https://doi.org/10.1007/978-981-99-2695-4_18

1 Introduction

No wonder R.E.M.'s 1987 song "It's the End of the World as We Know It (And I Feel Fine)" re-entered the top 100 on the USA iTunes list in the early days of the COVID-19 pandemic [1]. We have witnessed radical changes affecting our daily lives from employment to health and education. Urban mobility and transport systems are no different. Since the early days of the global lockdowns, city authorities have innovated new solutions to address their citizens' mobility needs. In this period, we have seen significant initiatives from cities like Paris and Barcelona leading other cities towards a more sustainable mobility future. Although there is much debate and ongoing speculations about whether COVID-19 will have lasting impacts on the behaviour of consumers and commuters [2], as well as the decisions of politicians, the cry for more sustainable and resilient mobility futures has been loud and clear for a very long time (e.g. [3]). Despite the calls from academia, supranational organisations and NGOs to reduce transport burdens on the environment and social and economic life, urban mobility and transport systems have not yet successfully met the sustainability targets [4].

Building on the key debates on urban mobility systems, which are expected to guide readers' approach to emerging mobility systems, technologies, and services, as well as the new governance models best suited to manage this changing urban mobility landscape in favour of sustainability, this chapter discusses how an effective governance framework may be established to meet the growing sustainable mobility challenges. Whether providing clean technologies for a low-carbon future [5] or developing governance frameworks for street reallocation [6], COVID-19 has provided a platform to elaborate on these existing challenges for much-needed socio-spatial and political transformations.

We frame the underlying challenges of sustainable urban mobility under three themes: movement, representation and practice, inspired by Cresswell [7]. We argue that an effective and just governance system needs to be constructed upon an understanding of these domains as they provide new trajectories to elaborate further on socio-spatial and ecological injustices which are deeply buried in day-to-day and long-term transport policies. The ongoing debates around movement, representation and practices are first outlined. We then provide a framework for the governance of urban mobility. Reviewing the urban mobility literature under these three themes critically through an inequalities lens, we outline the most prominent challenges brought by the COVID-19 pandemic to urban mobility systems and travel behaviour in Sect. 2. We then give an overview of recent advancements in urban mobility in Sect. 3 before underlying the key principles for sustainable mobility governance in Sect. 4. Here, the principles are discussed within a nested approaches framework suggested by Sheller [8] and a new scheme is proposed considering the relationships between the key mobility actors and stakeholders. In the final section we draw some conclusions based on the discussion in the chapter.

2 Rethinking Urban Mobility Through Movement, Representation and Practice

This section introduces key concepts shaping the current landscape in sustainable transport and mobility research and practice under three themes: movement, representation, and practice.

Movement

The world is organised by movement and flows. As Urry [9] once famously illustrated, the world seems to be on a constant move. Until the shock of the COVID-19 pandemic on urban mobility, the worldwide trend was moving towards the increasing movement of people, goods, energy and other sources for the continuation of the existing economic and social systems. We have witnessed an incremental growth of airline passengers, maritime shipping and increased mobility in urban areas. ITF [10] predicted a significant growth in travel demand, especially in developing countries. Average distances travelled by people and freight would be on the rise based on the growth of disposable income, an important determinant of trade volumes (ibid.). Passenger air travel was showing an upward trend before the pandemic and there were predictions for a further growth in 2020 [11].

Population growth has been one of the key determinants of increasing movement of people and freight as well as travel demand in cities. As urban areas accommodate a growing number of people residing and working in cities, mobility demand increases. Before the pandemic, this growth was predicted to be higher in non-OECD countries in 2050 [10].

This constant mobility growth comes with its burdens, especially on the environment. For instance, while all other sectors have managed to reduce their CO_2 emissions, transport sector has maintained an increasing contribution to the greenhouse gases. In 2016 transport related emissions in OECD countries was 30% of all emissions. In non-OECD countries, this ratio was 16% [10]. More than 70% of all CO_2 emissions from transport are generated in road transport [12] and an important share of it is produced in cities.

Contrary to the future scenarios for urban mobility that included a significant decrease in CO_2 emissions in urban areas by 2050 due to increasing shared mobility, COVID-19 has been a potential key disruption for future scenarios. A study that tracked mobility level changes since the beginning of lockdowns in Switzerland has demonstrated that populations moved away from collective mobility options towards private cars, walking and cycling [13]. Although car sales were below previous estimations worldwide in 2020 due to economic uncertainties in the beginning of the pandemic, country-wide statistics showed recoveries in later periods [14]. Especially, in developing countries such as China, India and Turkey, passenger car sales were higher compared to previous periods [15, 16, 17]. Moreover, although working from home has become an option, this is not common across sectors, populations and geographies.

Growing urban mobility levels pose an important challenge for local and national governments across the world. On the one hand, the increasing demand for fast, reliable and comfortable mobility options requires new infrastructure developments in cities. On the other hand, this situation points out an increasing concern for meeting the emission reduction targets in the next decades for local and national governments. Another key challenge is the financing of public transport, particularly for those vulnerable populations who have been negatively impacted by radical changes during and after the COVID-19 pandemic. As passenger-km travelled by public transport declines so do the revenues of public transport organisations. These challenges need to be addressed in the governance of urban mobility systems in the near future and further discussed as key dimensions in our proposed governance framework.

Representation

Without a doubt, mobility trends are not all even across populations and countries. Particularly in the last four decades, scholars have accumulated a wealth of knowledge on transport-related inequalities [18]. As Banister [19] and Sheller [8] discuss separately, two research strands provide important debates in deepening our understanding of transport-related inequalities: one sees mobility as a capability, a capital, intangible commodity individuals may or may not possess based on gender, age, class, education, income and other socio-economic and demographic factors (e.g. [20, 21]), while the second approach follows a more conventional accessibility framework (e.g. [22]). Although both approaches have influenced transport research agenda considerably, we consult to the former in order to discuss representation, power and mobility, yet, without disregarding the burdens of "inaccessibility built into our cities" on "diverse kinds of bodily mobility" [8, p. 54].

As Kern [23] rightly puts it "all forms of urban planning draw on a cluster of assumptions about the "typical" urban citizen: their daily travel plans, needs, desires, and values. Shockingly, this citizen is a man. A breadwinning husband and father, able-bodied, heterosexual, white, and cis-gender" (p. 34). This statement not only brings a gender-lens to urban and transport planning but also a comprehensive understanding of differences in the city. Mobility is political as much as it is spatial. A power dimension based on social relations is embedded in mobilities that are "both productive of such social relations and produced by them" [7, p. 21]. These social relations based on dimensions such as culture, nationality, income, ethnicity and religion as well as bodily differences such as age, gender and ability define the ways in which some may be more mobile than others as well as who are represented in the existing mobility system.

It is not the intention of this chapter to present a comprehensive debate on the ways in which the alliance of patriarchy and neoliberalism resulting in the neo-colonial city dictates how individuals move around and in-between cities. Yet, such a debate brings an important perspective in redesigning our mobility governance systems around the questions of who and what is represented, even more than before if we are to achieve low-carbon, sustainable mobility systems that address the ever-growing challenges our cities are facing.

In this sense, an important question is asked: **who** decides for **whom** to have the **choice** to be mobile or immobile? Such decision-making power for one's mobility may be embedded within family- and community relations as well as within the regulations of local and national governments, and the relations between nations. This question, also proposed by Akyelken [24] with regard to women, when expanded to include other social, economic, demographic dimensions becomes a catalyst for the right to mobility—a collective right, borrowing the term from Harvey [25], to change the mobility systems in the city collectively so that they accommodate diversity and provide a dignified mobility system for all, and in turn to change the power relations between individuals.

Discussions on unequal mobilities questioning the control of some powerful individuals/groups over others' mobility patterns gain further importance in the time of the COVID-19 pandemic as the news are full of "kinetic elites" enjoying luxurious hotels and rental housing during long weekends [26] while the majority of the public are stranded in their homes, immobile. Furthermore, within the working population, only a handful of individuals are able to work remotely, while for others such option is not remotely possible [27]. Therefore, the representation of mobilities in the age of immobilities prove to be an important dimension for rethinking urban mobility governance.

Practice

Cities are on the verge of a mobility revolution. Much of the improvements to urban mobility for the past half century have been gradual and incremental, and focused on increasing the capacity of the existing transport infrastructure to meet growing demand, which has been especially the case for the cities of the developing world. However, two factors are tremendously impacting the urban mobility landscape: radical new technologies and COVID-19. Electric vehicles are replacing fossil-fuel powered cars, autonomous vehicles are soon expected to coexist with vehicles with drivers, developments in the usage of data (e.g. mobility as a service, shared vehicles, etc.) have been improving the way mobility services are designed, planned and delivered.

These fundamental changes in urban mobility systems have been unprecedented in such a short period of time and across numerous technologies. An example of such a profound impact on urban mobility was the advent of affordable automobiles in the mid-twentieth century which radically changed the urban mobility landscape of cities. As a result, there was a decline in public transport use, cycling and walking as well as an increase in traffic congestion, greenhouse gas emissions and air pollution.

There has been an optimistic outlook, though severely diluted by the COVID-19 pandemic, on the impact of new technologies on urban mobility. The new transport technologies are expected to make journeys faster, safer, easier, more comfortable and affordable and make cities quieter and less polluted. They are also expected to provide the option to have mobility as a service, integrated and accessible to all. Therefore, the benefits of a new urban mobility system are greater than just transforming how to travel from A to B. Mobility entrepreneurs are developing new mobility options, hence contributing to delivering high quality jobs, increased productivity, and economic development. Innovations in the urban mobility area harness the domestic and global expertise to provide a cleaner, safer and more efficient urban mobility. The role of local governments is becoming key in this changing mobility landscape not just to regulate the existing transport modes in a city but also to support mobility entrepreneurs, innovators, and businesses as well as engaging with various mobility groups, NGOs, and academics. Local governments in cities are seeking to engage with the technology providers to invest in the development of urban mobility systems.

Developments in data science, artificial intelligence, and sensing technology have contributed to mobility innovation with cleaner mobility options, automation, new business models and new modes of travel. If managed and regulated successfully within an urban environment, they have the potential to deliver great benefits for society, the environment and the economy, the three pillars of sustainability.

However, if this new urban mobility ecosystem is not effectively managed by a competent local authority, they could have unintended impacts, such as increasing traffic congestion or preventing sustainable mobility. For example, urban policy-makers are not yet well prepared to manage the potentially disruptive impact of autonomous vehicles with clear policy strategies [28]. This is why developing an effective governance framework and regulatory policies is a central part of incorporating the changes in the urban mobility.

3 Recent Transformations in Urban Mobility Landscape

Urban mobility has always been at the center of disruptive changes to the cities. Evolution is part and parcel of the urban mobility landscape. The rapid technological developments, social changes, and post-pandemic mobility landscape are the drivers for change for the urban mobility. For example, increasing availability of data and enhanced connectivity are improving trip planning for customers. Vehicle-to-vehicle and vehicle-to-infrastructure connectedness provide information to network operators and passengers in real-time so that the transport fleet is used in an optimum way. Machine learning, a branch of artificial intelligence, helps improvements in areas of autonomous vehicles, identification of areas having more traffic congestion and forecasting bus arrival times more accurately. Improved sensing technology, computing power, and software engineering are helping the introduction of autonomous vehicles in urban mobility [29]. The autonomous vehicle projects usually involve multiple mobility actors from different sectors, local governments, and academy.

These landscape changes as well as developments driving changes in the regime and niche levels for urban mobility, could be better illustrated by the schematic below (Fig. 1), drawing on the socio-technical transition and multi-level perspective literature [30, 31, 32]. The schematic seeks to indicate the multi-dimensional and complex dynamics acting on the urban mobility landscape.

Developments in zero-emission or alternative fuel vehicles such as battery electric, fuel cell, hydrogen, etc. have the potential to provide a cleaner mobility environment

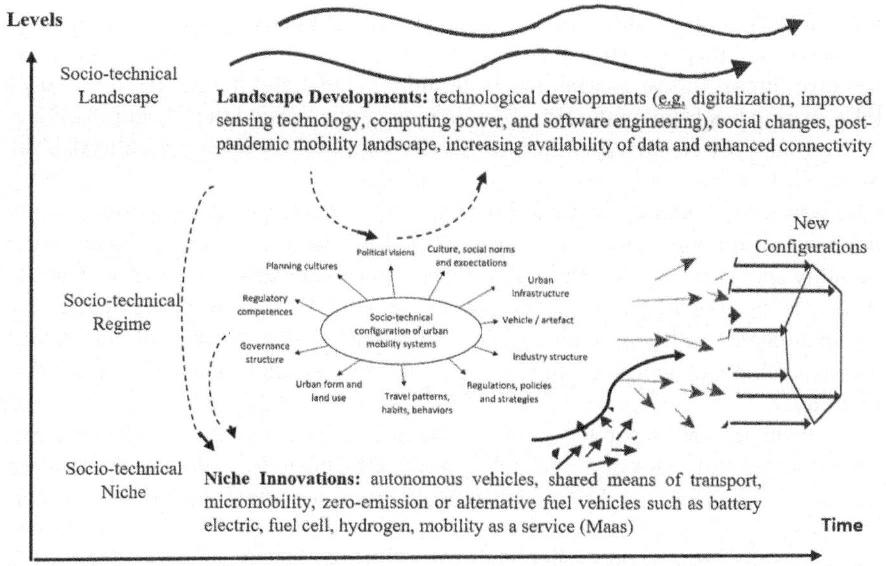

Fig. 1 Multi-level factors impacting sustainable mobility transitions based on Geels's [31] and Canitez's [30] socio-technical transition perspective

for cities, as they have no greenhouse gas or air pollutant tailpipe emissions. More than 1 million electric vehicles were sold in China in 2018 [33]. Cities and countries are setting targets for full electrification, to tackle climate change impacts and reach targets set in this area. The main impact on the local governance of these zero-emission vehicles is the increasing need for adequate charging stations across the city. Local governments play a major role in providing spaces for charging stations around their cities.

Improved batteries enable the proliferation of new means of micromobility not only for carrying passengers but also goods such as electric scooters, electrically assisted pedal cycles (e-bikes), and e-cargo bikes. Micromobility vehicles have the potential to complement the conventional public transport system in cities, where rapidly increasing transport demand, especially in developing countries and cities, exerts pressure on the existing public transport capacity. Many e-scooter operators, for example, have been cooperating with local governments in an effort to help achieve the transport accessibility goals. Providing as a first and last mile solution, micromobility services also offer an alternative to short car trips.

Although conventional public transport modes are still the basic form of shared mobility, new mobility means based on sharing economy are on the rise, facilitated by increasing digitalization [34]. This can take various forms including sharing journeys such as ride-sharing and demand responsive transport, or sharing access such as car sharing and bike sharing [35]. However, COVID-19 negatively impacted the use of shared mobility, especially during the initial phase in 2020 [36]. Unless recovery plans that consider social and environmental factors are put into practice, these negative impacts are expected to linger on for a longer time [37]. Not all types of shared mobility have been affected by the same degree, however. Whereas the negative impact on car-sharing has been more evident and longer, the demand for bike-sharing has been on the rise again in many cities around the world [38, 39].

New mobility technologies and services are expected to increase the affordability, availability, and accessibility of urban mobility, and reduce the existing inequalities in accessing to mobility and right-of-way in cities. Increased competition in the provision of mobility services is expected to drive down travel costs, while shared mobility and on-demand mobility models provide greater efficiency and availability. Additionally, mobility as a service (MaaS) apps make it more convenient to pay for mobility as well as travelling across different modes. MaaS can help reduce car ownership and, therefore, facilitate transitioning to active and sustainable urban mobility.

By reducing the dependency on car ownership, new mobility models increase vehicle utilization rates and road space usage efficiency. By reducing the need for parking spaces, new mobility models unlock spatial opportunities for more sustainable usage of urban space. However, this raises some other problems, such as mixed use of the road space by different mobility modes such as cycling, walking, and public transport. Local governments now play a more critical role in regulating the use of road space in a fair and equitable way across different mobility modes. For example, pop-up cycle lanes installed during the pandemic are expected to remain even after the pandemic. Local governments restrict access to certain roads and streets or introduce speed limits to facilitate sharing road space across different mobility modes. Therefore, at this juncture, it is critical for local governments to steer the course of the transition to sustainability by putting forward a holistic framework that would help capitalize on the dynamics shaping the urban mobility landscape. New mobility services, technologies as well as models, presented in this section offer an unprecedented opportunity for the local transport authorities, as critical agents of realizing the 'public good' in urban areas and cities. A sustainable governance framework, explicated in the next section, would help these critical actors 'seize the moment'.

4 Defining a Sustainable Governance Framework for Urban Mobility

As new developments for urban mobility occur and change the urban mobility landscape, the lines between the conventional public transport modes and new mobility models become blurry. This also challenges the existing governance frameworks and regulatory structures and systems. New modes of mobility facilitating a shift towards sustainability need to be clearly defined, as well as where they can be used (right-of-way) and the governance structure for this new urban mobility ecosystem. In this context, the role of local governments should be reconsidered. The key governance challenges surrounding the new mobility systems include fragmented mobility institutions [40], lack of future-oriented strategic planning [41], and opacity of public–private partnerships [42]. In addition, the lack of formal engagement structures or networks among the local policymakers and planners, mobility innovators, and various mobility groups is also a noteworthy challenge for a robust governance framework for the new mobility landscape. One of the gravest challenges, however, came in with the COVID-19 pandemic: the severe impact on local budgets. Increasing financial pressure limits the capability of local governments.

In this new urban mobility landscape, local governments need to provide a suitable environment for mobility innovation, enabled by a flexible, responsive, and participatory regulatory system. Transforming the former car lanes into cycle lanes, widening pedestrian walkways, installing EV-charging infrastructure, setting out regulations for AVs are just some examples of how crucial the role of local governments is in this new urban mobility landscape. Moreover, in order to address the representation issues regarding the needs and mobilities of diverse groups, the governance of mobility has to be designed inclusive of civil society.

The trend towards the public–private partnership models in recent decades has led to a compromise in the sustainability of the urban mobility systems [43]. The new mobility landscape, on the other hand, necessitated reconsidering the key role of local governments, or transport authorities, as protector of the social outcomes in this new mobility environment. As opposed to the economic partnership schemes that in the long term help private companies and financial institutions to gain profit while funneling public money into guarantees for e.g. bridge crossings which only a small group of the population use, the new mobility landscape should prioritize everyone's needs and also protect non-users. In this sense, the first and overarching principle of the proposed governance model would be to eliminate inequalities. Second, the regulatory environment needs to consider the developments in micromobility vehicles, zero-emission vehicles, autonomous vehicles, mobility as a service, open data, and private hire vehicles legislation. Third, the local governments should also have a suitable regulatory framework which would allow them to be flexible, innovative, and sustainable. On the one hand, the local governments should support an innovative mobility ecosystem by engaging with mobility entrepreneurs such as technology start-ups specializing in mobility. On the other hand, they should adopt a participatory and inclusive approach to establishing their mobility ecosystem by engaging with different mobility groups and NGOs, such as cyclists' groups, disabled people associations, etc.

Figure 2 illustrates the proposed tripartite governance framework built upon the three dimensions, movement, representation and practice, that would facilitate the introduction of new mobility models as well as sustainable urban mobility in cities. The proposed framework is inspired by Sheller's [8] "nested approaches" to ensure a transition to just decision-making systems for urban mobility (pp. 30–35). In this sense, **distributive** processes refer to the essential accessibility opportunities provided to communities by local and central governments and are highly covered in the conventional accessibility-based approach to transport (e.g. [44]). **Deliberative** processes propose recognising civil society, communities and groups with vulnerabilities and how their needs may be acknowledged in mobility governance.

In this sense, the bond between local governments, professional chambers and civil society would be more robust compared to the bond between these groups and the central governments (as shown as a dotted line). For example, the Department for Transport (DfT) in the UK updated the 'Inclusive Mobility' program in December 2021, acknowledging the changing needs of individuals with disabilities as well as engaging with their representatives. The guidance published by DfT [45] describes features that need to be considered in the provision of inclusive mobility related to disabling barriers, the use of technology, maintenance, awareness of the needs of disabled people, and engagement. Like deliberative processes, **epistemic** processes also refer to mechanisms for participation and include "proactive knowledge production to fill in knowledge gaps" that other scientific and deliberative processes may not cover [8, p. 33]. Participation mechanisms can be best established with the efforts of local governments, NGOs and other civil society agencies. Moreover, start-ups and entrepreneurs would play a key role in the constant production of knowledge using state-of-the-art techniques, tools and platforms for co-producing data and transferring this into new products. For example, Open Street Map (OSM) initiative is a collective project to create a free editable geographic database of the world. The data from OSM can be used in various ways, including route planning, geocoding of address and place names, etc. Real-time traffic information by Google Maps helps people plan their travels, hence helping to reduce traffic congestion. Therefore, it is key for actors generating and collecting mobility data to set up platforms for sharing this data with these third-party developers. For example, London TfL's Open Data initiative, whereby all public TfL data is freely released for developers to use in their own software and services (e.g. presenting customer travel information in innovative ways) is a good example of a collaborative approach in this regard. In our framework, both deliberative and epistemic processes include bidirectional relationships between different actors.

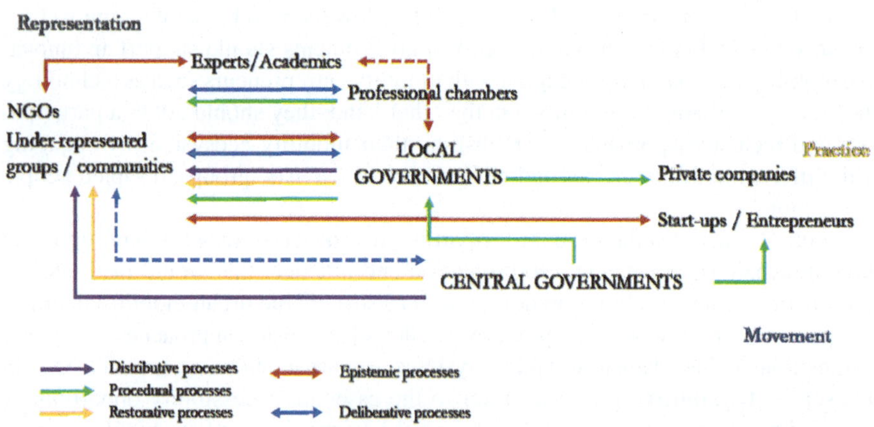

Fig. 2 Key actors and their relationships within the proposed governance model (Authors' own elaboration based on Sheller's [8] nested approaches)

Sheller [8] explains procedural processes as the conditions for meaningful participation of the affected population in the governance systems. Here, we describe **procedural** processes as the whole regulatory and informative procedures as well as mechanisms that empower communities to be able to participate in decision-making systems meaningfully. These regulatory frameworks for participation are often described by local and central governments and usually overseen by professional chambers. For example, 'Sustainable Mobility Plans (SUMP)' across European cities set out procedures for ensuring active support from the public and stakeholders. The engagement of citizens and stakeholders is regarded as crucial to creating a positive foundation for collaborative planning, improving the knowledge base, considering new ideas and opinions and increasing the overall quality of decision making [46].

The final component of our sustainable mobility governance framework is the restorative processes [8]. **Restorative** processes cover the reparations made by local and central governments to those who may be subject to more vulnerabilities compared to others due to policy change and/or disruptions in the system. In relation to the mobility system, disruptions due to the COVID-19 pandemic have forced governments to compensate vulnerable populations as investment was diverted to the neighbourhood level for healthier and sustainable mobility options. For example, the City of Seattle has put equity at the forefront of their urban mobility policies to eliminate the disproportionate impact of the pandemic on ethnic groups as well as low-income communities. Towards this end, Seattle launched street reallocation programs to address mobility injustices as well as setting out a general 'Transportation Equity Framework'.

This framework places a central role on the local governments as the main actor in the urban mobility landscape. In addition to the processes described above, local governments can play a key role in mobility innovation by fostering experimentation and trialing. By creating standards and platforms, they can also encourage the sharing and harnessing of mobility data. Partnership with universities and industry to support the research and development of low carbon technologies is also key for enabling the new mobility forms to get implemented in cities. Therefore, building local capability and governance is crucial in designing and allocating the urban space for the new urban mobility environment as well as supporting proactive knowledge production with citizens and NGOs [8]. The potential of political will in inducing the socio-technical changes quickly, once deemed as requiring a long and gradual transition process before the pandemic, has been made evident during the pandemic [47]. This arguably paves the way for a more favorable political context where local governments can mobilize resources and measures quite easily and quickly in initiating policies to enable sustainability-oriented pathways.

One of the major barriers to effective and robust decision-making to enable new urban mobility systems may also arise from the structures of the local governments due to the entrenched bureaucratic way of working. Building future-oriented, flexible and agile thinking into the local governments' decision-making can be achieved by providing strategic roadmaps and guidance for mobility modelling. Local governments also need an evidence base for evaluating the impact of new mobility technologies as well as conducting analysis and research. They should also continuously monitor and understand the public perceptions and expectations in a dynamic mobility environment through engagement with the public, NGOs, mobility groups as well as innovators, entrepreneurs, and academics.

Ensuring the safety of an increasingly connected, integrated, and digitalized mobility network against cyber-attacks requires the protection of technological systems, which can be achieved by the local governments. Ensuring the privacy of users' data in such a connected and digitalized mobility system is also crucially important. Local governments, with the help of national governments, need to play an active role in this area as well. Another risk that may arise from emerging mobility models is that mobility providers may not be willing to share the travel data with local governments to protect their business interests. However, this may limit the ability of local governments to plan and manage the mobility ecosystem adequately. Therefore, local governments should have a role in limiting the emergence of a dominant mobility provider, which can potentially reduce the affordability of services by capitalizing on their market dominance. Local governments have a greater role in ensuring that mobility data is received from service providers for better planning and protection of customer privacy. They should set out clear regulations to mitigate potential tensions that may arise, damaging the commercial opportunities of mobility entrepreneurs or the social benefits of mobility groups. Local governments, or transport authorities, need to develop data sharing regulations, standards, and formats so that the value gained from the mobility data is beneficial both for passengers and entrepreneurs.

New mobility models may negatively impact the use of conventional public transport modes such as metros and buses, if not managed well by a local transport authority. This, in turn, may reduce the revenues of local governments and public transport operators, which already suffer from a severe drop in demand and revenues during the pandemic. For example, autonomous cars may be more attractive to passengers compared to public transport, cycling, and walking, and this may lead to a reduction in demand for more sustainable modes. This can further aggravate inequalities in access to transport, as lower-income groups rely more on conventional means of public transport. Therefore, coordinating the mobility ecosystem through a local authority is also key for mitigating transport inequalities.

In terms of managing micromobility in cities, local governments should consider the right-of-way for these vehicles, and provide a suitable environment to allow testing for these vehicles and ensure the safety of these vehicles on the road as well as the safety of other road users (especially the elderly, children and individuals with disabilities). They should identify the key parameters for the safe design and road allocation of these vehicles such as e-bikes, and enable trials of innovative ideas in this area [48]. For example, local governments in some cities have helped micromobility service providers to improve safety, by providing helmets or modifying vehicle designs to take into account the uneven pavement [49].

Using the tripartite governance framework proposed above, the local governments should both engage with the entrepreneurs developing, for instance, e-bikes or e-scooters by facilitating the update of innovative models in this area and also engage with e-bike or e-scooter mobility groups to better understand and respond to their needs. Therefore, the innovation would have dual benefits: both in technology development and use **(practice)**, and in fostering community engagement **(representation)** while the local and central governments are providing infrastructure for **movement**. In relation to the e-bike/e-scooter example, the local governments should also consider their impacts on the design of cities. In this sense, they should set out necessary measures and regulations for the reallocation of streetscape. For example, London TfL launched e-scooter rental trials in June, 2021 (expected to run until November 2022) in collaboration with three different e-scooter operators: Dott, Lime, TIER. The operators were chosen after an open and competitive process to assess their ability to meet strict safety requirements and high operating standards. (For example, the scooter batteries can be monitored to ensure they meet fire safety regulations.) Participating boroughs have designated no-go areas where e-scooters cannot be ridden and where the rental e-scooters will come to a safe stop. There are also go-slow areas where the speed of e-scooters will be reduced to 8 mph. Therefore, in this new urban mobility environment, local governments should be more proactive in enabling emerging mobility technologies and models.

Another example of the implications of new mobility models on governance is from mobility as a service (MaaS) model. MaaS mobility model requires a seamless integration of various modes of mobility in terms of scheduling, ticketing, and interoperability of data from different service providers. Both central and local governments can assist in ensuring the integration and interoperability of these services to achieve a socially optimal outcome. The governments, both central and local, have a greater role in making sure that new mobility services enable accessible, equitable, safe, and inclusive mobility services. The Department for Transport (DfT) in the UK has carried out a public consultation to gather evidence on a code of practice for MaaS. DfT recognizes the potential of emerging business models such as MaaS to make the planning and payment of trips easier for consumers. The consultation report, published in February, 2022, aims to provide clarity by highlighting the roles and responsibilities of various actors in the MaaS ecosystem, provide guidance on central government objectives and highlight areas of best practice for MaaS solutions [50].

The role of the local governments is not limited to regulating the mobility landscape. They might be direct providers of some of the emerging mobility models. For example, many local governments around the world (e.g. Copenhagen, Amsterdam, New York, etc.) are installing bike sharing schemes in their cities to promote sustainable urban mobility. COVID-19 has accelerated the implementation of these schemes as the implementation and operational costs of these schemes are quite low besides their health, social, and environmental benefits. The role of local governments is crucial also for laying out the supporting infrastructure (e.g. cycle lanes) for the deployment of bike-sharing schemes [51].

5 Conclusions

The new mobility models and services presented in this chapter have brought about new challenges for urban mobility governance. These services are essential for cities to achieve their transition to sustainable urban mobility that is also fair and inclusive. Therefore, developing effective governance and regulatory structures to accommodate these models has been crucial to aligning these services with the sustainable urban mobility goals of the cities. Local governments must go beyond just reconciling the conflicting interests among the new urban mobility models and shape these dynamics towards sustainable urban mobility. The critical dynamics that local governments should consider while managing this transition are urban development and land use, sharing economy, digitalization, e-commerce, teleworking, electrification, and automation.

Having examined the current mobility landscape with the new and ongoing challenges through the entanglement of movement, representation and practice, we have been able to highlight the critical relationships between different stakeholders and actors for mobility governance. Adding the "nested approaches" [8] helped us configure the new dynamics within this governance and trajectories each relation should follow to ensure a transition to just decision-making systems for urban mobility.

The development of new urban mobility models may play out differently in different contexts, particularly in more developing settings. The key concepts and governance challenges identified in this chapter provide specific insights into the governance and policy framework needs for deploying new urban mobility models. Adopting an equality lens to urban mobility governance requires reassigning roles to the actors within governance networks. This facilitates the redistribution of power and transition to a sustainable and resilient governance system. The governance framework proposed in this chapter addresses those needs, considering the unique governance contexts of cities. However, this chapter does not aim to provide immediate solutions for the issues arising out of the new mobility landscape in cities but rather seeks to introduce the new governance challenges with a multi-actor framework providing conceptual tools to understand the new urban mobility landscape.

COVID-19 pandemic has reemphasized urban mobility as a unified system rather than a loose aggregation of individual mobility modes. Therefore, the role of local governments and transport authorities is now ever more critical in accelerating a shift towards a sustainable urban mobility system. Uncertainty is now the parcel of the current mobility environment. No one could have imagined how urban mobility would change in 2020 with COVID-19. The evolution of urban mobility will continue, with different impacts on the current state of "movement, representation, and practice" in urban mobility. Setting out scenario plans, roadmaps, and alternative futures for the new urban mobility models helps local governments to steer this evolution.

References

1. Burrieza J, Djukic T, Aifadopoulou G, Salanova Grau JM, Masegosa DA, Rojo MJ, Fabianski C (2020) New mobility options and urban mobility. Challenges and opportunities for transport planning and modelling. Momentum
2. Van Wee B, Witlox F (2021) COVID-19 and its long-term effects on activity participation and travel behaviour: a multiperspective view. J Transp Geogr 95:103144
3. Banister D (2008) The sustainable mobility paradigm. Transp Policy 15(2):73–80
4. Holden E, Banister D, Gössling S, Gilpin G, Linnerud K (2020) Grand narratives for sustainable mobility: a conceptual review. Energy Res Soc Sci 65:101454. https://doi.org/10.1016/j.erss.2020.101454
5. Jeyaseelan T, Ekambaram P, Subramanian J, Shamim T (2022) A comprehensive review on the current trends, challenges and future prospects for sustainable mobility. Renew Sustain Energy Rev 157:112073
6. Vecchio G, Tiznado-Aitken I, Mora-Vega R (2021) Pandemic-related streets transformations: accelerating sustainable mobility transitions in Latin America. Case Stud Transp Policy 9(4):1825–1835
7. Cresswell T (2010) Towards a politics of mobility. Environ Plan D: Soc Space 28:17–31
8. Sheller M (2018) Mobility justice: the politics of movement in an age of extremes. Verso Books
9. Urry J (2007) Mobilities. Routledge
10. ITF (2019) Transport outlook 2019. International Transport Forum. https://www.oecd-ilibrary.org/sites/transp_outlook-en-2019-en/1/2/1/index.html?itemId=/content/publication/transp_outlook-en-2019-en&_csp_=1b3375008054c148f41fef71cd42b552&itemIGO=oecd&itemContentType=book. Accessed 21 Feb 2021
11. IATA (2019) Passenger market analysis October 2019. International Air Transport Association. https://www.iata.org/contentassets/43f535d0113646ebbca928acce402f84/passenger-analysis-oct-2019.pdf. Accessed 21 Feb 2021
12. IPCC (2014) Climate change 2014: mitigation of climate change chapter 8: transport. Intergovernmental Panel on Climate Change. https://www.ipcc.ch/report/ar5/wg3/transport/. Accessed 21 Feb 2021
13. MOBIS (2021) COVID-19 mobility behaviour in Switzerland - coronavirus study. https://ivtmobis.ethz.ch/mobis/covid19/reports/latest#21_Updated_conclusions. Accessed 21 Feb 2021
14. Statistica (2021) International car sales since 1990. https://www.statista.com/statistics/200002/international-car-sales-since-1990/. Accessed 15 Feb 2021
15. Ergocun G (2021) Turkey: auto sales up over 61% in 2020. Anadolu Agency, 6 Jan 2021. https://www.aa.com.tr/en/economy/turkey-auto-sales-up-over-61-in-2020/2100452. Accessed 15 Feb 2021
16. Kubota Y (2020) China car sales rise again, cementing pandemic recovery. The Wall Street Journal, 8 Dec 2020. https://www.wsj.com/articles/china-car-sales-rise-again-cementing-pandemic-recovery-11607427907. Accessed 15 Feb 2021
17. WARC (2021) India's used-car sales boom amid pandemic with consumers now ready to wheel and deal online. WARC, 7 Jan 2021. https://www.warc.com/newsandopinion/news/indias-used-car-sales-boom-amid-pandemic-with-consumers-now-ready-to-wheel-and-deal-online/44536. Accessed 15 Feb 2021
18. Lucas K (2012) Transport and social exclusion: where are we now? Transp Policy 20:105–113
19. Banister D (2018) Inequality in transport. Alexandre Press, Oxfordshire, UK
20. Beyazit E (2011) Evaluating social justice in transport: lessons to be learned from the capability approach. Transp Rev 31(1):117–134
21. Kaufmann V, Bergman MM, Joye D (2004) Mobility as capital. Int J Urban Reg Res 28(4):745–756
22. Pereira R, Schwanen T, Banister D (2017) Distributive justice and equity in transportation. Transp Rev 37(2):170–191
23. Kern L (2019) Feminist city: claiming space in a man-made world. Verso Books

24. Akyelken N (2020) Transport for women: who decides what women need? Transp Rev 40(6):687–688
25. Harvey D (2003) The right to the city. Int J Urban Reg Res 27(4):939–941
26. Sputnik (2020) Bodrum'da oteller doldu, kiralık ev kalmadı: Yılbaşı için geliyorlar. 21 Dec 2020. https://tr.sputniknews.com/turkiye/202012211043441823-bodrumda-oteller-doldu-kiralik-ev-kalmadi/. Accessed 15 Feb 2021
27. Gould E, Shierholz H (2020) Not everybody can work from home. Economic Policy Institute, 19 Mar 2020. https://www.epi.org/blog/black-and-hispanic-workers-are-much-less-likely-to-be-able-to-work-from-home/. Accessed 15 Feb 2021
28. Faisal A, Kamruzzaman M, Yigitcanlar T, Currie G (2019) Understanding autonomous vehicles. J Transp Land Use 12(1):45–72
29. Department for Transport (2019) Future of mobility: urban strategy, Moving Britain Ahead. https://assets.publishing.service.gov.uk/government/uploads/system/uploads/attachment_data/file/846593/future-of-mobility-strategy.pdf
30. Canitez F (2019) Pathways to sustainable urban mobility in developing megacities: a socio-technical transition perspective. Technol Forecast Soc Chang 141:319–329
31. Geels FW (2002) Technological transitions as evolutionary reconfiguration processes: a multi-level perspective and a case-study. Res Policy 31(8–9):1257–1274
32. Geels FW (2012) A socio-technical analysis of low-carbon transitions: introducing the multi-level perspective into transport studies. J Transp Geogr 24:471–482
33. Bloomberg New Energy Finance (2018) Electric vehicle outlook 2018. https://about.bnef.com/electric-vehicle-outlook/. Accessed 2 Feb 2021
34. Kent JL, Dowling R (2018) Commercial car sharing, complaints and coping: does sharing need willingness? Urban Policy Res 36(4):464–475
35. Curtis C, Stone J, Legacy C, Ashmore D (2019) Governance of future urban mobility: a research agenda. Urban Policy Res
36. Shaheen S, Wong S (2021) Future of public transit and shared mobility: scenario planning for COVID-19 recovery
37. Shokouhyar S, Shokouhyar S, Sobhani A, Gorizi AJ (2021) Shared mobility in post-COVID era: new challenges and opportunities. Sustain Cities Soc 67:102714
38. Hu JW, Creutzig F (2022) A systematic review on shared mobility in China. Int J Sustain Transp 16(4):374–389
39. Teixeira JF, Silva C, e Sá, FM (2021) The motivations for using bike sharing during the COVID-19 pandemic: insights from Lisbon. Transp Res Part F: Traffic Psychol Behav 82:378–399
40. Rye T, Monios J, Hrelja R, Isaksson K (2018) The relationship between formal and informal institutions for governance of public transport. J Transp Geogr 69:196–206
41. Legacy C, Ashmore D, Scheurer J, Stone J, Curtis C (2019) Planning the driverless city. Transp Rev 39(1):84–102
42. Ashmore DP, Stone J, Kirk Y (2019) The need for greater transparency when assessing the performance and prospects of Melbourne's rail franchise contracts. Urban Policy Res 37(1):82–96
43. Currie G, Truong L, De Gruyter C (2018) Regulatory structures and their impact on the sustainability performance of public transport in world cities. Res Transp Econ 69:494–500
44. Geurs KT, Van Wee B (2004) Accessibility evaluation of land-use and transport strategies: review and research directions. J Transp Geogr 12(2):127–140
45. Department for Transport (2021) Inclusive mobility: a guide to best practice on access to pedestrian and transport infrastructure. https://assets.publishing.service.gov.uk/government/uploads/system/uploads/attachment_data/file/1044542/inclusive-mobility-a-guide-to-best-practice-on-access-to-pedestrian-and-transport-infrastructure.pdf
46. Lindenau M, Böhler-Baedeker S (2014) Citizen and stakeholder involvement: a precondition for sustainable urban mobility. Transp Res Procedia 4:347–360
47. Manderscheid K, Cass N (2022) A socio-ecologically sustainable mobility regime: can we move beyond the car? Editorial for the special issue "Shapes of socio-ecologically sustainable mobility regimes". Appl Mobilities 1–14

48. ITDP (2019) The electric assist: leveraging e-bikes and e-scooters for more livable cities. Institute for Transportation and Development Policy. https://www.itdp.org/wp-content/uploads/2019/12/ITDP_The-Electric-Assist_-Leveraging-E-bikes-and-E-scooters-for-More-Livable-Cities.pdf. Accessed 2 Feb 2021
49. Zarif R et al (2019) "Small is beautiful Making micromobility work for citizens, cities, and service providers." Accessed on June 2022 https://www2.deloitte.com/us/en/insights/focus/future-of-mobility/micro-mobility-is-thefuture-of-urban-transportation.html
50. Department for Transport (2022) Consultation on mobility as a service code of practice. https://assets.publishing.service.gov.uk/government/uploads/system/uploads/attachment_data/file/1055553/consultation-on-mobility-as-a-service-code-of-practice.pdf
51. Burriez, J, Djukic T, Aifadopoulou G, Salanova Grau JM, Masegosa DA, Rojo MJ, Fabianski C (2020) New mobility options and urban mobility. Challenges and opportunities for transport planning and modelling. Momentum
52. Fashingbauer Cooper G (2020) Coronavirus puts R.E.M.'s It's the End of the World As We Know It on charts. CNET, 16 Mar 2020. https://www.cnet.com/news/coronavirus-has-r-e-m-s-its-the-end-of-the-world-as-we-know-it-climbing-charts/. Accessed 15 Feb 2021

Rethinking the Sustainable Housing Concept in the Post-pandemic Era

Osama Omar

Abstract There was a surge of change during the Coronavirus period—even in the way we thought about architecture principles to improve sustainability. This period of quarantine gave scientists time to reconsider the main principles behind theories of sustainable housing because many new factors had emerged, such as the health and wellbeing of occupants, the science of providing a clean indoor environment without the contamination of coronavirus, etc. People eventually adjusted to the prevailing situation despite a lot of discrepancies initially, allowing them to continue functioning. A temporary change is one type of change, while a permanent change is another. After the outbreak of COVID, residential architecture has also seen many changes in priorities, just like any other industry. Changes in lifestyles and habits have influenced home designs. A sustainable housing environment needs to be addressed in the design and construction sectors. With this disease affecting our design and construction methods, as well as standards and regulations, how should we adjust and review them? In what ways can new technologies in construction and design make a difference? Architects face a huge challenge now, compared to before the Coronavirus period, with new houses since there is no guarantee that the indoor spaces will be free from Coronavirus contamination. Creating a safe, clean, and sustainable dwelling in the post-pandemic age is the goal of this chapter. It explains the principles that can assist in creating a clean environment in the house. One of the most significant contributions of this chapter is the formulation of new theories and principles for sustainable housing in the post-pandemic era. The development of guidelines or a framework for architects intent on designing sustainable housing concepts worldwide is also a contribution.

Keywords Post-pandemic era · Sustainability · Rethinking of housing concept · Coronavirus contamination · Clean and safe indoor environment

---

O. Omar (✉)
Associate Professor of Architecture, Faculty of Architecture—Design and Built Environment, Beirut Arab University, Beirut, Lebanon
e-mail: o.omar@bau.edu.lb

Z. Allam (ed.), *Sustainable Urban Transitions*, Urban Sustainability,
https://doi.org/10.1007/978-981-99-2695-4_19

1 Introduction

According to world health organization in world health day 2021, which mention in Health equity and its determinants. Through the global "stay at home" response, COVID-19 has increased people's exposure to health risks associated with poor housing. The housing, water and sanitation challenges in urban slums make over a billion people highly susceptible to COVID-19. Additionally, COVID-19 has demonstrated the impact housing can have on resilience, well-being, and access to wider social networks [1].

As a result of the COVID-19 pandemic, cultures around the world have changed their simple assumptions regarding human life. As a result of the quarantine, scientists have been able to study how the pandemic impacts health, personal control, planning, family relationships, education, employment, and mobility and migration within and across these core areas [2].

The outcome of this crisis indicates that we will not return to normal afterward, and one of the most positive expectations is that we will begin anew. A shift is occurring in beliefs, traditions, and homes as a result of this. In terms of social isolation, the theory of arranging people in a box-shaped complex building does not follow the current practices of the house and is in contradiction with its fundamental purpose. In addition, houses are safer than apartments from the perspective of public health. Studies have revealed that homes play a significant role in the decline of disease, mental health, and social security in families [3].

In his book *Designing Disorder: Experiments and Disruptions in the City* [4], Richard Sennett suggests that in the future a new concept of building creation will arise and the movement towards larger neigbourhoods will enable people to socialize without being filled in like sardines, taking into account the healthier foundations of construction design [4]. This is also consistent with a study by Sennett 2020 [3].

Social behavior indicates that contemporary urban dwellers spend a great deal of time working at home. The design of housing directly impacts their health, and if it is bad, it can negatively impact public health. House closures and isolation steps disrupted the conventional idea of houses as places for sleeping, working, and playing. There may be a need for regulations to be amended and studied in the future in order to take into account the suggested "social distances". Indoor design should also be more focused on the changing responsibilities that people are taking on at home. A workplace, for example, can be a well-organized, fully equipped private room. As a result, this concept encompasses not just private homes, but also all types of public buildings, such as schools, waiting rooms, and so on [3].

Designers might return to nature when remodeling our homes or adopt a biophilic design approach to general housing efficiency as a consequence of household separation, which results in carefully developing dwellings to increase their quality. Using natural components may be good in reducing alienation and other psychological consequences, which is consistent with a prior paper that advocates reconsidering and developing roofs in undeveloped regions. Furthermore, it is critical to preserve verandas as an outside area, building ties with nature, the city, and green spaces, as well as social relationships among neighbors [3].

Housing should create a healthy and therapeutic environment for its people. A variety of technical, social, planning, and policy issues can all have an impact on physical, mental, and social well-being. These elements may be expressed in terms of basic human needs and so integrated into housing standards, regulations, and attainment targets that are crucial for each specific country's needs, resources, and ambitions. It is not always feasible to interpret healthy housing, but normal criteria may serve as the basis for the variables needed. Unfortunately, many individuals throughout the world are housed in shelters that do not provide even their most basic requirements. In reality, a number of accessible shelters not only fail to protect against but also expose occupants to totally avoidable health hazards. Adopting safe housing concepts will assist countries, neighborhoods, and families in defending themselves against the hazards involved [5].

Rehabilitation for compassionate purposes is commonly believed to be acceptable because it increases the community's physical, emotional and social wellbeing for the greater good. The assumption in the linear association between housing circumstances and health status can be crudely articulated as:

$$\begin{aligned} \text{Better housing} &= \text{Better health} \\ \text{Poorer housing} &= \text{Poorer health} \qquad [5] \end{aligned}$$

"Healthy building" is defined mainly by the German Sustainable Building Council as an engineering approach based on European standards, including EN15251: 2007 "Indoor Environmental Input Criteria for Design and Energy Assessment of Buildings Addressing Indoor Air Quality" [6]. In the world, the World Health Organization estimates that 30% of new buildings or renovations may be possible carriers of sick building syndrome. It has been observed in many enclosed office buildings since then that sick buildings syndrome is a common occurrence. Many diseases are suspected to be caused by poor indoor environmental quality, but the exact cause is difficult to establish empirically or categorize [7]. Those living in poor living conditions, for instance, are more likely to contract respiratory diseases because of the lack of decent patios and ventilation. In 2004, Yuguo Li studied SARS transmission simulation in a suburban area of Hong Kong while researching fluid mechanics and disease transmission through ventilation yards [8].

There has been evidence that indoor concentrations of several organic substances are up to five times higher than outdoors, and in some cases they can reach 1,000 times higher. A proper ventilation system achieves adequate indoor air quality by providing fresh air or air quality. By opening doors and windows, natural ventilation can be achieved. It is rare to find hot, moist air in the summer with cooler temperatures. The frequency of this is higher. Mechanically ventilated spaces are typically equipped with a series of ducts that supply air. In natural and mechanically ventilated buildings, mixed airflow systems are provided. In order to reduce the percentage of poor air in a building, fresh air must be brought into the building in greater quantities. There is no requirement for adequate fresh air in the current minimum building code, which is insufficient to ensure the wellbeing of inhabitants—in fact, sick buildings would not have occurred if the minimum code did [9].

It supports the search for ideas, patterns, and hypotheses that can be used to prevent viral infection and continue to add layers to our built environment's security framework. In order to deal with the present pandemic, professional architects, urban planning firms, and architectural firms have already begun to imagine the post-pandemic age.

However, not enough study is being conducted to anticipate what the anti-virus world would look like. To address this deficit, this research examines architectural and home creation during the previous centuries. It then examines the scientific fields affected by the COVID-19 pandemic and highlights the challenges that are important to them. We examine social separation and quarantine as design issues in the post-pandemic period. Some pandemic lessons were afterwards offered in order to conceive and communicate the study's perspective on the world formed by coronavirus.

2 Vision of Healthy Housing According to the World Health Organization

Several years ago, housing environments were identified as significant factors affecting human health. Many factors affect residents' health, including their living and housing conditions [10]. The most significant threats to health in homes are the quality of indoor air, the protection at home, noises, dampness, indoor temperatures, asbestos, lead, and radon, as well as improper sanitation and hygiene. It is well known that living conditions affect physical, mental, and social health, however there have not been clear mechanisms developed to counteract these impacts. Residents' physical, psychological, and social wellbeing is affected by the local living environment. Health and quality of the environment are often linked, but we are not fully aware of exactly how these effects work [11].

Housing and health are interconnected from a public health perspective, despite the lack of definite interventions. Through deductive calculation, it is possible to assume that there is a strong relationship. A residential environment has an impact on health as a whole, since it consists of several elements that can adversely affect health [5].

Housing, more than just a roof, should provide privacy, physical and psychological wellbeing, development, and social inclusion—a vital component of human wellbeing. A holistic definition of "healthy housing" takes into account a range of factors that contribute to housing quality. In addition to being a residential area able to meet the needs of residents, a healthy home is not a specially constructed building. Over centuries of housing development, standards of housing and best practices have been agreed and accumulated, demonstrating that applying established expertise and validated concepts is a key component of constructing a "healthy home" [11].

In the second HABITAT conference in Istanbul [12], one of the United Nations member states defined housing in the following terms:

> A decent shelter is more than a roof above the shoulders. Suitable confidentiality, suitable space, physical mobility, proper occupant safety, tenure security, structural stability and reliability, adequate lighting, heating and ventilation, appropriate basic facilities such as water, sanitation and waste disposal, environmental and health considerations and adequate, and environmentally friendly position; Adequacy for the individuals involved should be calculated, taking into consideration the possibility of progressive growth. Suitability also depends on specific cultural, religious, social, environmental and economic influences from country to country. Gender and age-specific factors, including sensitivity to both kids and women to hazardous chemicals, should be taken into account in this sense.

The relationship between houses and health clearly exhibits a dynamic combination. Additionally, it gives national authorities, local governments, and the general public the opportunity to improve health conditions, whether physically or socially [11].

It is important to remember that housing is not limited to its physical structure. A four-layer model of housing is used by the World Health Organization to define the concept of housing, which considers the physical structure of your home, the dwelling of your home (for each family member and each individual), the external dimensions of the house as they relate to its immediate neighborhood, and all of your neighbours as a community [11] (Fig. 1).

Based on Fig. 1, there are several effects for each dimension that can be expressed either directly or indirectly as health effects or as a restriction of residents' quality of life.

Psychosocial benefits can be derived from a comfortable, private home. As an individual or a member of a family, it can provide a safe haven from the outside world. Making room for oneself is thus possible. The house becomes less psychologically and socially effective when there is any interference from outside forces or stressors.

Health may be adversely affected by inadequate living conditions. Infestations, inefficient heating and insulation systems, mould growth, indoor air pollution, and lack of hygiene and sanitation equipment are the most common challenges. Whether a dwelling is structurally sound, socially accessible, and able to accommodate residents with disabilities is determined by how it is designed. A dwelling's design and layout also play a role in crowding and noise emission, since they are both influenced by other variables outside of a home [11].

In the community, there are a number of variables that are unrelated to housing conditions that affect health. A socioeconomic factor can have a negative impact on health. There are many factors that determine the effects of neighbourhood groups

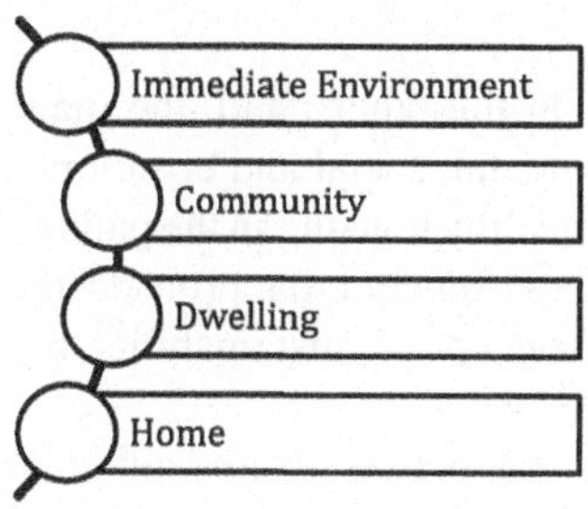

Fig. 1 The four dimension of housing (modified by author)

and city quarters, including education, socioeconomic characteristics, and ethnic composition. There is sufficient evidence from several studies that neighbourhood quality plays a role in facilitating or discouraging social activities by facilitating or prohibiting the presence (or absence) of various public facilities and places. The study provides sufficient evidence that community social cohesion, confidence and mutual effectiveness depend on neighbourhood quality [13].

Additionally, the immediate housing environment has an impact on wellbeing through the quality of its urban design. A lack of public facilities and greenery, parks, children's playgrounds, and walking areas in poorly developed suburban areas, which often lack greenery and greenery, has been linked to a decline in exercise, an increase in obesity among children, and a decline in socializing abilities. In addition to litter, pollution and graffiti, negative neighbourhood factors affect residents through segregation, loitering and increased insecurity. Urban development trends such as urban spread will lead to increased traffic and noise pollution as well as a greater reliance on individual transportation. This will place vulnerable communities such as children, the elderly or people with functional disabilities at risk and leave them isolated [11].

It is important to highlight that each of the four housing dimensions has the capacity to influence individual health status via physical, mental, or social mechanisms. As a result, suitable housing is contingent on acceptable amenities and circumstances in all four sectors. If this requirement is satisfied, housing provides individuals with a foundation for their bodily and emotional well-being. Therefore, it is clear that housing conditions play a relevant role for individuals as well as for public health:

- Health is a state of complete physical, mental and social wellbeing and not merely the absence of disease or infirmity.
- Housing is the conjunction of the dwelling, the home, the immediate environment and the community.
- The role of public health is to provide the circumstances under which people can be healthy [11].

In fact, the meaning of these three assertions is that housing is an issue that health ministries, as well as environmental, social, and building ministries, can address.

3 Rethinking the Sustainable Housing Concept

On the other hand, the main pillars of the sustainable housing concept—environmental, social and economic (Fig. 2)—do not provide the essential requirements for healthy housing in the pandemic era because they miss out the role of a healthy indoor environment that protects against different viruses. This leads to a new approach and concept of sustainability by adding two other pillars, namely health and science (Fig. 3).

A period of quarantine during the Coronavirus pandemic led to the addition of these two pillars (health and science), which represent what the entire world has

Fig. 2 Sustainable pillars [14]

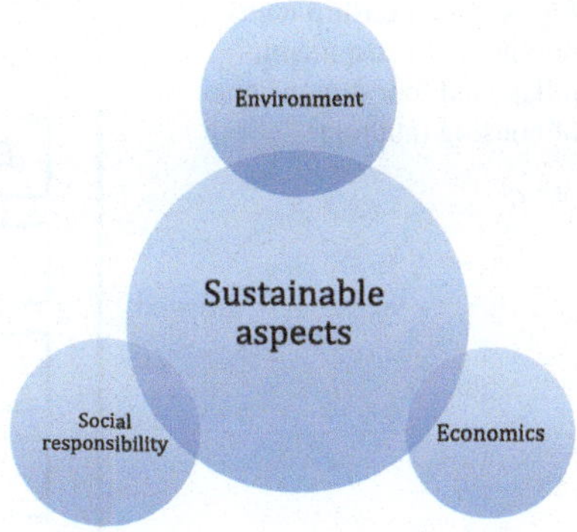

Fig. 3 Rethinking of sustainable pillars (author)

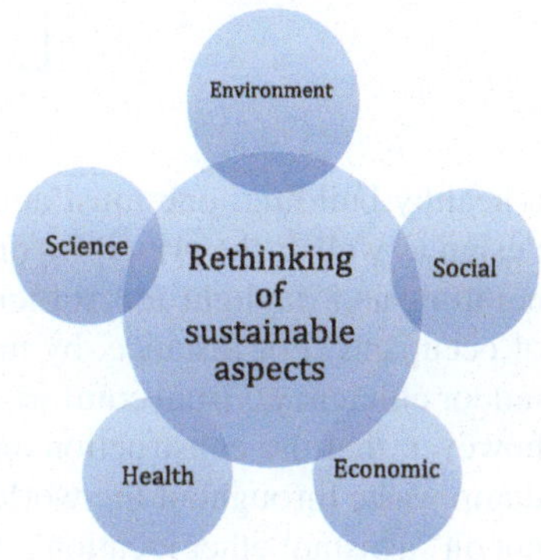

discovered: in 2019, the increase in respiratory infections caused by extreme acute respiratory coronavirus syndrome (COVID-19) caused both scientific and public concern regarding the prevention and regulation of the disease. It is imperative to also address other less well-known pathways of transmission to avoid further spread of respiratory viruses even though several precautions have been implemented that are standard for preventing the spread of respiratory viruses. Throughout history, environmental media have served as routes for pathogens from other houses, particularly hospitals. Several recent studies have examined how pathogens are transmitted, spread, and are present in the built environment, including the number, abundance, and diversity of species. The study of pathogen exchange has revealed many common pathways and mechanisms, which can reveal potential methods for transmitting viruses like COVID-19 via construction sites [15].

From previous studies and from what has been mentioned above, the matrix shown in Fig. 4 can easily explain the relationship between a new concept of sustainable pillars and the four dimensions of housing. Each pillar of the new concept has a direct impact and effect on every category of the four dimensions of housing.

As an alternative, a healthy building should be capable of meeting future needs, adaptive to "new drivers" like climate change, sensitive to changing environments, and adaptable to the preferences and needs of occupants (Table 1). In order to build

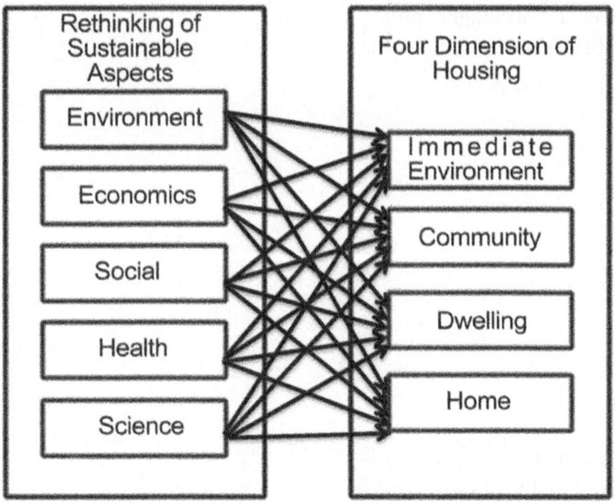

Fig. 4 Relationship matrix: rethinking of sustainable pillars and four dimensions of housing (author)

a healthy building, one must adhere to the required principles of architecture and design as well as the principles of performance. As a result, indoor efficiency, visual comfort, and daylight are sufficient without sacrificing the health and well-being of occupants. For instance, by meeting various technical specifications effectively, indoor efficiency, visual comfort, and daylight can be achieved [7]. Rees [16] argues, however, that the construction and design industries, which draw capital from and dump waste throughout the world, focus entirely on occupants (building users) and not on "a distant other location". It is this ecological reality that emphasizes the need for improved building industry health.

4 The Role of the Built Environment in the Transmission of COVID-19

The built world includes houses, vehicles, highways, public transportation, and other environments built by humans. The built environment plays a major role in 90% of human activity, so it is essential to understand how human behavior, spatial dynamics, and building operating factors can facilitate and mitigate the spread and transmission of COVID-19 in the built environment [15]. Through close human interactions, the use of fomites (infectious objects or materials), viral exchange, and airborne transmission, the built environment can serve as a vector for COVID-19 transmission [17, 18] (Fig. 5).

Buildings with high occupant density facilitate the accumulation of microorganisms associated with people due to type, schedule, and indoor operations. Through direct contact between people as well as communication with the environment, higher occupant density and higher indoor activity levels usually result in an increase in

Table 1 Timeline of building design and development from sustainable, intelligent and healthy design perspectives [7]

1999–2000	2000–2010	2010–2015	2015–2018	2018–future
Sustainable buildings	Intelligent buildings	Sustainable + intelligent buildings	Healthy buildings	Digitally sensible and fully responsive healthy building
Gradual attention to sustainable buildings	Increasing attention to green buildings	Green buildings are globally widespread	Increasing attention to health and wellbeing in buildings	Buildings to become fully sensible and responsive
USGBC establishment	Building automation as a core focus	Low-energy and BIM application	Sick building syndrome becomes a crucial issue	Digital design for enhancing health and productivity
Economic use of energy waste, water and pollution	More about directions towards intelligent buildings	Intelligent buildings incorporated with sustainable design	Healthy buildings as part of a smart city	Health as the focal point of future buildings
Global environmental movement + sustainable development movement	Technological dimension of building	Gradual shift to the social dimension of buildings	Emphasis on productivity and health design	

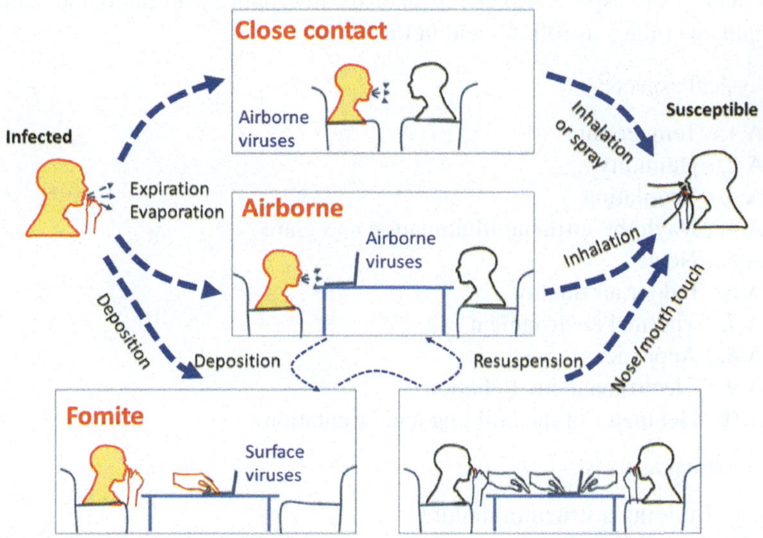

Fig. 5 An illustration of various possible transmission routes of respiratory infection between an infected and a susceptible individual [18]

social interaction, connectivity and the contact with abiotic surfaces. By demonstrating COVID-19's high transmissibility in these cases, the confined spaces within the built environment demonstrate its high transmissibility. There are both direct and indirect contacts with the surrounding surfaces during manufacturing processes. Footfall, walking, and warm human body thermal feathers can cause indoor turbulence that precipitates viral particles directly, resuspended by normal airflow patterns, artificial airflow patterns, or other causes. It is likely that the viral particles will then be reinstalled in the fomites and redeployed again. As a result of touching a surface, viruses are transferred from the person to the surface. Symptoms of COVID-19 infection are followed by the release of viral particles in the body. Abiotic BE items can serve as reservoirs for viral transmission when these viral particles are deposited on them. According to the evidence, fomites can be contaminated by bodies secretions, such as saliva, nasal fluid, soiled hands, and large droplets of aerosolized virus from people infected (SARS-CoV-2). Despite the fact that COVID-19 is only transmitted through respiratory gout and not via fomite deposition, washing and disinfecting all possible sources (SARS-CoV-2) are necessary to prevent an active virus from being transmitted by abiotic surfaces. There is a need to be cautious of the possibility that the infection will be spread through aerosols and surfaces [15] (Fig. 6).

5 Guidelines for a Sustainable Healthy Housing Concept

The guidelines for a sustainable healthy housing concept are based on five major aspects and 34 sub-aspects, which reflect on the five main categories of the rethinking of sustainable pillars as follows and in Fig. 7:

A. Physical aspects:

   A.1. Temperature
   A.2. Humidity
   A.3. Ventilation
   A.4. Daylight, artificial illumination and glare
   A.5. Noise
   A.6. Indoor air quality
   A.7. Thermal environment
   A.8. Acoustics
   A.9. Electromagnetic radiation
   A.10. Geometry of the building and orientation

B. Biological aspects:

   B.1. Building's structural failure
   B.2. Poor air ventilation
   B.3. Improper maintenance of building
   B.4. Local odour of mould
   B.5. Exposure to fungi

Fig. 6 SARS-CoV-2 deposition conceptualization. **a** Viral particles collect in the lungs and upper respiratory system after a person has been infected (SARS-CoV-2) [19]. Droplets and infectious airborne spores, through mechanisms such as coughing, sneezing and voice, and non-routine incidents such as vomiting, are released from the body by everyday work and spread to adjacent ecosystems and persons. **c** and **d** Mouth and nose viral spores are frequently detected on hands [20] and regularly touched objects (**d**), such as laptops, glasses, taps and countertops. Currently, fomite-to-human transmission is not verified, however, viral particulates are present on abiotic built surfaces [15]

C. Chemical aspects:

   C.1. Fine dust
   C.2. Volatile organic compounds (VOCs)
   C.3. Microbial origin (MVOCs)
   C.4. Airborne levels of bacteria
   C.5. Formaldehyde and the plasticizer Texanol

D. Psychosocial aspects:

   D.1. Anxiety
   D.2. Depression
   D.3. Environmental discomfort
   D.4. Job strain
   D.5. Reducing the occupants' performances
   D.6. User stress (systemic stress and psychogenic stress)

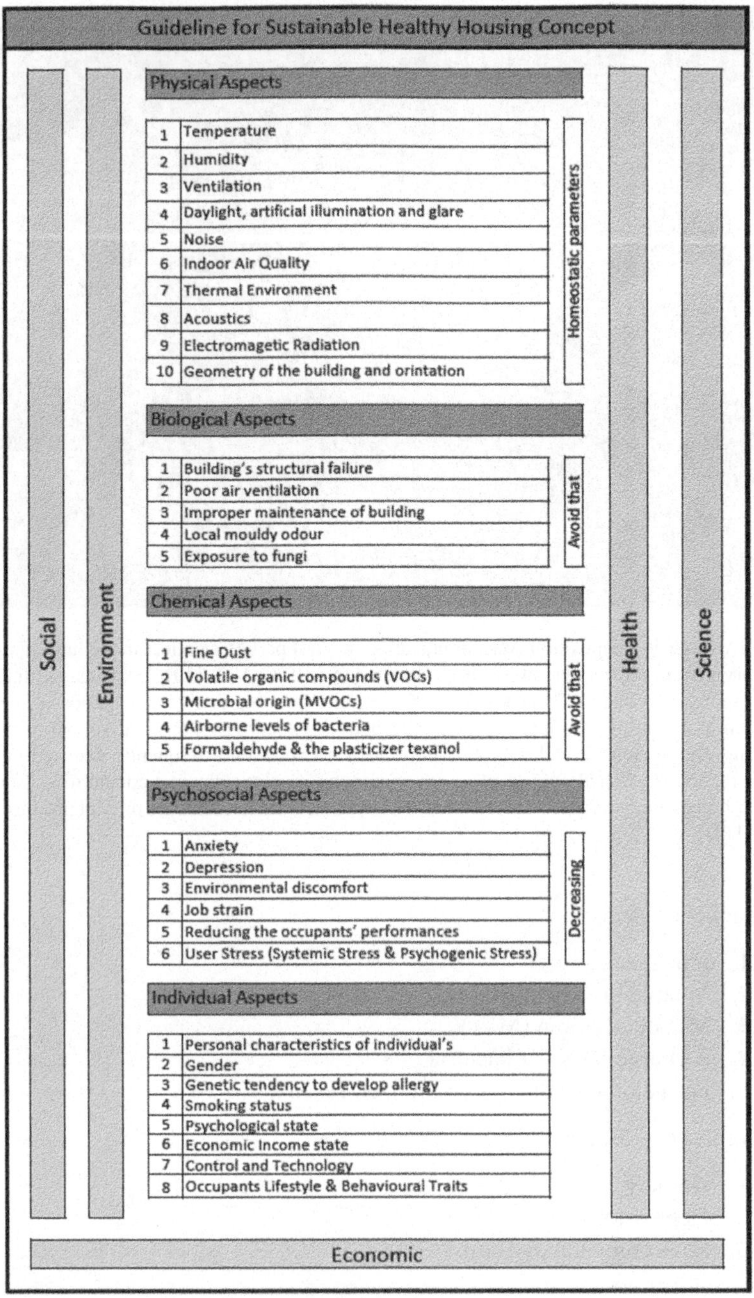

Fig. 7 Guidelines for sustainable healthy housing concept (author)

E. Individual aspects:

   E.1. Personal characteristics of individuals
   E.2. Gender
   E.3. Genetic tendency to develop allergies
   E.4. Smoking status
   E.5. Psychological state
   E.6. Economic income state
   E.7. Control and technology
   E.8. Occupants' lifestyle and behavioural traits

It's important to mention that these five major aspects—physical, biological, chemical, psychosocial and individual—have been mentioned in several studies [6, 7, 9, 18, 21] using varied terms and orders.

As a result of the Coronavirus pandemic and spread, the guidelines are rethinking the sustainable pillars, which is largely based on the direct and indirect lines of relationships and reflections. Several aspects and sub-aspects are tangible, while others are intangible, and future research will address those intangible parameters.

On other hand, there is one intangible parameter that draws huge attention in many studies, namely "occupants' lifestyle and behavioural traits" [6, 7, 22]. The health pillar of rethinking sustainable housing is greatly affected by this parameter. There are four dimensions to housing that can be reflected in these aspects. In addition, cultures and countries differ in how they define this parameter. Several countries have applied this guideline, so its validity has been proven by trial and error. A similar principle can be applied to the climate parameter, which affects all aspects and sub-aspects. It is therefore primarily applied to indoor environments in the research scope.

Moreover, neuroaesthetics research suggests that different types of spatial configurations affect the brain differently, under the sub-aspect of "home geometry and orientation". The perception of fun is significantly higher in curvilinear spaces. In areas with curvilinear contours of the brain, emotional responses are more likely to occur. It is more likely that people will find a room beautiful if it has a high ceiling that activates visual and spatial structures [6].

6 Conclusion

There is, therefore, a major asymmetry in our perception of the one thing that is limited in our awareness of and knowledge of the indoor climate. This is both physiologically and psychologically, as well as the environment itself. Obtaining comprehensive knowledge of all aspects of user behavior is crucial to meeting the growing demand for buildings driven by requirements such as information-based design and reduced CO_2 levels. This includes the psychological and physiological effects of an indoor environment.

To help architects and engineers design indoor spaces, these guidelines should provide a clear vision. Providing sustainable and healthy homes for their users is the result of this process. The architects and engineers can be more confident about their designs when these requirements are taken into consideration during the pre-design phase, through simulation and prediction programmes.

Finally, there is a need for more specific research by scientists to clarify the human interaction with the indoor environment. To be confident about a realistic result, the research must be interdisciplinary.

References

1. Orgnization WH (2021) It's time to build a fairer, healthier world for everyone, everywhere. World Health Day 2021 Health equity and its determinants, 5
2. Settersten RA Jr, Bernardi L, Härkönen J, Antonucci TC, Dykstra PA, Heckhausen J, Kuh D, Mayer KU, Moen P, Mortimer JT, Mulder CH, Smeeding TM, van der Lippe T, Hagestad GO, Kohli M, Levy R, Schoon I, Thomson E (2020) Understanding the effects of Covid-19 through a life course lens. Adv Life Course Res 45:100360. https://doi.org/10.1016/j.alcr.2020.100360
3. Sara Eltarabily DE (2020) Post-pandemic cities—the impact of COVID-19 on cities and urban design. Arch Res 10(3):70–84. https://doi.org/10.5923/j.arch.20201003.02
4. Sendra P, Sennett R (2020) Designing disorder. Experiments and disruptions in the city. J Urban Des 25(5):665–667. https://doi.org/10.1080/13574809.2020.1794803
5. Ranson RP, WHO (1988) Guidelines for healthy housing
6. Brunsgaard C, Fich LB (2016) 'Healthy buildings': toward understanding user interaction with the indoor environment. Indoor Built Environ 25(2):293–295. https://doi.org/10.1177/1420326X16636489
7. Ghaffarianhoseini A, AlWaer H, Omrany H, Ghaffarianhoseini A, Alalouch C, Clements-Croome D, Tookey J (2018) Sick building syndrome: are we doing enough? Arch Sci Rev 61(3):99–121. https://doi.org/10.1080/00038628.2018.1461060
8. Allam Z, Jones DS (2020) On the coronavirus (COVID-19) outbreak and the smart city network: universal data sharing standards coupled with artificial intelligence (AI) to benefit urban health monitoring and management. Healthcare 8(46):1–9. https://doi.org/10.3390/healthcare8010046
9. Osama Omar NA-S (2019) Impact of contaminated interior finishing materials on the educational buildings. BAU J - Creat Sustain Dev 1(1)
10. Jackson RJ (2003) The impact of the built environment on health: an emerging field. Am J Public Health 93(9):1382–1384
11. Bonnefoy X (2007) Inadequate housing and health: an overview. Int J Environ Pollut 30(3):411–429
12. NATIONS U (1996) REPORT OF THE UNITED NATIONS CONFERENCE ON HUMAN SETTLEMENTS (HABITAT II). Retrieved from https://documents-dds-ny.un.org/doc/UNDOC/GEN/G96/025/00/PDF/G9602500.pdf?OpenElement
13. Altgeld T (2004) Gesundheitsfördernde Settingansätze in benachteiligten städtischen Quartieren. Expertise E&C: Entwicklung und Chancen Junger Menschen in sozialen Brennpunkten
14. Council USGB (2016) LEED Green Association exam preparation study guide LEED v4 edition
15. Leslie Dietz PFH, Coil DA, Fretz M, Eisen JA, Van Den Wymelenberg K (2020) 2019 novel coronavirus (COVID-19) pandemic: built environment considerations to reduce transmission. Appl Environ Sci 5(2):1–13

16. Rees WE (1999) The built environment and the ecosphere: a global perspective. Build Res Inform 27(4–5):206–20
17. Adams RI, Bhangar S, Dannemiller KC, Eisen JA, Fierer N, Gilbert JA, Green JL, Marr LC, Miller SL, Siegel JA, Stephens B, Waring MS, Bibby K (2016) Ten questions concerning the microbiomes of buildings. Build Environ 109:224–234. https://doi.org/10.1016/j.buildenv.2016.09.001
18. Raymond Tellier YL, Cowling BJ, Tang JW (2019) Recognition of aerosol transmission of infectious agents: a commentary. BMC Infect Dis 19(101):1–9. https://doi.org/10.1186/s12879-019-3707-y
19. Ibrahim MA, Hanafi MA, Omar OM (2012) Nanoarchitecture and global warming. Int J Sci Eng Res 3(10):1–9
20. Peña-García A, Salata F (2020) The perspective of total lighting as a key factor to increase the sustainability of strategic activities. Sustainability 12:2751. https://doi.org/10.3390/su12072751
21. Megahed NA, Ghoneimb EM (2020) Antivirus-built environment: lessons learned from Covid-19 pandemic. Sustain Cities Soc 61. https://doi.org/10.1016/j.scs.2020.102350
22. Omar O (2020) Near zero-energy buildings in Lebanon: the use of emerging technologies and passive architecture. Sustainability 12. https://doi.org/10.3390/su12062267

Transforming Obsolete Spaces into Vital Places: Historic Italian Villages as Laboratories of Sustainability

Rossana Galdini, Silvia De Nardis, and Nicoletta Cerreti

Abstract Today, Italian villages, the so-called "borghi", are the focus of political, urban, economic and social attention. As specific "Italian urban models", the small historic villages seek to reinvent their image without losing identity and sense of place, creating a solid synergy between environment, economy and society. These actions safeguard natural and cultural resources and contribute to social well-being. Communities, institutions, entrepreneurs and associations are encouraging a return to the Italian hinterland for better living conditions. Bringing these realities back to life means adopting far-sighted policies focused on innovation, strategic investments and creativity. Our chapter analyses two Italian case studies. The first is in Abruzzo, in Santo Stefano di Sessanio, which has chosen to preserve and enhance its cultural heritage and biodiversity to counter depopulation, strengthening the recovery of its traditions and sense of community. The second is in Liguria, in Bussana Vecchia, where an abandoned village was given a new lease of life by several artists who upgraded abandoned spaces, transforming them into residences and ateliers. The two selected cases show how complex regeneration processes can transform some villages with unique features into laboratories of sustainability, capable of suggesting attractive urban solutions that take up traditions, innovate them, providing alternative answers to emerging needs and aspirations, contributing to a cultural, environmental and *people-centred* transition.

Keywords Natural · Cultural resources · Urban regeneration · Sustainability lab · Proximity hub

R. Galdini (✉)
Department of Social and Economic Sciences, Sapienza University of Rome, Rome, Italy
e-mail: rossana.galdini@uniroma1.it

S. De Nardis · N. Cerreti
Department of Social and Economic Sciences, Sapienza University of Rome, Rome, Italy
e-mail: silvia.denardis@uniroma1.it

N. Cerreti
e-mail: n.cerreti@cgil.it

S. De Nardis
Cgil Nazionale, Rome, Italy

Z. Allam (ed.), *Sustainable Urban Transitions*, Urban Sustainability,
https://doi.org/10.1007/978-981-99-2695-4_20

1 Introduction

In the socio-economic and health crisis affecting urban spaces' formal aspects, uses, and general organisation, contemporary cities and areas face urgent environmental, economic, and social challenges. Finding unambiguous and sectoral answers to the questions posed by urban complexity has become necessary. This means new policies, alternative development models and a new cultural approach that knows how to redesign spaces and transform them into living places by directly involving citizens. Environmental issues, increased focus on resources, and the quality of urban life pose challenges that need addressing globally. Change happens locally and becomes meaningful for people. New urban regeneration strategies shift the focus from spaces to people [20], from large-scale to small municipalities, and recently to small historical villages. "Ghost towns", "dust cities" and "lost villages" are terms that describe a phenomenon of abandonment and depopulation resulting from multifactorial decline processes. Rural villages and medieval centres are witnessing a process of demographic crisis, while a vast set of "endangered architectures" risk losing their historical identity and that of their inhabitants [17]. A valuable *genius loci* risks being destroyed alongside urban branding, commodification and standardisation. However, these millenary places are bearers of a historical-architectural, landscape and artistic-cultural heritage recognised as a strategic resource for economic and social development [14]. For this reason, new strategies are being considered to favour their rebirth through sustainability. Today, Italian villages, where 17% of Italians live, i.e. about 10 million inhabitants [2], are the focus of political, urban, economic and social attention.

The Next Generation EU strategy implemented by the European Union to exit the pandemic aims to transform the continent through regeneration, sustainability and innovation. The National Recovery and Resilience Plan (NRRP) implementing the EU programme prepared by Italy includes the regeneration of small historic villages through integrated local cultural-based projects. In Italy's polycentric regional system, these small historic centres embody those landscapes of everyday life that develop a more distinct diffused local quality over time [28]. As specific "Italian urban models", the small historic villages seek to reinvent their image without losing identity and sense of place, creating a solid synergy between environment, economy and society. These actions safeguard natural and cultural resources and contribute to social well-being. Communities, institutions, entrepreneurs and associations encourage a return to the Italian hinterland for better living conditions. Bringing these realities back to life means adopting far-sighted policies focused on innovation, strategic investments and creativity. This implies rethinking these villages as proactive communities, potentially able to offer solutions to urban complexity by creating proximity hubs linked to neighbouring cities. These aspects require an alternative living system [6] and identifying ample space for significant public works [4].

Our chapter analyses two Italian case studies, the first in Abruzzo, in Santo Stefano di Sessanio, which has chosen to preserve and enhance its cultural heritage and biodiversity to counter depopulation, strengthening the recovery of its traditions and sense of community. An Italian-Swedish entrepreneur regenerated the village according to the "where it was and how it was" principle. Keeping the original architecture, the careful use of materials, restrictions on constructing new buildings, respect for history, and an unconventional type of tourism have given this village a new life and vocation, avoiding distorting the environment.

The second case is in Liguria, where, since the 1950s, the abandoned village of Bussana Vecchia was given a new lease of life. Several Italian and foreign artists redeveloped the less run-down buildings and regenerated the abandoned spaces by transforming them into artists' residences and ateliers: a workshop of creativity known as the "artists' village." This experience was focused on a creative solution to counter depopulation, reinforcing the recovery of traditions and a sense of community based on a spontaneous participatory process.

Regardless of the structural and organisational differences in the redevelopment methods, these communities have revived these historic villages, albeit facing difficulties and obstacles, enhancing their natural, artistic and architectural heritage through a strategy based on environmental respect and rediscovering the *genius loci* and art. These two selected cases show how complex regeneration processes can transform some villages with unique features into laboratories of sustainability, capable of suggesting attractive urban solutions that take up traditions, innovate them, providing alternative answers to emerging needs and aspirations, and contributing to a cultural, environmental and people-centred transition.

2 From Lively Spaces to Ghost Towns and Vice Versa

A country like Italy comprises an urban fabric of a few large cities, a single metropolis and many medium and small municipalities, villages and minor historical centres. There are approximately 5,591 municipalities with a population of 5,000 or less (69.9% of the total). 45% of these have a population of between 1,000 and 3,000 inhabitants [1]. This phenomenon occurs mainly in the Italian "inland areas", which is 60% of the country [29]. Villages represent the country's armour and identity, they are historical and geographical frameworks with crucial resources and cultural values for the development of these places and the ecosystem. Historical centres play a decisive role in rebalancing the region by reactivating local economies. "The civilisation of the villages and their original historical nuclei", as a set of cultural, economic, social and political aspects specific to a community in a given era, has generated features revealing local typical identity elements, dubbing Italy the "Country of historical centres" [11].

As Sau [28] stated,

> Italian villages "are the outcome of the dynamics of the various and differentiated natural systems and the anthropisation of peculiar and secular processes. They present some common features such as the significant distance (remoteness) from the main centres supplying essential services (education, health and mobility), the presence of critical environmental resources (water, agricultural systems, forests, natural and human landscapes) and cultural resources."

A common aspect of these areas is the degradation of buildings, partly caused by depopulation and abandonment affecting some vital centres due to the loss of economic power and the low quality of life linked to a lack of services. This trend has led to legislative and administrative measures allowing citizens to reconstitute their resident communities, starting by enhancing the existing structures.

Since the middle of the last century, interest in inland areas has produced attempts such as the Gubbio Charter of 1960, promoted by the National Association of Historic-Artistic Centres, to safeguard and redevelop existing settlement structures, defining the constraints for suspending building.

Following a 2001 Anci investigation, the "Borghi più belli d'Italia Association" included 329 villages in the 2022/2023 Guide, selected based on architectural, cultural and landscape quality parameters. In 2021, the municipalities belonging to the network recorded more than 3.3 million arrivals and 12.1 million presences [22], encouraging the rediscovery of these areas.

A significant organic attempt to enhance these small villages is in the National Strategy for the Development of Inland Areas (Italian SNAI), deployed between 2012 and 2014 promoted by the Department for Development Policies in partnership with ISTAT and the Bank of Italy. The Strategy was an effort to map the most vulnerable and marginal areas identified based on their distance from the main cities and the absence of primary services. A second relevant action is put in place by the Minister for Cultural Heritage and Activities with Law no. 158 of 2017, the "Salva Borghi" law, containing measures to support and enhance small municipalities with fewer than 5,000 inhabitants and provisions to redevelop and recover historical centres [28]. Measures for the repopulation of old villages, the redevelopment, reuse and reactivation of Italian wasteland, defined as potential "local civic garrisons", are incentivised. In this case, the redevelopment will encourage settlement in these areas and stimulate the influx of tourists in a circular economy perspective through the coexistence of measures to improve the quality of services and life. Among the measures put in place the "reuse of the existing building heritage to connect historical-architectural and environmental values with the economic and climatic potential of the places," promoting "less soil consumption and resource-saving" [23, p. 118] appears particularly relevant. More recently, the National Recovery and Resilience Plan (NRRP), which implements the European Green Deal, allocated €1 billion for the economic and social development of small Italian villages in the most vulnerable areas. This Plan is divided into two different lines. Line A is "a social experiment," which establishes new functions, infrastructures and services in the field of culture, tourism, social or research, such as schools or academies of arts and cultural professions, Albergo Diffuso, research centres and university campuses, health and welfare residences. Line B is geared towards developing non-profit and for-profit commercial, craft or cultural enterprises in the selected villages or that wish to establish there. The "Borghi" (Villages) plan's objective under the National Recovery Plan is to create sustainable and quality growth and distribute it throughout the country.

The enhancement projects must include measures to improve the village's image through renovation projects and the regeneration of the cultural, social, and economic fabric receiving investment.[1]

Despite what is defined as wishful thinking or linked to an idealised idea of these places, small villages need plans, policies, strategies, and tools to overcome their fragmentation related to their marginal location and lack of connections with neighbouring towns. Measures to combat abandonment and desertification due to human actions or the absence of adequate procedures and policies can concern urban accessibility and mobility, and adequate public and private infrastructure and services. They can also enhance intangible aspects related to the identity values based on their unique features [28]. Villages require planning but presuppose an idea of community which can slowly be reconstructed by preserving the concept of "staying" and the will to generate a new sense of place [30]. This idea also means stimulating the revival of these places, traditional activities, and tourism responsible for a process that regenerates and gives new life, function and meaning to the existing historical, landscape and cultural heritage [19].

Cervellati [13] defines minor historical centres as mistreated entities since they are *encapsulated* in building expansion and industrialised agriculture, *abandoned* in uncultivated land or *transfigured* by the standardising recovery of tourism. The challenges of preserving, protecting and enhancing historical centres are linked to the transformations induced by the industrial model and digital technologies that change production processes and people's behaviour. The presence of new financial players and diversified tourism, the economic tertiarization, and the lack of adequate investments for maintenance and management exemplify the issue of inland areas. They are often penalised by the same isolation that in the past determined their emergence, linked to defensive reasons, their preservation and that today excludes them from the active life cycle. There is a need for enhancement processes that include new development models and green solutions capable of favouring new activities and the revival of past forms of craftsmanship, tourism and enterprise.

Policies and regulatory measures combine the need to protect the identity of these realities within the cultural landscape in which they are set. This must consider citizens' legitimate expectations for the area's economic development. The case studies presented below, Santo Stefano di Sessanio in Abruzzo and Bussana Vecchia in Liguria, are two positive examples of processes built over time with tenacity and determination to redevelop their image, generating a new economy and shared identity values among the local inhabitants. These results were reached by integrating past craft activities, new artistic expressions, innovative agriculture, the recovery of traditional products, hospitality, unconventional forms of tourism, and strengthening the social cohesion of residents. Like several other similar Italian and international cases, these aspects refer to good strategies and actions to implement an ecological and a new cultural transition. These cases highlight the need to include sectorial and micro- actions in an integrated design that promotes sustainable innovation for

[1] https://www.ilsole24ore.com/art/per-borghi-storici-miliardo-euro-pnrr-cultura-AEwj6v5.

inland areas, with considerable attention to the development and recovery of natural spaces built from a new solidarity dimension typical of small communities.

Several Italian ghost villages undergo processes in which old ruins, abandoned houses and landscapes immersed in silence have been the subject of recovery projects to preserve and revive a "minor heritage." The skilful recovery of homes and spaces of their history, area, and old traditions allows some villages to challenge a destiny that for many was already written and to write something new. It is the case of Santo Stefano di Sessanio, which tells a different story.

3 The Santo Stefano di Sessanio Case. From Roots to the Future: Urban Regeneration and Rediscovery of the *Genius Loci*

The Italy of small villages struggles to survive, and Santo Stefano di Sessanio represents it with great pride. Located in the Gran Sasso and Monti della Laga National Park mountains, this architectural jewel rises 1251 m above sea level in the Abruzzo region, near L'Aquila. This small city is sadly known for the 2009 earthquake that caused hundreds of casualties, €15 billion in estimated damages and the disappearance of 80% of its artistic heritage. On that tragic day, L'Aquila, a city of approximately 70,000 inhabitants, ceased to be a city and became "the crater", a term that denied the area its identity value and the people of L'Aquila their sense of community. The village of Santo Stefano, formerly known as "*Sextantio*", began as a small Roman settlement, developed in the Middle Ages between the eleventh and twelfth centuries, and between 1579 and 1743, reaching its peak under the ancient and powerful Tuscan Medici family. The commercialisation of "carfagna" wool made Santo Stefano di Sessanio known throughout Europe. The Seigniory of Florence identified the village as a centre of operations to produce this high-quality yarn that, due to its dark colour, was used to make military uniforms and the monks' robes. Over the years, the village experienced a gradual decay process that coincided with the abrupt end of the millenary activity of transhumance. This is an ancient practice of sheep farming that saw numerous flocks of sheep migrate seasonally from the Apennine mountain range to the Adriatic Sea. Transhumance was included in the UNESCO Intangible Cultural Heritage List in 2019, recognising its

> cultural standing of a tradition that has shaped the relationships between communities, animals and ecosystems, giving rise to rituals, festivals and social practices that punctuate summer and autumn, a recurring sign of a practice that has been repeated for centuries with the cyclical nature of the seasons worldwide.[2]

The depopulation of the village accelerated in the second half of the twentieth century. The economic boom that followed the Second World War meant that those

---

[2] Unesco, "La transumanza" (Transhumance), https://www.unesco.it/it/ItaliaNellUnesco/Detail/820.

who lived in villages such as Santo Stefano di Sessanio closed their doors and moved to the city in search of work and better living conditions. This trend coincided with the gradual abandonment of agricultural and pastoral activities in favour of industries. For many families in Abruzzo, already used to looking elsewhere for work to make a living, it was not difficult to decide to emigrate. This migratory flow took them to internal Italian destinations, such as the large cities that in those years expanded their borders by constructing new districts, and external destinations, such as European states.[3] The population of Santo Stefano di Sessanio followed this migration process. If we analyse the demographic evolution from the end of the nineteenth century to the present day, we can easily trace a downward curve: in 1861, there were 1,315 inhabitants the peak was reached in 1901 with 1,489, today the village has just 116 residents.[4]

Tracing back over the broad economic and social history of Santo Stefano di Sessanio, we noted how it is an urban settlement that has its roots in pastoral traditions. Sheep farming, together with typical mountain crops—lentils (now included among the Slow Food presidia[5]), saffron and potatoes—have represented a valuable economic and social driving force for the town and a fundamental basis that defined customs and traditions for constructing the area's identity. From this place's roots, traditions and culture, an attempt was made to give life to a new community that could animate the narrow streets of this medieval village. This aspect did not involve nostalgic feelings but looking ahead, using sustainable innovation.

> The little balconies where Florentine ladies looked out with their graceful, mullioned windows and tenderly sculpted doorways resonate the sweet motherly songs of this secret and solid land and make the hard water of the peaks flow to the fountain.[6]

This spirit gave life to '*Sextantio*', the Albergo Diffuso built thanks to the entrepreneurial skills of the Italian-Swedish Daniele Kihlgren and the support of the municipality and the Park Authority. This experience provided the impetus for small entrepreneurs to invest in the village, opening of handicraft workshops, relaunching local production and start-up of new accommodation facilities, restaurants and B&Bs. One fundamental guideline is the sustainable regeneration of spaces of inestimable urban, architectural and landscape value.

---

[3] A.S.E.I. Archivio Storico dell'Emigrazione Italiana (Historical Archives of Italian Emigration), "L'Abruzzo migrante dall'Unità d'Italia alla Grande Guerra", (Abruzzo migration from the Unification of Italy to the Great War) https://www.asei.eu/it/2013/11/storia-e-memoria-dellabruzzo-migrante-nella-seconda-meta-del-xx-secolo-2/.

[4] Istat, "Calculation of the resident population", http://dati.istat.it/Index.aspx#.

[5] Slow Food is a global, grassroots organisation founded in 1989 to prevent the disappearance of local food cultures and traditions.

[6] Most Beautiful Italian Villages, "Santo Stefano di Sessanio. Il nido d'aquila del Gran Sasso" (Santo Stefano di Sessanio. The eagle's nest on the Gran Sasso).

https://borghipiubelliditalia.it/borgo/santo-stefano-di-sessanio/#1480496820077-2b27c1ff-e93b.

Art, sustainable tourism, revitalisation of local production, and financial incentives provided by the municipalities—such as the "*Case a un euro*"[7] (houses at €1) programme, which has been a success, especially among foreign citizens—are just some of the incentives chosen to save Italian villages from abandonment. In Santo Stefano di Sessanio, the stimuli used to regenerate these places were quality cultural and nature tourism and the rediscovery of small and old local economies.

At the heart of the small village regeneration process, now inhabited by just a few souls, was the '*Sextantio*' Albergo Diffuso, created in 2001 by entrepreneur Daniele Kihlgren. According to Kihlgren

> Italy has suffered the amputation of its "minor historical heritage" and landscapes, which have been systematically violated.[8]

Santo Stefano di Sessanio, because of its gradual depopulation, has managed to keep its environmental and architectural values intact. This uniqueness, i.e. the integrity between the area and historical buildings, made the entrepreneur decide to implement a project of recovery and redevelopment of accommodation facilities without entailing the loss of local identity but enhancing the original urban configuration, local customs and traditions. Kihlgren's mission aimed to avoid the risks of transforming a village into a "mass holiday village", as unfortunately occurred in other municipalities that have chosen to bet everything on tourism to revive themselves, but without recovering the identity of the place or creating a new community.

In the 1990s, Kihlgren began to buy some empty properties and started a detailed recovery work to give meaning and identity to these urban spaces and fill them with purpose. He found the support of the municipality and the Park Authority, with which he signed the "*Charter of Values for Santo Stefano di Sessanio*". He began an important partnership with the Museo delle Genti d'Abruzzo (in Pescara) to materialise an old idea of his—restore a ruined medieval village philologically, to make the entire village into an Albergo Diffuso.[9]

An ambitious urban regeneration project started from a long ethnographic work, from the recovery of oral traditions to the study of historical documents and photos, such as those taken by the Swiss linguist Paul Scheuermeier who immortalised the agro-pastoral culture of Abruzzo between 1920 and 1930 with his camera. He recovered traditional weaving techniques and the recipes of the old folk tradition, reintroducing them into the cultivation of seeds that had disappeared entirely.

> The revival of local traditions contemplates the most minute details, towels, wool mattresses, sheets from wedding trousseaus, handmade blankets with wooden looms and natural colours. The private part of the project includes preserving the original domestic organisation.[10]

---

[7] This is an initiative that offers houses for sale for a symbolic sum to encourage the repopulation of villages.

[8] Report, "C'è chi dice no" (some say no) 2010/2011 TV season, https://www.raiplay.it/video/2011/05/Daniele-Kihlgren-5f760fb4-964c-4962-ae8d-bccebe2e183c.html.

[9] Sextantio, "Santo Stefano di Sessanio", https://www.sextantio.it/santostefano/abruzzo/.

[10] Sextantio, "La salvaguardia del patrimonio storico minore" (Preserving the minor historical heritage), https://www.sextantio.it/salvaguardia-patrimonio-storico-minore/.

4 The Bussana Vecchia Case. From Ghost Town to "Artists Village"

Bussana Vecchia is a small village of about 75 inhabitants belonging to the municipality of Sanremo, located in the hinterland of the Liguria region in Northwest Italy. Its historical, architectural and landscape features contribute to the design of Italy's "widespread cultural heritage." The first settlement dates to 1050, when the Counts of Ventimiglia built a castle. Over the centuries, it saw a gradually growing number of inhabitants, making it necessary to carry out recurring enlargements of the historic core.

Bussana Vecchia is a medieval village abandoned due to environmental seismic, and a new centre being founded. The triggering event was the earthquake that struck western Liguria on 23 February 1887, causing the destruction of many houses and the expulsion of the inhabitants outside the boundaries of the original settlement. In 1889, a new Bussana was rebuilt further downstream, while the government military authority ordered the area evacuation and demolished some buildings considered dangerous. As shown by the historical reconstructions of Calvini [10], the combination of these events condemned the village to abandonment.

The old village of Bussana was a ghost town until the early 1960s, when a group of artists fascinated by the spirit of the place squatted there, fostering its rebirth. The Turin ceramist and painter Mario Giani, with painter Vanni Giuffrè and poet Giovanni Fronte, had the idea of the "International Community of Artists", a self-established group to regulate life in the village. During this period, painters, musicians, writers and actors were assigned empty houses to set up workshops and exhibition spaces. Bussana witnessed the spontaneous creation of a workshop for experimentation and cultural research, allowing it to live again. The artists were financially and emotionally invested in renovating and recovering abandoned ruins, facing the absence of infrastructures and services. Despite the informal and spontaneous settlement process, they continued repopulating Bussana Vecchia with small physical adjustments and revitalising public spaces with artistic works and cultural events.

The site quickly became a popular tourist destination and still houses some of the first occupants referred to as "the resilient." Among the main attractions are the workshops and ateliers, set in the typical Mediterranean wooded landscape, surrounded by murals, sculptures and frescoes amidst largely unaltered medieval architecture. At the same time, artistic and cultural initiatives in a large pedestrian island inaccessible by cars are flanked by small refreshments and commercial hospitality businesses. However, some artists promoting the revival decided to leave the village, thinking it had lost the innovative character and identity that enabled its revitalisation in the first phase. In addition, the speculative period of the 1980s and 1990s saw the growth of a profitable property market, often in the shadow of regulations and property rights.

In 1977, the Borgo Committee was founded, a body that attempted to solve the problems caused by the legally complex nature of Bussana Vecchia. In 1983, the municipality of Sanremo launched an international competition for a detailed plan to clarify the legal status of property ownership and encourage the overall village recovery. Concluded in 1986, the competition awarded the "Città Invisibili" project drawn up by the group coordinated by architect Claudio Baracca. This proposal, which never came to fruition, involved leaving the portion already recovered unaltered, except for necessary structural and technological works, and recovering the buried part with earthquake-proof frames, required by the area's geological conditions. These would be assigned to the public body, which would take charge of their redevelopment. In 2004, the municipality of Sanremo assigned architect Carmen Lanteri the task of drafting and implementing urban planning tools to recover the Bussana Vecchia historic core. This provided a framework to identify functions, intended uses, and implementation methods for the renewal and enhancement.

In 1983, the village was included among the State's assets and the properties of the Agenzia del Demanio, an entity entrusted with public real estate management. It was during this period that the first actions to restore legality began. To continue their lives in Bussana Vecchia, artists are obliged to bear high compensation costs for the improper use of the houses. Between 2000 and 2003, the Ministry of Cultural Heritage and Activities confirmed the historic core of Bussana Vecchia as being of crucial cultural interest and subject to protection. In 2016, Bussana Vecchia returned to the availability of the municipality of Sanremo on a non-payable basis by implementing the Cultural State Property Federalism procedure under Legislative Decree no. 85/2010. Lastly, 2017 saw the approval of the Programme for the Enhancement of Bussana Vecchia, entrusted again to the Lanteri group, under which the municipality ensured its conservative restoration and safety, favouring the preservation of the environmental and architectural features of the old urban layout. With an expected investment of approximately €43 million over ten years, the plan was to carry out primary urbanisation, building recovery works, an earthquake research and study centre and a restoration school that should host laboratories, workshops, master's degrees, conferences, exhibitions and artistic performances. Other functions considered crucial for the tourist development of the village were identified. These include an Albergo Diffuso and a hostel. The residential units and the art atelier[11] were confirmed.

The housing ownership dispute between the State and residents is still ongoing and the illegal condition interfered with the redevelopment projects and especially accessing funds. The critical assessment forwarded by Bussana Vecchia inhabitants underlines the absence of an organic and inclusive regeneration strategy that preserves the life of artists in the village, regulates mass tourism, and protects medieval architecture by favouring gradual redevelopment principles.

[11] https://www.architettilanteri.it/.

5 Results

The analysis of these case studies highlights the multidimensional nature of the small village depopulation phenomenon. The observation of recent experiences shows that at the basis of the decline of small historical centres, natural and human causes often coexist. It highlights the variety of possible responses [21] and players involved, whether enlightened entrepreneurs or artists, with careful care for traditions. The challenge of the revival of abandoned Italian historical centres is still open. Protection, restoration and preservation actions are flanked by strategies of enhancement, revitalisation and regeneration [8].

The '*Sextantio*' Albergo Diffuso got it right—regenerating an abandoned village, and recovering the *genius loci* while respecting the past.

> A conservative approach to places that were born, lived, and sometimes died, abandoned because of the destiny of their inhabitants, economies and activities that sustained them, must consider extreme choices of preserving the traces of human experience. These include plasterwork that bears the signs of the suffering of the time, and the evidence of subsistence activities, which are an integral part of the identity of these places, and need to be protected.[12]

It is evident how the regeneration plan of Santo Stefano di Sessanio represented a reference point, a starting reference model. '*Sextantio*' is an experience that began 21 years ago and is highly avant-garde. This single accommodation project, unlike many other projects that have turned out to be pure tourist exploitation, has the merit of having relaunched the sector, encouraging the opening of several B&B restaurants, and craft workshops—such as aquiLANA, which raises sheep and produces spun wool according to tradition. The project respects the place's history and rediscovered its soul.

With the Albergo Diffuso, a regeneration based almost exclusively on the conservation and enhancement of historical and natural resources has been carried out, but to overcome what we can define as the "heritage paradigm",[13] often aimed exclusively at relaunching tourism, it is indispensable to reconstruct a "productive vision of the inland areas", a vision that knows how to make the knowledge of the past coexist with innovation, which cannot be left to free private initiative alone. This aspect is the weak point of the small-town revitalisation programme. Investments have been made almost entirely by private subjects. For the true rebirth of these realities, public investments are indispensable. They can re-launch private investments to guarantee services and employment for those who stayed and those who might arrive.

The experience of Bussana Vecchia provides an assortment of themes for reflection that embrace the restoration, conservation and adaptive reuse of cultural heritage, community-based urban informality and the regulatory role of institutions, culture

[12] Sextantio, "La salvaguardia del patrimonio storico minore", (Preserving the Minor Historical Heritage), https://www.sextantio.it/salvaguardia-patrimonio-storico-minore/.

[13] Montagna in rete, "La nuova centralità della montagna e dei territori di 'margine' Dialogo con Antonio De Rossi" (Mountain networking The new centrality of mountains and 'edge' areas. Dialogue with Antonio De Rossi), https://www.montagneinrete.it/testimoni-d-eccellenza/la-nuova-centralita-della-montagna-e-territori-di-margine-dialogo-con-antonio-de-rossi.

as an incentive for recovery and the strategic but problematic function of tourism. The Ligurian case is a historic village abandoned for environmental reasons that has undergone cultural and community-based reactivation. The spontaneous nature of the recovery provided virtuous outcomes but, at the same time, continues to pose questions on how to proceed with the village's sustainable transformation from a technical-structural and administrative point of view. For this reason, the revival of Bussana Vecchia seems to find a home in an ambivalent category, understood as the coexistence of non-excluding alternatives and a constitutive dimension of contemporary social phenomena [7]. The strengths that have allowed its repopulation are connected to some still unresolved critical issues that slow down the implementation of a unified regeneration strategy.

The potential for the village's restoration can be identified in the cultural revival implemented by what has been dubbed a "resilient community." In this case, the relationship between spontaneity and collective action highlights attractive social innovation opportunities [25]. The new inhabitants of Bussana Vecchia invested their energies and resources in a joint project under the banner of creative experimentation. The practice of the social and financial reactivation of an old abandoned village through arts and culture, weaving a cohesive network of neighbourhood relations, is embedded in the paradigm of innovation [24] and sustainability. The added value of this experience lies in its ability to activate a value circuit around cultural heritage, reusing empty spaces in the village and transforming them into common goods [16]. In this way, artists have contributed to the village's rebirth by preserving its architectural authenticity and fostered the sedimentation of a new place identity embedded in the cultural sphere. Today, the "artists' village" is a mosaic of experience, work and collective effort that gives back the image of a vital place of national and international interest.

However, some elements of weakness persist, for which a critical assessment is needed. The issue of informality is undoubtedly one of the main knots to be unravelled in Bussana Vecchia. The village is an aggregation of houses restored independently by artists without any structural verification, which needs urban reorganisation. The lack of rules has led to a change in the building structures obtained by merging different units to meet the space requirements of the new occupants. Dwellings are realised by uniting several house ruins. Alleys are narrow and made even more cramped by walls, external staircases and long stretches covered by vaults [26]. The illegality issue hinges on the economic dynamics. Towards the end of the 1960s, the visibility of the village and the associated profit opportunities led to the escalation of legal disputes between the historical owners, many of the residents of the new village downstream, who had abandoned the dwellings and now claimed ownership, and the occupants. They revived the village and rescued it from degradation with great effort. Finally, the experience of this small historic centre provides a general consideration of the relationship between tourism and the city. This is proved by the migration of an albeit small group of original occupants due to the tourism effects on the village.

6 Discussion

The two case studies reveal substantial differences and overall similarities in the process, which followed different directions and led them from a state of abandonment to a process that enhanced their natural and cultural resources.

Sextantio is an example of how urban regeneration can be achieved by rediscovering the *genius loci*—place and identity have returned to interact with each other. The spirit of the place was reawakened among the dwellings built of white limestone in this fortified village. This creates

> an identity that includes the historical/architectural heritage, landscape, area inhabitants with their customs, traditions, and specific physiognomy. In their indissoluble unity, these features characterise the area's unique, unrepeatable identity.[14]

Recovering a village, such as Santo Stefano di Sessanio, under earthquake-proof standards has meant protecting the spirit of the place from the risk of a new tear following the devastating earthquake that struck the area in April 2009. The Albergo Diffuso houses had already undergone a major renovation and withstood the strongest tremors. Earthquakes, floods and landslides often cause small towns partial or total destruction. This problem often leads to the "double villages" phenomenon [30] and the inevitable depopulation, as happened, most recently in 2016, to Amatrice, in the province of Rieti—Lazio.

Minor historical centres have a strong symbolic and identity dimension, a heritage of cultural biodiversity made of material and immaterial traditions that are not static but continuously changing [27]. This suggests a focused and situated enhancement that pays attention to the identity of the place. As Forbes and Coletta [18] pointed out, bringing citizens closer to cultural heritage implies that "care for the past is interlaced with care for the future". "Do not betray the spirit of the place" is a motto that draws attention to identity features without neglecting possible scenarios and the creative flow of human life. The sustainability of a recovery operation passes through a quality that is not only architectural and environmental but social [27]. In the Bussana Vecchia case, preserving the village's artistic image and structure as the result of the spontaneous architecture of "artist-builders" can be as crucial as the reconstruction and protection of historical buildings.

Local identity, which is often the expression of rhetorical authenticity, beauty and uniqueness, cannot avoid facing the performative role of residential communities, considering new tourism development models. The tourism dynamics of some Italian art cities activate processes of degradation, growth and hyper-growth [3]. The risk of a "village-commodity" [5] imposes the search for solutions geared toward sustainable tourism according to mechanisms that unite the area's spatiality with its material and immaterial heritage and its human dynamism [9]. Alongside citizen engagement actions, the theme of empowerment that places people and communities at the centre increasingly recurs. Dispelling the existing and more widespread idea of the village

[14] Sextantio, "La salvaguardia del patrimonio storico minore" (Preserving the minor historical heritage), https://www.sextantio.it/salvaguardia-patrimonio-storico-minore/.

as a fake, stereotyped "object" without context, inhabitants and human relations, Barbera et al. [5] require focusing attention on the "social infrastructure" made up of goods and services essential to developing quality everyday life. As stated by [12]

> Land is being continually re-inhabited and regenerated. But the word "inhabiting" must not be emptied of meaning and reduced to are location of people in search of a non-urban elsewhere, a pre-packaged naive tranquillity, a dormitory where they can recharge their "batteries" during the weekend.

The trajectory indicated by the authors contrasts with the extractive logic of capitalising on the small historical village to re-evaluate development models close to the environmental features.

It is time to increase the attractiveness of villages. An objective contained in the National Recovery Plan,[15] coherent with the Next Generation EU relaunches the economy of European countries after the crisis caused by the Covid-19 pandemic. Approximately 250 small towns and villages will receive funding to counter depopulation and stimulate lasting, inclusive and sustainable economic growth. Among these is Rocca Calascio, a village just ten kilometres from Santo Stefano di Sessanio, which has a similar past and can redeem its identity thanks to European resources. "Rocca Calascio luce d'Abruzzo" is the programme's title. It envisages the restoration and conservation of the site and the creation of an Albergo Diffuso in the closed and semi-abandoned buildings.[16] The hope is that the substantial public funding from the European Union will be used to consolidate virtuous experiences and launch new projects with a long-term vision. It is fundamental to set up "districts", a group of villages located in a circumscribed area, with a similar history that can pool their strengths to relaunch the social and economic fabric. Interconnecting similar realities, such as Santo Stefano di Sessanio and Rocca Calascio, fosters integrated projects.

Recent literature identified the possibility of starting a broader sustainable transformation process by enhancing the built heritage and new management models, integrated and creative ways of using spaces and technological innovation [16]. The emerging theme is the cultural, social, economic and environmental regeneration that inserts the village into a functional relations system with surrounding areas and stimulates innovative mechanisms of heritage administration with a view to integrated local development. It is a challenge linked to the conscious rediscovery of local Italian polycentrism and the search for the complex balance between village identity, tourist promotion and basic welfare. This implies adopting assessment, monitoring and tools that involve social, economic, environmental and historical-architectural components simultaneously [15].

---

[15] Ministry of Culture, "Piano Nazionale Borghi, Franceschini: 21 progetti pilota e 1.800 candidature per vincere la sfida della crescita sostenibile", (National Village Plan, Franceschini: 21 pilot projects and 1,800 applications to win the challenge of sustainable growth) https://cultura.gov.it/pnrr-borghi.

[16] Ministry of Culture, "Pnnr Borghi. 21 straordinari territori tornano a vivere, progetto 'Rocca Calscio luce d'Abruzzo'", (NRP for villages. 21 extraordinary locations come back to life, 'Rocca Calscio light of Abruzzo' project) https://media.beniculturali.it/mibac/files/boards/be78e33bc8ca0c99bff70aa174035096/Card/PNRR-Borghi/PNRRborghiA_Card_1.jpg.

7 Conclusions

The cases analysed solicit critical reflections on processes involving tangible and intangible measures and presuppose the joint action of different players, skills, and medium- and long-term strategies. In village revitalisation processes, it is necessary to involve those who, with determination, decided to stay and their participation is indispensable for fostering a new community. Leaving and staying are two extremes of human history. The right to migrate corresponds to the right to stay, building another sense of place and self. "Staying" means feeling anchored and, at the same time, lost in a place that needs to be protected while being radically regenerated [30].

Today, villages are places to stay temporarily or permanently where it is possible to create conditions that allow for a good quality of life.

The two cases analysed are stories with happy endings that enhance an essential heritage, doomed to oblivion. Good practices may be applied to many ghost towns affected by the force of nature or alternating development phases.

As the analysis of the cases reveals, a village's regeneration must be based on two principles: identity and community, which are essential but insufficient. To build prospects and not thwart resettlement attempts, it is crucial that the community has access to essential services, such as health, education, mobility, land protection and enhancement, and that investments are employed in creating new employment. Public policies that can multiply private initiatives with quality investments are indispensable. As in the cases of Santo Stefano di Sessanio and Bussana Vecchia, these must safeguard the historical, architectural and landscape heritage. Works that enhance the positive outcomes produced by communities are crucial, as can be seen by observing the value of the identity and cultural heritage resulting from the creative experimentation of the Bussana Vecchia artists. "Place-based" actions[17] that consider abandoned or semi-abandoned villages which are concentrated in the country's inland areas,[18] as a resource and not a problem. Villages are not just places with critical issues and they represent excellent opportunities considering the crisis in metropolitan areas. The idea is to suggest actions on a regional basis that do not contrast those carried out in metropolitan areas but are in close correlation. The emergency brought about by the pandemic, and increasingly widespread awareness of environmental issues have boosted the desire for nature, green spaces and less chaotic life. The possibility offered by smart working has rekindled interest in these areas on their potential benefits for the community.

[17] Minister for the South and Regional Cohesion, "Strategia Nazionale Aree Interne", (National Strategy for Inland Areas), https://www.ministroperilsud.gov.it/it/approfondimenti/aree-interne/strategia-nazionale-aree-interne/.

[18] Inland areas are significantly distant from the centres which supply essential services (education, health and mobility), rich in valuable environmental and cultural resources and strongly diversified by nature and as a result of secular processes of anthropisation https://www.miur.gov.it/documents/20182/890263/strategia_nazionale_aree_interne.pdf/d10fc111-65c0-4acd-b253-63efae626b19#:~:text=Wecall%20interior%20those%20areas%20significantly%20from%20secular%20processes%20of%20anthropisation.

Isolated or short-term measures are insufficient. They must be structural, with integrated policies to recover a cultural, environmental and economic heritage of inestimable value that represents Italian uniqueness.

These policies can effectively regenerate these places and create a new planning process based on sustainable development. Starting from these small villages, this approach will restore these areas by creating a renewed balance between individuals and the community.

References

1. ANCE (2017) I borghi d'Italia. Dalla visione alla rigenerazione. https://ance.it/wp-content/uploads/archive/30858-Paper%20Piccoli%20Comuni%20-%20Borghi%2015dic2917.pdf. Accessed 10 July 2022
2. ANCI (2019) Atlante dei Piccoli Comuni. https://www.anci.it/atlante-dei-piccoli-comuni/. Accessed 5 July 2022
3. ANCSA (2017) Centri storici e futuro del Paese. Indagine nazionale sulla situazione dei Centri Storici. http://www.cresme.it/doc/rapporti/Centri-storici-e-futuro-del-Paese.pdf. Accessed 4 July 2022
4. Arminio F (2020) Contro il coronavirus torniamo nei piccoli borghi. https://www.fanpage.it/cultura/franco-arminio-contro-il-coronavirus-torniamo-nei-piccoli-borghi/. Accessed 5 July 2022
5. Barbera F, Cersosimo D, De Rossi A (2022) Contro i borghi. Il Belpaese che dimentica i paesi. Donzelli Editore, Roma
6. Bascherini E (2020) Riabitare i borghi abbandonati. Nuove strategie abitative contro la crisi pandemica. FAMagazine 52/53:204–209. https://doi.org/10.12838/fam/issn2039-0491/n52-53-2020/518
7. Beck U (2000). L'era dell'e. Asterios, Trieste
8. Berizzi C, Rocchelli L (2019) Borghi rinati. Paesaggi abbandonati e interventi di rigenerazione. Il Poligrafo, Padova
9. Bizzarri C, Micera R (2021) The valorization of Italian "Borghi" as a tool for the tourism development of rural areas. Sustainability 13(12):6643. https://doi.org/10.3390/su13126643
10. Calvini N (1987) Bussana. Dall'antico al nuovo paese. Famija Sanremasca, Sanremo
11. Cerasoli M, Mattarocci G (2020) Un futuro per i centri storici minori. Scenari possibili nell'era post-covid. Aracne Editrice, Roma
12. Cersosimo D, Librandi F, Nisticò R (2022) Case a 1 euro: Critica dell'ideologia del borgo-merce. In: Barbera F, Cersosimo D, De Rossi A (eds) Contro i borghi. Il Belpaese che dimentica i paesi. Donzelli Editore, Roma
13. Cervellati PL (2009) La sorte dei piccoli centri storici: Abbandonati, trasfigurati, turisticizzati. Minori e maltrattati. Bollettino Italia Nostra, 445. https://www.italianostra.org/wp-content/uploads/Bollettino_445.pdf. Accessed 6 July 2022
14. Coletta T (2008) Il paesaggio dei centri abbandonati. Riv Int Cult Urban 2. Università degli Studi di Napoli Federico II
15. D'Andria E, Fiore P, Nesticò A (2021) Small towns recovery and valorisation. An innovative protocol to evaluate the efficacy of project initiatives. Sustainability 13(18):10311. https://doi.org/10.3390/su131810311
16. Daprà F, Fabi V (2017) Territories of culture between regeneration and social innovation. An Italian experimentation. TECHNE - J Technol Arch Environ (14):200–208. https://doi.org/10.13128/Techne-20819
17. Fineschi S, Rescic S, Riminesi C, Debenedictis D, Robador MD, Burzagli L, Reimao Costa M, Degrigny C (2022) Toward sustainable regeneration of historic endangered towns: strategies for increasing resilience. In: Hadda L, Mecca , Pancani G, Carta M

18. Forbes N, Colella S (2019) Embedding engagement: participatory approaches to cultural heritage. Sci Res Inf Technol 9(1):69–78. https://doi.org/10.2423/i22394303v9n1p69
19. Galdini R (2019) Urban re-use practices in contemporary cities: experiences in Europe. Cities 87:103–105. https://doi.org/10.1016/j.cities.2018.12.026
20. Galdini R (2022) Rigener-azione. Idee, pratiche e politiche per una città condivisa. Sociologia Urbana e Rurale 128. https://doi.org/10.3280/SUR2050-531001
21. Gizzi FT, Bentivenga M, Lasaponara R, Danese M, Potenza MR, Sileo M, Masini N (2019) Natural hazards, human factors, and "ghost towns": a multi-level approach. Geoheritage 11:1533–1565. https://doi.org/10.1007/s12371-019-00377-y
22. ISTAT (2022) Viaggi e vacanze in Italia e all'estero. Anno 2021. https://www.istat.it/it/files// 2022/04/Viaggi_vacanze_2021.pdf. Accessed 15 July 2022
23. Marchionni C, De Berardinis P, Bellicoso A (2014) The compatibility of off-grid technologies in the rehabilitation of energy network of minor historical centres. TECHNE - J Technol Arch Environ 7:118–124. https://doi.org/10.13128/Techne-14540
24. Moulaert F, MacCallum D, Hillier J (2013) Social innovation: intuition, precept, concept. In: Moulaert F, MacCallum D, Mehmood A, Hamdouch A (eds) The international handbook on social innovation: collective action, social learning and transdisciplinary research. Edward Elgar, Cheltenham, pp 13–24
25. Murray R, Caulier-Grice J, Mulgan G (2010) The open book of social innovation. NESTA, London, UK
26. Pirlone F, Spadaro I (2016) Borghi antichi abbandonati e rischio sismico: Aspetti di vulnerabilità ed esposizione. In: Pirlone F (ed) I borghi abbandonati. Patrimonio da riscoprire e mettere in sicurezza. Franco Angeli, Milano
27. Pistidda S, Bersani E (2020) Costruire accoglienza. Il patrimonio culturale dei centri minori come occasione di rigenerazione urbana e sociale. ArcHistoR 7:1459–1479. https://doi.org/10.14633/AHR277
28. Sau A (2018) La rivitalizzazione dei borghi e dei centri storici minori come strumento per il rilancio delle aree interne. Federalismi.it 3
29. SNAI (2020) Relazione annuale sulla Strategia Nazionale per le Aree Interne. Dipartimento per le politiche di coesione. https://www.agenziacoesione.gov.it/wp-content/uploads/2021/11/Relazione-CIPESS-2020_finale.pdf. Accessed 10 July 2022
30. Teti V (2022) La restanza. Einaudi, Torino

Augmenting Mobility Safety in Cities by Increasing Data Pools from Connected Urban Devices

Aditya Dixit and Zaheer Allam

Abstract While cities pursue transitions to 'smarter' pathways, the use of technology within urban quarters is gaining in popularity, including for rendering safer mobility. Specifically, Machine Learning (ML) is revolutionizing the automotive industry and the Artificial Intelligence (AI) being applied to almost every stage of development of automobiles. While development stages can be time consuming, the use of AI and ML powered simulations can help in increasing the efficiency of design stages, while reducing the amount of computational and human resource dependency, hence enabling faster results with more accuracy leading to faster development cycles and vehicle roll outs. Taking the idea further is the fact that Autonomous Vehicles (AVs), are currently restricted to using data sourced from the vehicles and omit the data-rich landscapes that surround them in within 'smarter' cities. Through this chapter, we argue that the development of AVs can gain from AI and ML tools, along with the development of tools that can integrate data sourced from third party service providers. Doing this will help in increasing efficiency of driving as well as the safety of passengers and urban dwellers.

Keywords Machine learning · Data · Artificial intelligence · Autonomous vehicles · Simulation · Smart devices · Mobility

---

A. Dixit (✉)
Department of Mechanical Engineering, Friedrich Alexander University Erlangen-Nuremberg, Schloßplatz 4, 91054 Erlangen, Germany
e-mail: aditya.dixit@fau.de

Z. Allam
Chaire Entrepreneuriat Territoire Innovation (ETI), IAE Paris—Sorbonne Business School, Université Paris, 1 Panthéon-Sorbonne, 75013 Paris, France

Live+Smart Research Laboratory, School of Architecture and Built Environment, Deakin University, Geelong, VIC 3220, Australia

Network for Education and Research On Peace and Sustainability, Hiroshima University, Higashihiroshima 739-8530, Japan

Z. Allam (ed.), *Sustainable Urban Transitions*, Urban Sustainability,
https://doi.org/10.1007/978-981-99-2695-4_21

1 Introduction

The fourth industrial revolution brought about diverse transformations especially in the urban planning arenas—with adoption of digital solutions is becoming predominant. This has been affirmed by increasing number of technologies (including Internet of Things (IoT), Artificial Intelligence (AI), Geospatial technologies, Cloud Computing and Big Data et cetera) currently being deployed in cities [1]. As a result of these technologies, the number of smart devices, sensors and other components have been increasing—currently estimated to be about 13.1 billion connected devices and projected to increase to over 29.4 billion devices by 2030 [2]. The influx of the devices and their subsequent connection into the digital networks have prompted an increase in the amount of data generated (increased from 17.3 Zettabyte reported in 2019 to projected 73.1 Zettabyte by 2025 [3]) and subsequently being relied upon in decision making on a wide range of issues [4]. Further, those devices have led to an unprecedented disruption on traditional urban landscapes—with elements such as street lighting, car parking, water metering, waste management and others transforming to adopting 'smarter' components. As such, urban areas are now transforming courtesy of increased efficiency, real-time response to issues, increased security, improved citizen collaboration and participation, better traffic flow and upgraded infrastructures et cetera [5]. In addition, the liveability status in cities that have embraced and incorporated vast technologies has improved, as expressed in the latest report by the Economist Intelligence Unit [6].

The vast amount of data being generated has however not been exploited to the maximum, as there are a number of urban realms that still require substantial transformation. For instance, the human-scale dimension has not been transformed to a satisfactory level and this could be pointed by factors like the increasing number of homelessness (As of 2019, approximately 2% or 150 million globally were reported to be homeless, and it is anticipated that the COVID-19 pandemic promoted an addition 100 million people [7]), the exclusivity scenarios, unemployment, and gender and racial inequalities, safety on roads, among others. However, proponents of 'smart' cities are upbeat that it is possible to expand the use of data from urban governance, urban morphology and private profiteering pursuits to the least considered dimensions, for the social goods [8, 9]. This can be explained by the gradual emergence of human scale oriented solutions such as liveable cities, like the 15-min City planning concept [10], and the increased interest in alternative urban mobility options [11–13] to address safety levels in transport networks.

The attention on urban mobility solutions, where alternatives, such as autonomous vehicles—pertinent to this chapter, is particularly paramount, as the increasing data from different dimensions can help accelerate the achievements of rendering safer cities. More importantly, data generated by diverse interconnected smart devices and tools such as sensors in vehicles, lighting systems, packing meters, charging ports et cetera could help increase the efficiency in the use of the AVs, which is currently being limited to its data generated from individual vehicles. Further, such data would also help in increasing the safety of road users, including those using alternative options like cycling or walking, who represent the largest demographics of road accidents. A case in point of how data from different urban sources could help reduce the number of urban fatalities, by adopting autonomous vehicles can be drawn from Hangzhou City, Zhejiang Province, China that relies on AI-driven system [14]. With real-time data from installed cameras and sensors, the city has managed to reduce accident by approximately 95%, and subsequently, benefitted the social dimension through reduced traffic build-up, saving travel time and reduced cost travel as energy usage is reduced [14].

In view of the above background, this chapter will delve into how rider experience and safety can be enhanced via autonomous vehicles, especially in urban areas hosting a wide range of smart devices.

2 Background

Globally, in the present-day, machine learning is being used to enhance safety and design in almost every industry. One of the more relevant cases of our study is in the enhancing the safety of an automobiles, by integrating them with a connected urban environment. The data could be collected from connect sensors present at traffic lights, road crossings, fuels stations, road bulwarks etc. This data can then be collected and processed by the AV to plan its path and reduce accidents because it would "know" the present dangerous and the position and speed of other vehicles.

One of the most promising approaches is to build a ML Prediction based model which is trained against a simulated connected urban AV system. Here we first build a simulation where the urban environment and the AVs are completely connected and the data is generated, data types such as crash, traffic, average acceleration, average velocity etc. The data synthesized from this simulation would then be used to train an ML Model, it can be used for future predictions of characteristics such as accident rates, casualties and can even be used to predict the amount of fuel emissions. This is important especially in view of the changing urban demand, where AVs are being sort following their potential to reduce fatalities, as well as promote the urban sustainability agendas.

Features and parameters from the geometric/simulation model through a process known as feature engineering. The data extracted from this feature engineering would then be cleaned and formatted. This data would then be visualized using plots and a general understanding of how the data obtained is related. This data is then divided into testing and training datasets and the mode is fit over various ML Models such as supervised learning methods such as Regression, Classification, Ensemble Decision K-Tree etc. this would then be validated on the test data set and the predictions would be obtained.

3 An Augmented Model for Autonomous Vehicles in Urban Areas

We propose that the data collection for AVs include data sourced from both the car and its surroundings, even if its from third party providers, acknowledging that this poses some limitations in the immediate future do the lack of standardization, access and disconnect between networks and protocols. While those hurdles will be addressed in the future, our model theorizes that the following approach can benefit both car development and the safety of passengers and urban dwellers. In principle, our approach works as represented in Fig. 1 below.

Handling such huge amounts of data is done in a standardized way. One of the more popular methodologies to approach in data science is the adoption of the CRISP DM—Which refers to Cross Industry Standard Process for Data Mining (CRISP DM). It mainly consists of six main steps and phases to target and solve the problem. This can be helpful in the development of simulations, as part of the proposal, and is thus described in the following sections.

3.1 Understanding the Business/Problem Statement

The need for ever increasing safer automobiles and urban spaces is increasing, hence it is necessary to develop a connected system where in the urban environment interacts with the AV to enhance the safety of both its passengers and the urban dwellers.

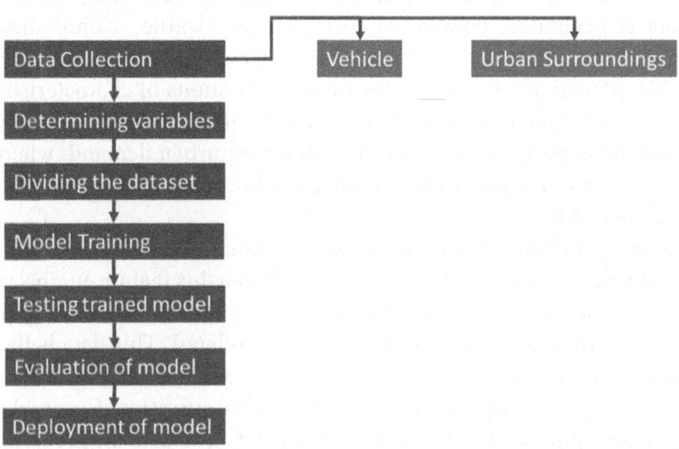

Fig. 1 Proposed model for simulating AVs to use data sourced from both vehicles and urban surroundings. Illustration from authors

3.2 Data Understanding—What Does the Data Indicate?

This stages expressly help in understanding the data that is to be extracted from the simulation model for the urban space. Here, feature engineering is employed to extract data points from the urban model. The data is described as in the data format, the field identities and the significance of these data points. It is also explored and visualized through plots and an understanding of relations among the data points is realized.

3.3 Data Preparation—Organizing the Data for Modelling

First is to determine which datasets are to be used and reasons must be given for inclusion/exclusion is noted. Then this selected data is cleaned from Null, NaN, repeated and other unwanted values from the extracted dataset. Re-formatting of the data may also be necessary as any string values which store numbers need to be converted into numeric values so that mathematical operations can be performed. This is then organized, reshaped etc. for the modelling stage. Here again the dataset is divided into the test and training (validation) datasets which is then fed into the model.

3.4 Modelling Stage—Which Model Fits Well?

This is one of the more crucial stages, where one needs to determined which ML model can be fit over the data. Here we start with a simple linear regression model and then increase the complexity of the model (increasing the index). We make use of a toolbox in MATLAB called Regression Leaner which can run multiple models on the dataset such as Robust Linear, Interactions Linear, Stepwise linear etc. CRISP DM suggests an iterative model-based approach until a strong model has been found. The out puts the evaluation metrics such as the Root Mean Square Error (RMSE) and RRMSE values which is assed in the next stage.

3.5 Evaluation of the Model

Here the results of the model are evaluated, does the model meet and provide a solution to the problem statement, and if it should be approved. The review process consists of summarizing the findings and correct anything if needed.

3.6 Deployment of the Model

The complexity of this phase varies considerably and depends on a major part on the type of the problem but it mainly consists of four sub stages. Plan deployment a model for deploying the built model, then a plan is established for reviewing and maintaining this model. This would first be deployed at a much smaller scale then expanded to accommodate a much larger landscape.

4 Method of Approach for Prediction Based Modelling—Feature Engineering Based or Design Variable Based Based?

Coming to our Urban Model, there are mainly two approaches in which a model can be built.

4.1 Feature-Based Machine Learning

The feature-based ML method is built upon a process known as "Feature Extraction". In the process, the features of the urban model are collected and their result to the required response would be evaluated and studied. For training, the feature-based model, these models must first undergo a feature extraction process, where in the only required features such as number of dwellers, traffic frequency, type of vehicles etc. are chosen to build the ML Model. The data collected from this is divided into test and training datasets, and the ML prediction model is trained. In the trained model when a sufficient Root Mean Square Error (RMSE) value is obtained the model with the lowest RMSE is chosen and used. To make a predictions on number of casualties, air pollution, traffic jams etc. which can be used to quantify the "connectivity" of the urban-mobile infrastructure—higher the score, lower is the connectiveness of the infrastructure. An interesting approach here is that the feature-based predictor first reads the new "unseen or unknowns" urban models, for example for a city where an urban model isn't built yet, after feature extraction, the ML Model based on its past training, provides a prediction of the required attribute values of the urban model under study.

4.2 Design-Variable Based Machine Learning

The Design-Variable based method is a very common process in data mining for collecting the design (input) and response (output) variable and then creating a predictive model. For this process, multiple urban models are required, and in this way the design variables are collected. And as for responses (outputs) during post processing can be retrieved at every iteration and thus forms the output matrix. With these variables, the data can be divided in test and training datasets and training can be performed to create a new DV based predictor. Once the ML Model has been trained optimally, it does not require any Urban Model to "extract" variables, it just need the DV Values which is present at the design stage, once these values are given the model predicts the responses (Switzerland).

4.3 Why We Choose Feature Engineering Over Design Variable Based ML

The main difference between these two approaches is that the FE based approach relies on the actual model geometry but whereas the DV-based model depends on the values of the deign variables, and the model learns from this. Here DVs are faster, they can be trained much faster and there is no feature extraction step, but this is not practical and optimal as the Urban Model doesn't necessarily match with the established design variables. But whereas, the FE Based approach may take much longer to train and predict due to feature extraction, but here input it the urban model itself and thereby it results in more practical and result oriented training.

5 Feature Based Training and Its Scenarios

5.1 Basic

This is the fundamental and the most basic process of building an ML Model. The basic principle is to split the test and the training according to a defined ratio which is usually a 70:30 split to prevent over fitting of the data (Xu, 2018). The model is then trained over a defined training dataset which is free of any errors or NaN values and is then tested and validated over a test dataset and the requires RMSE and Mean Average Error (MAE) are obtained and studied this is represented in the below figure. This Basic process is usually suitable for our Damper model study wherein there is little to no variation between the various damper models for vehicles.

5.2 Incremental Learning and Training

This case is suitable for our Urban model as the Urban model is gradually developed overtime, the feature engineering scripting extracts the data points from the Urban model and is continuously "incrementally" added to the training dataset, hence the ML model is always up to date with the Urban model being developed, this style of model building is very dynamic and is constantly in sync to the latest developments taking place. The already present predictor which is capable and making accurate predictions is selected and is then updated with new date obtained from the new Urban model. This new dataset is again split in testing and training datasets and this new data is passed on to the predictor to train.

A definite propagation ratio is set for this process. For example, a PR of 0% refers that the training will occur only on the new data and the old data does not have any contribution for training the predictor. A PR of 100% refers that both the previous old data and the new current data is used for the training of the ML Model this is usually preferred as this prevent the predictor model from overfitting and completely changing to only "suit" this newer dataset (Rong, 2018). The speed of the training is inversely proportional to the speed of training the ML Model. At the end the RMSE and the MAE are again compared with the older key parameters of the ML Model and is then deployed (Rong, 2018).

5.3 Retrain Phase

This is phase usually comes into picture when the existing ML Model is performing inefficiently and the required RMSE and MAE is very high and do not meet the allowable tolerance limits. The retrain scenario is very similar to the basic scenario. This uses all the data from the existing predictor and the additionally newly selected data from the new predictor. In contrast to the Incremental model, in which hyper-parameter tuning and model selection do not take place, in the retrain phase the all the models and the HPT are run again (Rong, 2018). This is especially suitable if the new data is mostly unrelated to the initial dataset and is completely new.

6 Discussions and Conclusion

The availability of vast amount of data in cities resorting from increased adoption and deployment of advanced technology is hailed for increasing efficiency, reducing costs and promoting operations. In the transport sector, data has played a significant role especially in facilitating the production and subsequent introduction of AVs on various cities across the road—with positive outcomes especially in reducing accidents and related fatalities. As discussed in this chapter, most of the fatalities from crashes are related to vehicles front end structures. However, it has also been established from the literature that over 90% of the accidents are as a result of human error [15]; hence, even with the right vehicle design, chances of crashes occurring still remain high. But with Autonomous vehicles, their reliance on data generated by diverse components such as sensors, cameras that allow them to communicate with other different urban components and other vehicles is argued to increase the safety capabilities in them. The safety features are particularly celebrated and emphasized noting that cities are becoming the priority habitats for a majority of the global population. Further, it is being championed as it would allow for an increased shift by a majority of urban dwellers to sustainable mobility options such as cycling and walking; which have been very risky in the traditional setup where automobiles have accounted for substantial accidents; some being fatal [16].

The discussion above showcases that it is possible to enhance the efficiency and reliability of AVs by integrating advanced technologies such as ML, which are modelled and trained such that they have capabilities to respond in real-time to diverse scenarios [17, 18]. However, it has been appreciated that increasing the safety in the AVs is a work-in-progress, occasioned by disparities in the amount of data available from different urban spheres. On this, it is anticipated that, as the number of smart devices and tools continue to increase, there would be an increase in the number of databases, and datasets that will eventually lead to better modelling and simulations.

Modelling and simulations, as highlighted in the study, has not been smooth and in tandem with the emergence of AVs due to challenges with the data and their data sources. Issues at the fore in derailing the modelling include lack of unified sources of data, as most operate in silos, where smart devices and component manufacturers pursue proprietary policies—that is, maintaining unique communication protocols for their products [19]. This way, most of the data produced have been in silos, and not provided in open- access platforms [20]. With data silos, seamless interoperability across systems has not been forthcoming, and this has led to limitations such as high upfront costs for manufacturers and startups deploying AVs [21]. Further, this has been a bane especially in respect to security and perception of AVs especially at the societal levels, leading to low acceptability by a majority of the population. Additionally, with insufficient data from different quarters, it has been argued that it is impossible to effect changes on available infrastructures to align them with the increasing demand for AVs.

Despite the above challenges, it is worth noting that there is evidence of strides being made in the realization of uniformity of data architectures, and it is expected that this will be realised in the near future. However, even before this is realised, the outcomes of this study points to the need to continue pondering on alternative ways that challenges with data can be overcome as it is obvious that the deployment of AVs in cities is eminent. This is true especially in view of emerging new technologies with capacities to stimulate even further success on the manufacturing and performance of AVs, coupled with the changing global landscape due to discourses on sustainability, liveability and need for responsible consumption and production.

7 Future Scope

This study is a steppingstone to a much wide scale application of AI and ML in the product development cycle, especially in view of emerging trends on the global scale where attention is focused on sustainable agenda. In particular, these are expected to have huge impacts in the adoption of alternative mobility option—more so the AVs. Thereby, warranting the need to explore how these two technologies could be deployed to influence robust rollout of new AVs designs that are responsive to emerging and anticipated urban changes without the need for simulations. Future works could also focus on how these could be integrated with emerging technologies such as Digital Twins, 5G and others, that are taunted to have potential to revolutionize different urban components, especially as data from smart devices and urban elements continue to increase exponentially.

References

1. Cugurullo F (2020) Urban artificial intelligence: from automation to autonomy in the smart city. Front Sustain Cities 2
2. Vailshery LS (2022) Number of IoT connected devices worldwide 2019–2030. Available online: https://www.statista.com/statistics/1183457/iot-connected-devices-worldwide/
3. Jovanovic B (2022) Internet of Things statistics for 2022—Taking Things Apart. Available online: https://dataprot.net/statistics/iot-statistics/
4. Jeble S, Kumari S, Patil Y (2018) Role of big data in decision making. Oper Supply Chain Manag: Int J 11:36. https://doi.org/10.31387/oscm0300198
5. Paiva S, Ahad MA, Tripathi G, Feroz N, Casalino G (2021) Enabling technologies for urban smart mobility: Recent trends, opportunities and challenges. Sensors 21, doi:https://doi.org/10.3390/s21062143
6. The Economist Intelligence Unit (EIU) (2022) Safe cities index 2021. Available online: https://safecities.economist.com/
7. Chamie J (2022) As cities grow, so do the numbers of homeless. Available online: https://archive-yaleglobal.yale.edu/content/cities-grow-so-do-numbers-homeless
8. Özdemir A, Kourtit K, Nijkamp P (2019) Social policy in smart cities: The forgotten dimension. pp 235–261
9. Rowe F (2021) Big data and human geography. In: Demeritt D, Lees L (eds) Concise encyclopedia of human geography. Edward Elgar Encyclopedias in the Social Sciences series, UK
10. Allam Z, Nieuwenhuijsen M, Chabaud D, Moreno CJTLPH (2022) The 15-minute city offers a new framework for sustainability, liveability, and health. 6: e181–e183
11. Bouton S, Hannon E, Knupfer S, Ramkumar S (2017) The future(s) of mobility: How cities can benefit. McKinsey
12. Camagni R, Gibelli MC, Rigamonti P (2002) Urban mobility and urban form: the social and environmental costs of different patterns of urban expansion. Ecol Econ 40:199–216. https://doi.org/10.1016/S0921-8009(01)00254-3
13. Jones P (2014) The evolution of urban mobility: The interplay of academic and policy perspectives. IATSS Res 38:7–13. https://doi.org/10.1016/j.iatssr.2014.06.001
14. Yijing H (2022) Smart cameras keep an eye out for traffic accidents in Hangzhou. Available online: http://chinaplus.cri.cn/news/china/9/20180412/116911.html

15. Haghi A, Ketabi D, Ghanbari M, Rajabi-Vardanjani H (2014) Assessment of human errors in driving accidents; analysis of the causes based on aberrant behaviors.Life Sci J 11
16. Budget Direct (2020) Car accident statistics 2020. Available online: https://www.budgetdirect.com.au/car-insurance/research/car-accident-statistics.html
17. Molinario G, Deparday V (2021) Demystifying machine learning for disaster risk management. Available online: https://blogs.worldbank.org/opendata/demystifying-machine-learning-disaster-risk-management
18. Elbattah M (2019) How can machine learning support the practice of modeling and simulation?—A review and directions for future research. pp 1–7
19. Madsen AK (2018) Data in the smart city: How incongruent frames challenge the transition from ideal to practice. Big Data Soc 5:2053951718802321. https://doi.org/10.1177/2053951718802321
20. Franke J, Gailhofer P (2021) Data governance and regulation for sustainable smart cities. Front Sustain Cities 3
21. Mulder T, Vellinga NE (2021) Exploring data protection challenges of automated driving. Comput Law Secur Rev 40:105530. https://doi.org/10.1016/j.clsr.2021.105530

Litoral Besòs, an Urban Sustainability Transition in the Barcelona Metropolitan Area

Nancy Andrea Ramírez-Agudelo, Joan de Pablo, and Elisabet Roca

Abstract Innovative solutions integrated into urban systems enhance cities' responses to societal challenges if created in collaboration with citizens. Nature-based solutions (NBS) are a comprehensive approach that places a territorial emphasis on the social, ecological, and technological systems involved. Transformative shifts in urban systems—which are widely recognized as intersectoral, multi-level, and complex—lead to transitions toward urban sustainability. As a result, NBS implementations are getting more support, because it has the potential to change different urban domains, while also offering several benefits and more inclusive living conditions. A case study of NBS implementations explores how nature endorses urban sustainability transitions in terms of place-based transitions, multi-actor dynamics, urban experimentation, problem-reframing, and visioning importance. The analysis of the *Litoral Besòs* in the Barcelona metropolitan area, an urban area bounded by two waterfronts, the Besòs river waterfront and the Mediterranean seafront, which have undergone significant change because of NBS implementations shows the extent to which urban sustainability transitions are useful conceptual lenses for this purpose. We conclude with a call to action for more neighborhood-focused initiatives because, with a greater understanding of NBS as—a continuous, broad, and local—process, we can make better decisions regarding NBS scalability.

N. A. Ramírez-Agudelo (✉) · J. de Pablo · E. Roca
Institute for Sustainability Science and Technology, Universitat Politècnica de Catalunya, Jordi Girona 1–3, 08034 Barcelona, Spain
e-mail: andrea.ramirez.agudelo@upc.edu

J. de Pablo
Department of Chemical Engineering Universitat Politècnica de Catalunya, Eduard Maristany 10–14, 08019 Barcelona, Spain

Z. Allam (ed.), *Sustainable Urban Transitions*, Urban Sustainability,
https://doi.org/10.1007/978-981-99-2695-4_22

1 Introduction

Cities are required to solve in an urgent and systemic manner issues related to environmental degradation, changes in climate, and associated risks, as well as the pressures of urbanization and population growth. As a response, cities are deploying innovations such as Nature-based Solutions (NBS), in order to address the complexities of these societal challenges, while delivering simultaneous and multidimensional benefits. By reintroducing nature in cities, NBS facilitate transformative changes in their socio-ecological-technical systems [19]. Therefore, urban implementations of NBS are increasingly promoted as sustainability actions in policy, science, and practice [8].

Nature-based solutions (NBS) represent alternative practices to socioecological adaptation and resilience that place equal emphasis on social, environmental, and economic domains [7]. NBS have been described in the agenda policy of the European Union as "actions that are inspired by, supported by, or copied from nature…" [2]. Moreover, NBS is a type of urban experimentation that reintroduces nature to address various societal challenges [8]. For this reason, NBS have a place-based transformative potential. Research on NBS has used the term as a green concept umbrella, including other concepts as green infrastructure and ecosystem services [5, 6, 11].

In policy, for instance, the European Commission officially introduced NBS into policy in the last decade, and as a result, it was initially regarded as a 'European' concept [5]. However, the United Nations Environmental Assembly agreed to formalize a definition of NBS recently, and now, it is being promoted globally as a sustainable development action [29]. Negative perceptions and conflictual concerns about NBS may emerge, which can be handled by increasing awareness and knowledge through citizen participation and engagement [24]. Hence, NBS is regarded as a comprehensive approach useful for integrating different governance levels, and the citizens, in order to promote a transition toward sustainability.

Transformative shifts in urban systems—which are widely recognized as intersectoral, multilevel, and complex—lead to transitions toward urban sustainability. In transitions management, urban sustainability transitions are characterized through five key aspects: Place-based transitions, multi-actor dynamics, urban experimentation, problem-reframing, and visioning importance [15]. Reflexive understanding of urban sustainability transitions by NBS interventions requires evidence to inform on the processes of change enabled on these key aspects to better inform urban management. The analysis of a case study allows for reflection on the coherence of NBS implementations by informing on how the innovative use of nature in urban areas enables a sustainable urban transition in its five key aspects.

This chapter focuses on the Barcelona metropolitan area in southern Europe, the *Litoral Besòs* that comprises an area within two waterfronts, the Mediterranean seafront and the Besòs riverfront. Currently, the seafront reconfiguration is under discussion as a brownfield redevelopment plan in which nature seems to be a common ground. However, its approval is being scrutinized from different standing points as the area is the final piece of a 'Metropolitan littoral puzzle' to be assembled. In contrast, the riverfront reconfiguration presents the consensus on what is considered a historical and successful experience of a riverfront restoration, which was motivated by social, technological, and ecological goals. The *Litoral Besòs* analysis aims to the understanding of NBS as a local process, because NBS is an action for sustainable development that is implemented in a local context, which is shaped under the framing of specific territorial and historical concerns.

The three following sections discuss to which extent the analysis of NBS implementations benefit from the conceptual lenses of an urban sustainability transition. First, the presentation of an overview of the unit of analysis in terms of the components of the *Litoral Besòs.* Second, the summary of the findings as place-based transitions, multi-actor dynamics, urban experimentation, problem-reframing, and visioning importance. Third, we raise a call to action for more local as neighborhood-focused strategies for urban NBS implementations as decision-making for NBS scalability can be better informed. In particular, by considering the perspectives of the residents, who seem to support transformative change while aware of the threat of gentrification. Finally, we conclude by arguing how NBS and sustainable urban transitions are useful concepts to increase precaution and awareness on the fundamental, systemic, and long-term changes in cities.

2 *Litoral Besòs*: Overview of the NBS Implementation and the Processes of Change

The '*Litoral Besòs*' is an urban area bounded by two waterfronts, the Besòs river waterfront and the Mediterranean seafront in Sant Adrià de Besòs (Fig. 1). Symbolically, the *Litoral Besòs* is the territory that corresponds to the influence of the Besòs river's restoration, which is currently considered as an a strategic 'axe' of green infrastructure [1]. In addition, the seafront of the Three Chimneys (3X) is the missing (and last) plot for the reconfiguration of a metropolitan littoral.

The metropolitan configuration reveals the difficulty of a 'greater city,' for example, small municipalities as neighboring settlements to Barcelona face unbalanced demands versus local capacities for dealing with them.

In a comparative view of the Barcelona metropolitan area, for instance, the urban vulnerability differences are shown in aspects such as the urban environmental quality such as green and open space, as well as air quality.

The local capacities to cope with the functional complexity are not enough to solve the traditional needs related to transportation, housing, and jobs.

Thus, the articulation through the metropolitan area's governance has been central to better addressing this 'imbalanced' situation, by the joint effort within the 36 municipalities of the Barcelona metropolitan area.

The *Litoral Besòs* is a context to be considered as an urban, and peri-urban area, as different activities are mixing [30]. This area is facing environmental degradation, including soil, air, nature, and water. The area's processes of change can be

Fig. 1 Location of the *Litoral Besòs*. The top image corresponds to the location in the Barcelona metropolitan area. The image below shows the two waterfronts: As the Besòs River axe, and the Mediterranean Sea, as the last 'puzzle-piece' for the configuration

described through: (i) evolution of the *Litoral Besòs*, as an anthropized landscape near Barcelona city; (ii) Besòs riverside waterfront, as the restoration through NBS as coping approaches to deal with the industrialization effects; and (iii) Mediterranean seafront, as the contemporary development of a littoral reconfiguration.

In terms of the evolution of the *Litoral Besòs* as an anthropized landscape in close proximity to Barcelona city. Initially, the River Besòs was a natural limit within these urban centers, in which the placement of industrial activity generated a human-made coastline and river waterfront. Together, these activities and the interest in configuring the metropolitan area motivated the development of massive transport infrastructures (Fig. 2).

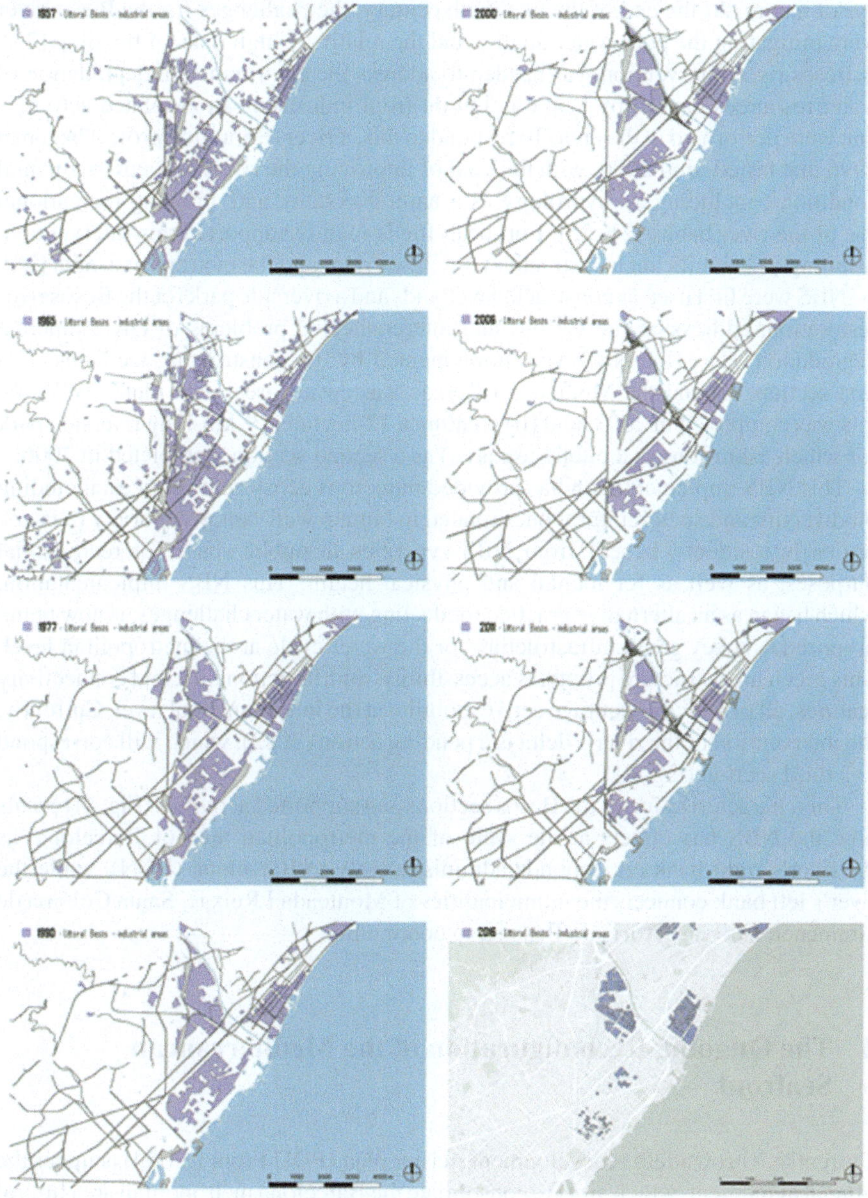

Fig. 2 Evolution of the industrial areas in the *Litoral Besòs*

3 The Historical Reconfiguration of the Besòs Riverfront

The analysis of the area of the Besòs riverside park, using a chronological approach, exposes two decades of changes, addressing environmental degradation through the use of nature. At the end of the twentieth century, the challenges for the Besòs river were mitigating the poor water quality and the relatively high risks of flooding [25]. A Besòs river intervention was needed to address the environmental degradation of water resources caused by heavy pollution from industrialization-related activities that were performed in the area. To respond to this, a river restoration project began in 1996 that lasted until 2006, with the goal of improving the riverbed's environmental conditions, including its hydrology as a natural system, and allowing recreational use of the river banks [21, 25]. European funds mainly supported this intervention, resulting in a significant investment in the Besòs river and the metropolitan area [18].

NBS were first used as constructed wetlands and a riverside park for the Besòs river restoration, addressing the various and interconnected problems of environmental degradation. Here, the NBS were implemented by: (i) constructing wetlands, as a first section, around the Montcada i Reixac wastewater treatment plant (WWTP); this was completed in 2003; and (ii) creating a 22-hectares, 9-km-long riverside park (of which 5-km-long is a public use area) as a second section, completed in 2006.

This NBS implementation has provided numerous ecosystem services, including biodiversity and socio-cultural ones related to human well-being. A variety of users, primarily residents, benefit from NBS synergies in public space for recreational purposes, as well as for mental and physical health. This NBS implementation, which began as an alternative practice for dealing with water challenges, is now being proposed as a key green infrastructure for the water cycle at the metropolitan level. This green infrastructure promotes accessibility, multifunctionality, and connectivity features, all of which encourage active mobility at the intermunicipal level. Currently, the intervention of the river's delta is a pending action (2022), which will correspond to a third section (Fig. 3).

This characterization of the Besòs sections and supported activities (Fig. 4) reveals how the NBS has effects at the scale of the metropolitan area of Barcelona, as the river's right bank corresponds administratively to Barcelona (BCN), while the river's left bank connects the municipalities of Montcada I Reixac, Santa Coloma de Gramenet, and Sant Adrià de Besòs as a continuum.

4 The Ongoing Reconfiguration of the Mediterranean Seafront

Currently, a brownfield redevelopment master plan (PDU Front Litoral) is under the official agreement, which aims to consolidate the Barcelona metropolitan area littoral (Generalitat [10]). In particular, this process has allowed to opening a non-binding participatory process during the plan's formulation phase, in which the citizens of the

Fig. 3 Sections along the Besòs River intervention. Section 1—Location of the constructed wetlands. Section 2—Besòs riverside park, Section 3 Restricted area

Litoral Besòs area were directly involved. Because of the economic, regulatory, and technical support needed for a seafront reconfiguration; multiple actors, at different governance levels and sectors, need to be involved. The master plan presents the redevelopment of 32 hectares of industrial land on the Mediterranean seafront, in which the multi-scale and multi-level conditioning are implicit.

In particular, the area was the location of the 'Three Chimneys' (3X), a coal-energy plant that is considered a twentieth century heritage at the municipal level. The 3X is an icon for the inhabitants, representing the link between energy production and the industrialization process of Catalonia (Fig. 5).

Originally, the location of the 3X allowed the settlement of highly polluting industry. These activities caused land and water contamination, and currently this environmental degradation is a central issue to be dialed with in the redevelopment. The vision of the 3X brownfield redevelopment appears to be informed by different perspectives of what is a sustainable built environment in a metropolitan littoral.

As the intervention is being planned to transform the industrial land use for a mixed used program, the decisions on how to promote the land occupation were the most controversial issues. For this purpose, the renaturalization of the seafront generates consensus, thus, NBS are considered to being implemented as an open space. Thus, the seafront is foreseen as a reconfiguration, in which NBS endorses the multifunctionality feature that integrates a natural coastal line, as well as the overlapping of mobility systems and land-uses. NBS is supported, as the approach

Fig. 4 Characterization of the Besòs sections and supported activities

that addresses the need for physical changes, emphasizing the aim of a coastal front in which different activities take place to respond to the metropolitan needs of an articulated littoral.

This participatory process has shown that a renaturalized seafront is considered as the option to ensure social practices that integrate actual residents and the new

Fig. 5 The Three Chimneys' in the Mediterranean seafront

users through public space. Although, the intention of housing development for higher income population has been critically received by the citizens who advice on the threat it poses to them as the residents perceive gentrification as an unintended result. In contrast, citizens support this vision of a seafront renaturalization, which as participants insisted on the experience gained through the Besòs river restoration. Thus, the river restoration seems to be a major intervention that exposes the potential of this territory for transformative change.

5 Five Key Aspects of Sustainability Urban Transitions

Research on urban sustainability transitions emphasizes on the shared characteristics along the processes of change; namely, place-based transitions, multi-actor dynamics, urban experimentation, problems reframing and visioning importance. From a geographical perspective, the 'place-based transitions' are the changes in the (urban) context, that is physical configurations of space, place, and scale [3, 9]. In addition, transition governance research proposes the shared characteristics of multi-actor dynamics, problem reframing, the importance of vision, and the relevance of experimentation [15]. The 'multi-actor dynamics' refers to the involvement and role of multiple actors in the transition process. 'Reframing the problem', explores the societal consensus and the actors' ways to understand the need for systemic change. The 'importance of visioning' denotes the collective value of believing in alternative futures as the potential to take aligned actions. The 'relevance of experimentation' is the open-ended exploration of ways to adapt, change, and transform existing dominant cultures, structures, and practices.

5.1 Place-Based Transitions

For a place-based- transition, the Littoral Besòs is a territory in which the physical and social dynamics have changed through the innovative use of nature. Hybrid implementations of grey and green infrastructures on the riverfront support NBS implementations. The Besòs riverfront hybridization has been a sequential process of coordinated actions, which has resulted in simultaneous and synergetic benefits. For a renaturalized seafront reconfiguration, the river experience of NBS implementation is currently highlighted as successful.

Two central features of the physical and social change are emphasized in both waterfronts, the first one related to the scale of the intervention and the second considering the multifunctionality of the place. The multi-scale condition of NBS exposes the consideration of the river Besòs as key green infrastructure at the metropolitan level [1]. In addition, the reconfiguration of the seafront as a Metropolitan littoral emphasizes the need for local capacities aiming to transformative change. The multi-scale feature as an open space, and place, in which different municipalities are physically connected in the case of the Besòs river, and a metropolitan littoral in the case of the Mediterranean Sea.

5.2 Multi-actor Dynamics

Multi-actor dynamics is an aspect of interest in understanding the involvement of multiple actors, the different roles played in innovation development, as they influence the speed and direction of a transition [15]. The role played by the non-usual actors, in particular the lay-citizens, is a complementary bottom-up driver demanding on the 'right to shape the city' [14].

Therefore, social changes are closely related to the physical changes promoted as place-based transitions, which are shaped by different governance schemes for NBS and for the urban systems supporting waterfronts. The waterfronts reconfigurations involve transformations in urban systems, which belong to different governance levels, including the metropolitan, regional, and national levels. On the Mediterranean seafront, the formulation plan for the urban reconfiguration follows a non-binding design, which is critically received by the citizens, who advice on unintended consequences, such as gentrification. In fact, these social concerns seem to restrict the acceptation of the physical changes planned for the area, and it exposes citizens' awareness and precaution as a valid input for decision-making.

In the Besòs riverfront, coordinated action as a top-down design and implementation of a river restoration [27], has shown to work for the delivery of intersectoral benefits as a riverside park, as well as constructed wetlands articulated to a wastewater treatment plant [23]. In particular, NBS maintenance as a biotic infrastructure [17], opens the opportunity for citizens' involvement as users can contribute to NBS management [22].

Therefore, the scalability of NBS can benefit from multi-actor dynamics, in which the role of citizens is a key local capacity. In both cases, the waterfront reconfiguration depends on higher levels of decision-makers, which relate to the need for transition management promoting vertical coordination [28].

5.3 Urban Experimentation

Experimentations are open-ended explorations that allow learning-by-doing and doing-by-learning, in order to create ways to adapt, change, and transform existing dominant cultures, structures, and practices [15]. However, urban experimentation is currently questioned as a means for broader transformative dynamics promoting a coherent sustainable urban transition [26]. Therefore, an open concern in urban sustainability transitions is the potential of cities to escalate (scale up) and replicate (scale out) their learnings and solutions into mainstream approaches. For this purpose, it has been considered beneficial to advance on integrating reflexive learning and higher awareness on the specific cities' capacities [16, 31]. This means that cities learnings and capacities address its societal challenges, in which the role of place shapes the urban transformative potential [20].

5.4 Problem Reframing

Problem reframing insists on the importance of societal consensus by establishing shared recognitions that can facilitate aligned actions for problem resolution [15]. Therefore, active participation is central in sustainability transitions, as it promotes the development of a (new) shared discourse for problem resolution. As a result, a comprehensive understanding of the different actors involved in urban systems is required for a systemic identification of the sustainability challenge in an urban reconfiguration as a socio-technical system [12]. In fact, a challenge pertaining to a specific urban system may impact other systems, thus, for reframing the challenge a better understanding of 'place' is fundamental [20].

In the *Litoral Besòs*, the ongoing process of change has exposed the initial concerns that promoted the NBS intervention, and which are now ensured results that demand to scale up and scale out. Despite the environmental concerns of the Besòs river as the water quality and quantity have been improved, new issues have emerged. By facilitating a more biodiverse waterfront, additional benefits and environmental services are being delivered, however, the aim for the waterfronts' integration is questioned by the kind of social benefits expected versus the need to maintain a risk management approach. Therefore, besides the context-specificities of a place, accountability is a critical feature to take into consideration for urban systems change. The effort for a renaturalized seafront can explicitly consider the users insight as a replicable process of awareness and knowledge, in which 'hybridization' promotes

NBS scalability [13]. Here, by higher communication on how the water associated changes are faced through NBS, the *Litoral Besòs* advances not only to avoid risks, but also to move towards new paradigms in urban systems [22, 23]

5.5 Visioning Importance

Visioning importance is the strategic aim of inspiring urban change as the actors' belief in futures of more sustainable urban living are useful for providing direction and speed of change [15]. The fundamental sustainability values that the actors involved aspire to realize with these alternative futures are the keystone for innovation and experimentation at all levels. If these values are 'non-prescriptive sustainability principles´, they are useful for determining 'what are the essential aspects of the ecological and social systems that need to be sustained' [4].

In the *Litoral Besòs*, the River waterfront did not include a visioning exercise. However, the Mediterranean seafront reconfiguration has shown that the participatory stage, which was part of the formulation plan for its redevelopment, allowed to expose the citizens standpoints of how a brownfield has the potential to advance towards a sustainable future. In this phase, their concerns on the seafront were beyond the physical changes, which are inputs for reframing the problem of a brownfield redevelopment, by including social concerns such as gentrification, and furthermore, a more just transition. In consequence, this ongoing seafront reconfiguration exposes the importance of visioning with different actors, as alternative versions of a sustainable urban future may emerge as confronting images, informing on the need for higher awareness.

In summary, this characterization of the Litoral through NBS implementation identified the extent to which the transformative shifts can be better analyzed in terms of place-based transitions, multi-actor dynamics, urban experimentation, problem-reframing and visioning importance (Fig. 6).

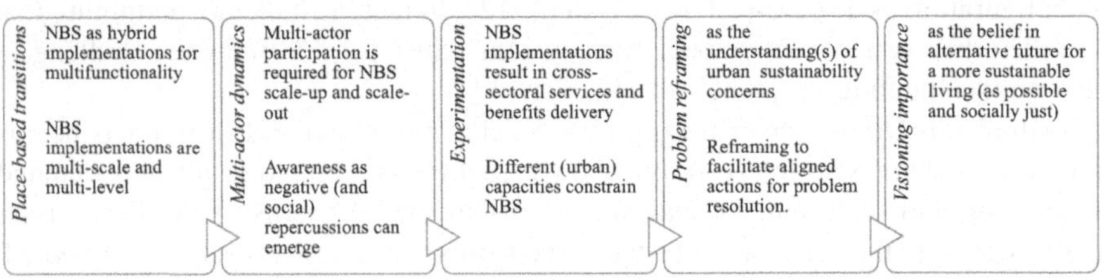

Fig. 6 *Litoral Besòs* and the urban sustainability transitions

6 Call to Action for NBS: A Neighborhood-Focused Process

This chapter has presented the analysis of the *Litoral Besòs* in the Barcelona metropolitan area, exposing a local experience of "how-to" promote transformative change through nature-based solutions. The *Litoral Besòs* demonstrated that NBS and the associated services and benefits can be identified as a sustainable development action by the use of the key aspects that characterize an urban sustainability transition. Therefore, this chapter contributes to advance our knowledge of NBS, by providing tools that benefit its comprehensive understanding as implementations that are deployed at the local level.

The *Litoral Besòs* case study, consisting of an urban area within two waterfronts, has characterized the processes of reconfiguration as a sustainable urban transition. The river reconfiguration, in which a polluted version of the Besòs river, dealing with water quality and quantity issues, has been radically transformed into a lively riverside park is validating NBS as a sustainable development action. The Besòs riverfront exposed how the riverside park and the constructed wetlands as NBS interventions have been beneficial for a reconfiguration that delivers several services and benefits in multiple urban domains. The riverfront reconfiguration has created local capacities to scale up at the metropolitan level, but also it seems practical, to scale out the learnings for a more sustainable Mediterranean seafront reconfiguration.

The ongoing debate over the plan for the Mediterranean seafront redevelopment highlights the challenge of integrating new urban developed areas in lower-income settings while avoiding negative consequences such as gentrification. Yet, a renaturalized seafront brings diverse viewpoints together in the interest of its reconfiguration as an open space designed in response to climate change pressures. Renaturalized waterfronts ensures public accessibility, connectivity, and multifunctionality, which in the *Litoral Besòs* is facilitating precaution and awareness of the citizens as users, pointing out potential threats. However, to deal with the societal challenges in an integrated way requires dialogue and collaboration for a common vision of the future, which favors the mainstreaming of NBS as a comprehensive approach to interact with a local context.

Overall, this analysis implies that if well implemented, NBS respond to a territorial and historical context, in which their potential endorses the change needed, such as an urban sustainability transition. By using conceptual lenses of 'urban sustainability transitions' the understanding of "how" the process of change -as fundamental, systemic, and long-term- is deployed by cities and with citizens at the local level is highlighted. The use of the NBS as a comprehensive concept in a specific case study serves to reframe urban problems, in which the vision of future alternatives is an opportunity to include the local concerns. Finally, a vision of a sustainable future entails the mediation of the different perspectives, guiding more resilient and inclusive urban reconfigurations.

Acknowledgements This work was supported by the Pect Littoral Besòs: Territori Sostenible project partially funded by the European Regional Development Fund (FEDER), and the NATWIP project—Nature-Based Solutions for Water Management in the Peri-Urban: Linking Ecological, Social and Economic Dimensions, partially funded by the Spanish Ministry of Science and Innovation (MCIU/AEI/FEDER) [PCI2019-103674, 2019].

References

1. AMB (2020) Avanç del Pla Director Urbanístic Metropolitá (Col·lecció). Àrea Metropolitana de Barcelona—AMB
2. Bauduceau N, Berry P, Cecchi C, Elmqvist T, Fernández M, Hartig T, Krull W, Mayerhofer E, Sandra N, Noring L, Raskin-Delisle K, Roozen E, Sutherland W, Tack J (2015) Towards an EU research and innovation policy agenda for nature-based solutions & re-naturing cities: final report of the horizon 2020 expert group on nature-based solutions and re-naturing cities. 76. https://doi.org/10.2777/765301
3. Binz C, Coenen L, Murphy JT, Truffer B (2020) Geographies of transition—From topical concerns to theoretical engagement: A commentary on the transitions research agenda. Environ Innov Soc Trans 34:1–3. https://doi.org/10.1016/j.eist.2019.11.002
4. Broman GI, Robèrt KH (2017) A framework for strategic sustainable development. J Clean Prod 140:17–31. https://doi.org/10.1016/j.jclepro.2015.10.121
5. Dorst H, Van der Jagt APN, Raven R, Runhaar H (2019) Urban greening through nature-based solutions—Key characteristics of an emerging concept. Sustain Cities Soc 49: 101620. https://doi.org/10.1016/j.scs.2019.101620
6. Escobedo FJ, Giannico V, Jim CY, Sanesi G, Lafortezza R (2019) Urban forests, ecosystem services, green infrastructure and nature-based solutions: Nexus or evolving metaphors? Urban For Urban Green 37:3–12. https://doi.org/10.1016/j.ufug.2018.02.011
7. European Commission (2021) Evaluating the impact of nature-based solutions : A handbook for practitioners. In: Dumitru A, Wendling L (eds) Publications Office. doi: 10.2777/244577
8. Frantzeskaki N, McPhearson T, Collier MJMJ, Kendal D, Bulkeley H, Dumitru A, Walsh C, Noble K, Van Wyk E, Ordóñez C, Oke C, Pintér L (2019) Nature-based solutions for urban climate change adaptation: Linking science, policy, and practice communities for evidence-based decision-making. Biosci 69(6):455–466. https://doi.org/10.1093/biosci/biz042
9. Fuenfschilling L, Frantzeskaki N, Coenen L (2019) Urban experimentation & sustainability transitions. Eur Plan Stud 27(2). Routledge. https://doi.org/10.1080/09654313.2018.1532977
10. Generalitat de Catalunya (2018) Pla Director Urbanístic d'Ordenació del Front Litoral en l'Àmbit de les Tres Xemeneies
11. Hanson HI, Wickenberg B, Alkan Olsson J (2019) Working on the boundaries—How do science use and interpret the nature-based solution concept? Land Use Policy, October 2018, 104302. https://doi.org/10.1016/j.landusepol.2019.104302
12. Hodson M, Geels FW, McMeekin A (2017) Reconfiguring urban sustainability transitions, analysing multiplicity. Sustain (Switz) 9(2):299. https://doi.org/10.3390/su9020299
13. Hoffmann S, Feldmann U, Bach PM, Binz C, Farrelly M, Frantzeskaki N, Hiessl H, Inauen J, Larsen TA, Lienert J, Londong J, Lüthi C, Maurer M, Mitchell C, Morgenroth E, Nelson KL, Scholten L, Truffer B, Udert KM (2020) A research agenda for the future of urban water management: Exploring the potential of nongrid, small-grid, and hybrid solutions. Environ Sci Technol 54(9):5312–5322. https://doi.org/10.1021/acs.est.9b05222
14. Hollands RG (2015) Critical interventions into the corporate smart city. Camb J Reg Econ Soc 8(1):61–77. https://doi.org/10.1093/cjres/rsu011
15. Loorbach D, Frantzeskaki N, Avelino F (2017) Sustainability transitions research: transforming science and practice for societal change. Annu Rev Environ Resour 42(1):599–626. https://doi.org/10.1146/annurev-environ-102014-021340

16. Luederitz C, Schäpke N, Wiek A, Lang DJ, Bergmann M, Bos JJ, Burch S, Davies A, Evans J, König A, Farrelly MA, Forrest N, Frantzeskaki N, Gibson RB, Kay B, Loorbach D, McCormick K, Parodi O, Rauschmayer F, ... Westley FR (2017) Learning through evaluation—A tentative evaluative scheme for sustainability transition experiments. J Clean Prod 169: 61–76. https://doi.org/10.1016/j.jclepro.2016.09.005
17. Marcus L, Berghauser Pont M, Barthel S (2019) Towards a socio-ecological spatial morphology: integrating elements of urban morphology and landscape ecology. Urban Morphol 23:115–139
18. Martín-Vide JP (2015) Restauración del río Besòs en Barcelona Historia y lecciones aprendidas. Ribagua 2(1):51–60. https://doi.org/10.1016/j.riba.2015.07.001
19. McPhearson T, Haase D, Kabisch N, Gren Å (2016) Advancing understanding of the complex nature of urban systems. Ecol Indic 70. Elsevier B.V. https://doi.org/10.1016/j.ecolind.2016.03.054
20. Peris-Blanes J, Segura-Calero S, Sarabia N, Ribó-Pérez D (2022) The role of place in shaping urban transformative capacity. The case of València (Spain). Environ Innov Soc Transits 42. Elsevier. https://doi.org/10.1016/J.EIST.2021.12.006
21. Pol Masjoan M, Alarcón i Puerto, A., & Puig i Pons, F. (1999) Recuperación medioambiental del tramo final del río Besòs Revista Del Colegio Oficial de Ingenieros de Caminos. Canales y Puertos I(46):80–85
22. Ramírez-Agudelo NA, Badia M, Villares M, Roca E (2022) Assessing the benefits of nature-based solutions in the Barcelona metropolitan area based on citizen perceptions. Nat-Based Solut 2:100021. https://doi.org/10.1016/J.NBSJ.2022.100021
23. Ramírez-Agudelo NA, de Pablo J, Roca E (2021) Exploring alternative practices in urban water management through the lens of circular economy–A case study in the Barcelona metropolitan area. J Clean Prod 329:129565. https://doi.org/10.1016/J.JCLEPRO.2021.129565
24. Raymond CM, Frantzeskaki N, Kabisch N, Berry P, Breil M, Nita MR, Geneletti D, Calfapietra C (2017) A framework for assessing and implementing the co-benefits of nature-based solutions in urban areas. Environ Sci Policy 77:15–24. https://doi.org/10.1016/j.envsci.2017.07.008
25. Santasusagna Riu A (2019) La gestion des cours d'eau dans la Barcelone métropolitaine (Espagne): les enjeux de la valorisation des espaces fluviaux du Llobregat et du Besòs. Sud-Ouest Européen [En Ligne] 47:11–23. https://doi.org/10.4000/SOE.5133
26. Torrens J, von Wirth T (2021) Experimentation or projectification of urban change? A critical appraisal and three steps forward. Urban Transform 3(1). BioMed Central. https://doi.org/10.1186/S42854-021-00025-1
27. Tort-Donada J, Santasusagna A, Rode S, Vadrí MT (2020) Bridging the gap between city and water: A review of urban-river regeneration projects in France and Spain. Sci Total Environ 700:134460. https://doi.org/10.1016/j.scitotenv.2019.134460
28. Turnheim B, Asquith M, Geels FW (2020) Making sustainability transitions research policy-relevant: Challenges at the science-policy interface. Environ Innov Soc Trans 34:116–120. https://doi.org/10.1016/j.eist.2019.12.009
29. 5th Resolution UNEA (2022) (testimony of UN Environment Assembly)
30. Wandl DIA, Nadin V, Zonneveld W, Rooij R (2014) Beyond urban-rural classifications: Characterising and mapping territories-in-between across Europe. Landsc Urban Plan 130(1):50–63. https://doi.org/10.1016/j.landurbplan.2014.06.010
31. Wolfram M (2016) Conceptualizing urban transformative capacity: A framework for research and policy. Cities 51:121–130. https://doi.org/10.1016/j.cities.2015.11.011

Learning from Experience: Reflections on the Delivery of Nature-Based Solutions as Part of a Multi-Partner Collaborative Project

Ian Mell, Sarah Clement, Fearghus O'Sullivan, Juliet Staples, Christine Derbyshire, Paul Nolan, Clare Olver, and Stella Shackel

Abstract The delivery of the EU Horizon 2020 funded project *URBAN GreenUP* required 27 institutions in 8 countries to work collaboratively to design, develop, deliver, and evaluate a portfolio of small-scale Nature-Based Solutions (NBS) interventions. The 6-year project focussed on understanding the role that NBS could play in future-proofing cities against climate, demographic, and economic change. Throughout its lifespan the project engaged extensively with public, private and community stakeholders to identify areas of need, and match them with innovative urban solutions. This chapter focusses on the Liverpool (UK) component of the URBAN GreenUP project reflecting on the challenges and solutions witnessed to deliver NBS. We discuss how the complexity of local government, the planning and development structures in Liverpool, as well as those issues related to land ownerships and project focus, led to difficult choices being made regarding how best to integrate nature into urban areas. Moreover, we focus on a range of tensions between stakeholders at the local and the overarching project scale related to the delivery of innovation addressing issues of climate change, health and well-being and economic enhancement. Consequently, we identify the following areas: *effective governance and communication between partners, flexibility in the approach to design and delivery, complementarity between the skills and experiences of project partners*, and a *willingness to challenges decisions* as essential criteria for the successful delivery

I. Mell (✉) · F. O'Sullivan
Department of Planning & Environmental Management, University of Manchester, Manchester, UK
e-mail: ian.mell@manchester.ac.uk

S. Clement
Department of Geography, School of Social Sciences, University of Western Australia, Perth, Australia

Department of Geography & Planning, University of Liverpool, Liverpool, England

J. Staples · C. Derbyshire
Liverpool City Council, Cunard Building, Brunswick Street, Liverpool L3 1AH, England

P. Nolan · C. Olver · S. Shackel
The Mersey Forest, Risley Moss, Ordance Avenue, Birchwood WA3 6QX, England

Z. Allam (ed.), *Sustainable Urban Transitions*, Urban Sustainability,
https://doi.org/10.1007/978-981-99-2695-4_23

of a multi-partner and multi-location NBS projects. The chapter concludes with a reflection on how we select, focus, deliver, and monitor NBS investments as part of large consortium project to provide insights into how to meet strategic and local objectives.

Keywords Nature-Based Solutions (NBS) · Governance · Multi-partner collaboration · Climate · Stakeholders

1 Introduction

As cities continue to grow their reliance on environmental resources, i.e., for water, food production or to support public health, compromises the functionality of the landscape to meet current and future socio-economic and ecological needs. In many cases this undermines the delivery of the UN Sustainable Development Goals and calls for all stakeholders involved with the management of urban areas to consider their impact on the environment [21]. Although the exploitation of ecological resources is not a new phenomenon—the development studies literature has examined this issue for decades (cf. [7])—the rate at which 'natural' landscapes and their environmental capacity are being exceeded has led many academics, politicians, and environmentalists to call for immediate action to halt, and indeed reverse the trend of land conversions from green space to built [6]. The visibility of the climate and biodiversity crises and associated public health issues are just three examples of this in practice.

Within this space Nature-Based Solutions (NBS) have been proposed as a mechanism to deliver more sustainable and resilient forms of urban management. By focussing on the composition and interaction of ecological resources within an urban location, NBS advocates ground nature as a central tenant of city debates [13]. Moreover, the literature on NBS argues that by working with nature to *compliment* or *mirror* its systems and processes, we can increase the proportion of urban landscape that are considered "natural"; we help to renature cities in order to ensure their long-term functionality [18]. Within the following discussion NBS are defined as follows:

> …living solutions inspired by, continuously supported by and using nature, which are designed to address various societal challenges in a resource-efficient and adaptable manner and to provide simultaneously economic, social, and environmental benefits.
>
> (European Commision [14])

However, the pathways to this achieve these ambitious solutions are fraught with political, physical, and technological barriers linked to existing disciplinary practices, siloed urban management processes, and competition for space for investment that limits opportunities to deliver an increased proportion of nature in urban areas [22, 28]. Consequently, NBS are being proposed within environmental discussions as a way of challenging the orthodoxy of carbon intensive urban development, and alternatively to place nature as the centre of design, implementation, and management discussions.

To examine the potential for NBS to address such complicated urban issues we present a case study of Liverpool (UK), and its implementation of a programme of NBS interventions as part of the 6-year EU funded Horizon 2020 URBAN GreenUP project. By exploring how stakeholders in Liverpool engaged with, and found solutions to, the need for effective governance and communication between partners, flexibility in the approach to design and delivery, complementarity between the skills and experiences of project partners, and a willingness to challenges decisions, we outline current thinking supporting how best to deliver NBS effectively [25].

Moreover, the chapter highlights tensions between stakeholders at both the local and the project scale related to variation in the focus of the project and its approach to innovation in addressing climate change, health and well-being and economic enhancement issues. The value of this discussion is to identify where pinch points exist that influence delivery [22]. This is especially critical if or when a project has multiple outputs, Key Performance Indicators (KPIs), partners, and oversight bodies, i.e., the public, local political systems and/or the EU. The outcome of this discussion is an assessment of how real-world problems can be solved to support the successful delivery of multi-partner and multi-location projects.

2 URBAN GreenUP

URBAN GreenUP (https://www.urbangreenup.eu/) is 6-year EU funded Research and Development (R&D) project that aimed to design, test, and report on the value of integrating innovative NBS in urban areas. The project is made up a consortium of 27 institutions based in cities, universities and research institutions, Small-Medium Enterprises (SMEs) and the environment sector located in eight countries (China, Columbia, Germany, Italy, Spain, Turkey, the UK, and Vietnam). URBAN GreenUP comprises three front-runner cities—*Liverpool* (UK), *Izmir* (Turkey) and *Valladolid* (Spain), who have led on the application of new NBS technologies and techniques and five follower cities (Chengdu in China, Ludwigsburg in Germany, Mantova in Italy, Medellin in Columbia, and Quy Nhon in Vietnam), who will test the innovations developed by the front-runner cities.

Each city (and their supporting partners) brings a wealth of local government, environmental and technological expertise to the project, which has been integrated into its work programmes and reporting to support the evaluation of a range of socio-cultural, economic, and ecological problems faced by urban areas. The three frontrunner cities developed portfolios of innovative NBS interventions to address these issues, which have subsequently been expanded into as a series of renaturing strategies for cities. The project has thus facilitated a holistic understanding of the complexity associated with NBS delivery, as the project has worked across geographic, climatic, political and governance systems. This has been achieved via the establishment of a series of complimentary approaches to implementation developed by the frontrunner cities.

URBAN GreenUP does not sit in isolation but is one of three EU Horizon 2020 funded projects focussed on NBS R&D. Its sister projects *Connecting Nature* (https://connectingnature.eu/) and *Grow Green* (https://growgreenproject.eu/) are comparable to URBAN GreenUP in terms of their composition of local government, SMEs, environment sector and academic partners drawn from across a range of locations in the EU and internationally. Variations are visible though in how each project has set its strategic objectives with Grow Green's work in Manchester focussed on a single large neighbourhood-scale project, *West Gorton Sponge Park*,[1] rather than the acupuncture style interventions seen in URBAN GreenUP. The Glasgow (UK) based projects associated with Connecting Nature included a range of small and medium sized interventions. The breadth of delivery options embedded in each project, and the subsequent cross-examination between projects, has offered important insights for the EU to examine what can work, how it works, and where barriers remain to successful NBS interventions.[2]

Over its lifespan[3] URBAN GreenUP has worked extensively with local government, the environment and business sectors, technology start-ups/companies, engineers, landscape architects, and local communities to ensure that (a) the right NBS is located in the most appropriate place, (b) that local stakeholders are made aware of the NBS interventions and the associated socio-economic and ecological benefits, and—potentially—most importantly (c) that the range of benefits associated with each NBS project are grounded in robust evidence and publicised widely to all relevant stakeholders. This approach has been crucial in gaining a more detailed appreciation of what NBS interventions are needed, how they address local needs, and if and/or how they support local government policy mandates to address climate change, health and well-being, and economic issues associated with urban areas. The following sections outline the outcomes of these discussions and highlight the approaches taken in Liverpool to deliver its components of the URBAN GreenUP project.

3 Environmental Politics in Liverpool

Liverpool is subject to an ongoing political debate regarding its environment. There is a visible and significant engagement with environmental issues within some parts of the city—noticeably wealthier parts of the city and locations with limited provision of greenspace—that have been heightened due to proposed losses of green and open space to facilitate house building [24]. Within discussions of the environment in Liverpool issues of spatial parity between communities, environmental quality, and the variability and functionality of amenities have been debated as being of paramount importance to local communities. Moreover, the proposed sale of public green/open space for housing in locations in north and south Liverpool led to vocal opposition, protesting of the city council, and the subsequent consultation on the Liverpool

Green & Open Space Review (LG&OSR) [20]. The outcome of which has been a clear indication from the public that (a) all green, blue, and open spaces in Liverpool are valuable, (b) there is a perception that parks are unevenly located and funded between the north/south of the city, (c) and that greater investment in environmental resources is needed to support the city's efforts to address climate change, biodiversity loss, sustainable transport issues, and health and well-being.

One direct consequence of the LG&OSR has been a foregrounding of environmental issues in local media, community consultations, and in the subsequent approach taken by elected members of Liverpool City Council to addressing environmental issues within development. The LG&OSR review built on the city's Green Infrastructure Strategy [16] and the emerging Local Plan [10, 23] and identified specific issues associated with health, the quality and provision of green space, and climate change that needed to be addressed via strategic policy and local project work. Specially issues of flooding in north Liverpool, improving access to high-quality green space across the city, and addressing health inequality via the promotion of alternative lifestyles that utilise parks and green space as the location for sports, recreation and well-being activities were all identified as key challenges [20]. The increased visibility of these issues was part of the rationale for engaging with the EU's Horizon 2020 call for partners to join NBS R&D projects and facilitated Liverpool's inclusion in the URBAN GreenUP project.

A political willingness to engage with environmental issues had thus been developed over several years and had been linked to responses to the strategic development objectives of the city. Consequently, senior city officials were supportive of attempts to align more effective planning for climate change mitigation and adaptation with economic development [26]. Key to this was the ongoing relationships that Liverpool City Council had developed with regionally innovative environmental partners, i.e., The Mersey Forest, and with academics working on environmental issues at the University of Liverpool and Liverpool John Moores University. These institutions have provided evidence and helped to engage local expertise with the City Council to shape environmental thinking. The integration of such institutional knowledge provided the technical, academic, and practical experience used to ground URBAN GreenUP conceptually and promote innovative practice.

However, despite the ongoing relationships between the city, the environment sector, developers, and the public there remain potential issues associated with the development and application of Liverpool's environmental policies. These critiques continue to focus on the equitability of green and open space provision, the downplaying of environmental issues when considered against economic development objectives, and a perceived lack of accountability of the city from some communities who view all development as negatively impacting the natural environment [27]. The equitability, transparency within decision-making, and the long-term resilience of the city's environment have, as a consequence, continued to be queried over a number of years [8, 24, 26]. The URBAN GreenUP project aimed to redress a number of these issues via the implementation of a programme of NBS interventions and subsequent dissemination work of the added-value associated with investing in nature.

4 Effective Governance and Communication Between Partners

As noted previously the development of URBAN GreenUP could be seen as a direct response to the critiques of the city's environment policy-making. The delivery of the LG&OSR could be viewed as a catalyst for URBAN GreenUP, as it helped cement a view within the city's decision-makers that environmental issues were of paramount importance to Liverpool's business development and quality of life for residential communities. Moreover, the historical partnership successes between the city, The Mersey Forest—most noticeably via their reporting in the Green Infrastructure Strategy [16], and with the city's universities laid the groundwork for the political acceptance of a collaborative partnership. These relationships were used to shape the city's thinking regarding whether they should:

(a) Engage with the developing URBAN GreenUP consortium to bid for funding due to the financial commitments of doing so, and
(b) Whether there was sufficient personnel, technical capacity, and financial support within the city to deliver the project.

The emergence of support from within Liverpool City Council regarding both issues, and especially the acknowledgment that capacity existed within city council officers to address the time, skills, and technical skills needed to deliver the project, were key factors in generating support for URBAN GreenUP bid.

Translating the support of city leaders into a successful implementation plan was subject to a range of complex economic, governance and real-world planning/delivery issues. It was therefore imperative that the project team, and specifically the city council officers, were able to lead engagement with relevant political, business, environmental and public stakeholders to generate support for the project. Additionally, it can be argued that the delivery of the project hinged on the political entrepreneurship of city officials [9]. They acted as conduits promoting the portfolio of URBAN GreenUP interventions to the city's leaders, i.e., those responsible for finance, and the delivery teams in Highways, Community Services, and Parks/Open Space Management. Without the innate knowledge of the city's political structures, existing barriers to implementation, i.e., in procurement, ownership/location issues, and knowledge of alternative mechanisms to support maintenance costs, it could be argued that the project would not have received the same level of political support. Furthermore, the expertise held by the project team regarding project timeframes, the acceptable scope of investment, and the immediate and longer-term benefits of delivery of the portfolio of NBS interventions was considered critical to implementation. Across each the three EU funded NBS projects these factors were noted as being signifiers of effective delivery [12, 15, 18]. Although considerable time and technical constraints were placed on the project because of intra-institutional practices and/or delivery objectives the political awareness of officers was a key influence in the successful roll out of the NBS programme in Liverpool.

In addition to the entrepreneurship shown by city council officers in managing the political considerations of the URBAN GreenUP project, the ability of the wider project team to work with businesses, landowners and local communities to evaluate the success of each NBS intervention provided the project with real-time feedback on their added-value. This included immediate feedback from public consultation events and more considered evaluations of the NBS intervention via ex-ante and ex-post postal and online surveys. Moreover, due to the legacy of project work by project partners there was a perception that investments were appropriate and focussed on addressing the key climatic and socio-economic issues facing the city [16, 24]. It is therefore important to acknowledge that the development of trust between project partners, delivery agencies and local communities is critical to the acceptance of any interventions. Within the academic literature the creation and maintenance of trust between stakeholders is noted as being pivotal to the long-term support of urban and environmental development (cf. [17, 33]. Sustaining trust remains an ongoing issue in Liverpool due to the variation in perceptions of the city council's legacy of urban planning. However, such variation in viewpoints was accepted and addressed, in part, via the broader make-up of the URBAN GreenUP project team. We also argue that the project's academic partners were able to add value to the evaluation of URBAN GreenUP specifically because they were not perceived to be the "city council" by the public. Thus, the surveying work undertaken was viewed by some respondents as being "independent" of the city and therefore the outcomes of the work were not predetermined. This reflects the view of some residents within the city that the city council have historically failed to effectively deliver environmental enhancement works and should not be charged with leading such projects [20, 28]. Surveying people as the university (although attached to the URBAN GreenUP project and the city council) provided an increased validity to some respondents and helped to move responses from a critique of Liverpool City Council to a critical reflection of the NBS interventions.

The final aspect of good governance witnessed in Liverpool was the development of an effective communication strategy between partners. Although the project work packages (WP) were assigned to specific organisations the technical, delivery, and evaluative approaches taken were created collaboratively. These reflected to overarching aims, KPIs and financial constraints of the project but were subject to extensive discussion to establish an acceptable delivery programme and framework for monitoring/evaluation that could be agreed upon by all partners.

The success of these processes was partially related to the established trust between partners generated on previous projects but also reflects an acute awareness that each organisation brought specific technical, financial, communication and engagement, and evaluative skills to the project. Thus, where knowledge gaps were evident in organisational skill sets, another partner could deliver that specialism. Examples of this include the city council's ability to procure access to land from landowners to deliver NBS interventions, the Mersey Forest were able to use their contacts and experience to lead on innovative NBS designs, and the universities led the ecological/biophysical and socio-economic evaluations of each intervention. In many projects a lack of such awareness would lead to delays, as additional skills would need to be procured [11, 31]. For URBAN GreenUP the project team was able to acknowledge its weaknesses and identify the complimentary skills needed to deliver the portfolio of NBS interventions and undertake its monitoring, as required by the EU funding agreement.

5 Flexibility in the Approach to Design and Delivery

As the delivery of URBAN GreenUP has progressed its focus has diversified in terms of the portfolio of projects implemented, its strategy for consultation, the timeframe for delivery, and how each intervention has been received locally. Such issues are common across projects with longer implementation timeframes, as they respond to changes in financing, political support, procurement regulations, and in the case of URBAN GreenUP the impacts of Covid-19 and the associated lockdown protocols in the UK.[4]

The changes required to deliver the URBAN GreenUP programme highlighted the need for a level of flexibility to be embedded within all project work in terms of design, procurement, locating and implementation of NBS in Liverpool. Reflecting on the six-year project illustrates a series of issues that arose hindering the delivery of NBS, and included but were not limited to:

1. Working with highways specialists to address potential damage issues relating to street tree root systems on sidewalks and road carriageways (Fig. 1).
2. Working with local law enforcement officials to make sure that any new NBS interventions did not disrupt the sightlines of surveillance equipment.
3. Engaging with landowners and businesses to ensure access to land for implementation and maintenance work for new NBS, i.e., the Parr Street green wall (Fig. 2).
4. Engaging with local communities to communicate the rationale for locating specific interventions, i.e., ecological islands in the Wapping Dock area and Sefton Park.
5. Working with data analyst and providers of technology/monitoring equipment to identity the most appropriate place to locate equipment.
6. Working with utilities providers to ensure that water and electricity services were not disrupted by NBS interventions, i.e., via de-culverting works and/or the installation of pollinator lampposts (Fig. 3).
7. Addressing the variation in costs associated with procurement, delivery, and installation/maintenance of innovation equipment over a five-year programme period (from the outset of the project to the implementation of NBS).

The project team, and specifically Liverpool City Council, were responsible for addressing the majority of these constraints and worked extensively over a prolonged timeframe to ensure that the portfolio of NBS associated with URBAN GreenUP was delivered. This drew extensively on knowledge of local environmental conditions to ensure that interventions were sited to maximise their ecological/climatic benefits.

Fig. 1 URBAN GreenUP interventions—street tree interventions on The Strand

Furthermore, they were able to call on existing working relationships with landowners to deliver NBS and to engage communities of interest (including residents) in understanding the added-value of investing in NBS. They also led discussions of the financial rationale for delivery to ensure that the technology and ecological resources, i.e., trees, floating islands, and varieties of plant, used were cost-effectively sourced and delivered within budget. All of which required a constant dialogue between stakeholders in the public and private sector, as well as an acknowledgement from within the project team that delivery would change depending on the intersection of these issues.

Moreover, the project team had to make significant changes to the NBS investment and evaluation programme due to Covid-19, and the associated lockdowns protocols in the UK. Due to the restrictions placed on people's movement the delivery of URBAN GreenUP slowed. Delivery continued though as restrictions were eased. Additionally, the evaluation of the implementation programme, especially regarding its socio-economic values, was modified to move away from a predominately face-to-face form of surveying to online and postal approaches. This may have provided less detailed commentary compared to face-to-face work but was considered to deliver a sufficient response from residents and businesses to evaluate if, and if so how, the interventions associated with URBAN GreenUP were successfully delivering benefits to society and the environment. The project was also subject to staffing changes leading to minor delays in its development and delivery, however, the core

Fig. 2 URBAN GreenUP interventions—Parr Street Green Wall

team of individuals (and their organisations) have remained in place throughout the project providing continuity over the five-year project period. The outcome of which has been a focussed approach to the delivery of NBS interventions that although modified due to circumstances outside of the project's team authority have delivered on the core aims, KPIs and types of NBS set out in the project bid.

6 Willingness to Challenges Decisions

With all multi-partner, and specifically multi-city projects, potential conflicts arise regarding what, where and how implementation should occur [8]. The willingness to challenge the strategic objectives of a project and an acknowledgement that these may be at odds with more localised focus of a project is paramount to ensuring the right intervention is located effectively to address both strategic issues, i.e., air pollution or biodiversity loss, as well as meeting the needs of the local/host population [34]. Within URBAN GreenUP such issues arose with the siting of specific projects, which were challenged locally as being biased towards more affluent areas in the south of the city despite evidence suggesting that interventions should be located in northern neighbourhoods [16]. The project team thus had to consider whether the initial locations identified for NBS interventions, as outlined in the project proposal,

Fig. 3 URBAN GreenUP interventions—Pollinator lamp posts in the Baltic Triangle

were fit for purpose in the city where climate justice is a key issue in several parts of the city. Moreover, we can also identify issues related to which interventions were located in which location. Due to the breadth of NBS types delivered by URABN GreenUP there was on ongoing discussion of whether green walls, floating ecological islands, pollinator lamp posts, de-culverting works, street trees, and/or moveable forests were being located in the most appropriate places [25]. In part the choices made reflected perceptions of where interventions would have the most impact but also reflected the outcome of negotiations within landowners and businesses regarding who was willing to host a new NBS.

Within URBAN GreenUP these discussions perpetuated an ongoing dialogue between the city council, the project team, other landowners, and local stakeholders. The result of which was a constant need to reappraise what could be delivered and where. However, the relationships developed by the project team, and specifically the city council and Mersey Forest, were noticeable in addressing potential restrictions placed on the locating of NBS interventions. Their working relationships with local partners in conjunction with their perceived expertise in delivering NBS was used to reassure stakeholders regarding the technical aspects of locating NBS on specific sites, i.e., ecological value of floating islands or potential damage to services/utilities associated with street trees. Consequently, the reluctance of some stakeholders to participate in the project was minimised as questions over costs, maintenance, and structural issues were answered. However, despite the positives associated with the project team's relationship with local stakeholders there were instances where NBS interventions needed to be relocated due to technical issues associated with specific sites [32]. The ability to adapt the delivery programme was considered critical to effectively delivering URBAN GreenUP and illustrates an acknowledgement that work programmes are subject to externalities which cannot be controlled by the project team [19]. These externalities include landowners withdrawing support for the project, changes in availability of specific NBS technology, rising costs of labour, additional procurement and legal issues, and in the case of URBAN GreenUP the impacts of Covid-19.

Successful projects are therefore acutely aware of the potential impacts of each of these factors and integrate contingency plans to address these issues if they arise. Within this discussion it is important to reflect on the integration of optionality in delivery, an issue debated extensively within the construction and project management literature. Within this work an emphasis is placed on the examination of potential cost and/or implementation issues that need to be considered to promote best practice options [1, 2, 5, 29]. In practice URBAN GreenUP has worked successfully to address these issues. It could be argued that the 6-year timeframe of the project was a key factor in this, as the extended delivery period absorbed these challenges.

Multi-partner projects also need to navigate the diversity of thinking between project managers and local delivery agents [13, 30]. Those organisations charged with managing continental scale projects for the EU are subject to extensive scrutiny and auditing and thus need to ensure conformity to an agreed programme of works [11]. However, this can fail to account for the climatic, political, or socio-economic variation that influences local decision-making. Thus, the strategic objectives of a project, as set out at its inception may need to be challenged as the project progresses. This was the case in URBAN GreenUP, as each of the frontrunner cities of Izmir, Liverpool and Valladolid faced additional complexity unforeseen at the project's inception due to changes in local context. Staffing changes for partner organisations, a growing diversity of political support, changes in local environmental context, as well as variation in economic costs and global political issues, i.e., Covid-19, all required each partner to renegotiate elements of the project with the coordinating body.

Finally, URBAN GreenUP needed to constantly appraise whether it was delivering the most appropriate benefits to Liverpool's population. This centred on how the project managed the need for innovation in delivery (as required by the EU funding agreement) compared to what investment Liverpool needed, as shown in the LG&OSR and the Liverpool Green Infrastructure Strategy [16, 20]. Unpacking the variation in these positions remained central to project delivery, as the project team attempted to deliver the portfolio of NBS whilst being mindful of the evidence outlining where interventions should be located. The outcome of this process was deemed unacceptable to some stakeholders due to the locating of interventions in areas identified as strategically important rather than in areas of greatest need. It was acknowledged within the project though that it was not always possible to successful deliver on both points simultaneously. However, the project's aims, and indeed those of the wider EU Horizon 2020 NBS funding programme were structured to showcase innovation [3, 32]. Consequently, strategic decisions were taken to deliver NBS interventions that maximised the visibility of the URBAN GreenUP project, and which could subsequently be used as a template for further work in Liverpool and the project's follower cities. The NBS interventions associated with URBAN GreenUP thus acted as a set of exemplars that could be replicated. This was perceived to be of critical importance as it provided an evidence base (and associated financial costing) to enable Liverpool City Council to address the issues of climate change, health inequality and lack of green space provision in other areas of the city. Whilst some stakeholders did not support this approach it did generate a robust and detailed evidence base that is now being used to shape future NBS interventions across the city.

7 Conclusion

The approaches taken to delivering NBS in Liverpool have been shaped by disciplinary and local geographical context (in conjunction with socio-economic and political considerations). The variation in disciplinary thinking and associated skills between local government, the environment sector and academic research components of the Liverpool URBAN GreenUP team need to be taken into consideration when developing a portfolio of NBS interventions. In practice the additional complexity associated with the management of the wider project—enforced via multi-city KPIs and delivery mandates—led to challenges in terms of what NBS could be delivered, how they could be implemented, and the added socio-economic and ecological values associated with interventions. The delivery of NBS therefore continues to be a complex process shaped by the various socio-economic, ecological, and political factors visible in urban areas. What, how, where and for whom we develop, deliver, and manage NBS therefore continues to vary between locations. Consequently, attempts to find a unifying solution to urban problems remains elusive, and the URBAN GreenUP project, as well as its sister Horizon 2020 projects, remain works in progress. However, analysis of each project highlights a series of issues which need to be considered if NBS are to be applied effectively in divergent climatic, urban, and political contexts. These considerations include reflections on the local governance context in terms of capacity to develop and deliver NBS, the willingness of local politicians to support investment, the allocation of capital and revenue financing to support investment, the identification of the most appropriate location and type of investment and the pathways to communicate the added value of delivering NBS to range of community, government, and development/environment stakeholders.

Acknowledgements This project has received funding from the European Union's Horizon 2020 Research and Innovation Programme under grant agreement no. 730426. Support was also received from the University of Manchester via their Open Access funding agreements.

Notes

1. https://www.groundwork.org.uk/groundbreaking-green-space-west-gorton-community-park/, accessed 1st August 2022.
2. A fourth project Naturvation (https://naturvation.eu/about.html) focussed more of setting a foundational baseline for NBS interventions. This looked at case studies to identify a set of governance, NBS typologies, stakeholder characteristics, and working approaches that can utilised to support investment in NBS (Almassy et al., 2018).
3. Due to Covid-19 the project was extended for a further year to enable additional monitoring of the NBS interventions to take place.
4. Lockdown protocols were in place across several European countries leading to delays across EU funded projects. Consequently, extensions of approximately 12 months were permitted to a series of projects to ensure the work programmes and associated monitoring were completed effectively. Lockdowns were in place twice in the UK in 2020.

References

1. Ahmed MO, Abdul Nabi M, El-adaway IH, Caranci D, Eberle J, Hawkins Z, Sparrow R (2021) Contractual guidelines for promoting integrated project delivery. J Constr Eng Manag 147:05021008. https://doi.org/10.1061/(ASCE)CO.1943-7862.0002173
2. Ahmed S, El-Sayegh S (2021) Critical review of the evolution of project delivery methods in the construction industry. Buildings 11:11. https://doi.org/10.3390/buildings11010011
3. Al Sayah MJ, Versini P-A, Schertzer D (2022) H2020 projects and EU research needs for nature-based adaptation solutions. Urban Clim 44:101229. https://doi.org/10.1016/j.uclim.2022.101229
4. Almassy D, Pinter L, Rocha S, Naumann S, Davis M, Abhold K, Buckeley H (2018) Urban nature atlas: A database of Nature-Based Solutions across 100 European cities. Durham University, Durham
5. Atkinson G, Doick KJ, Burningham K, France C (2014) Brownfield regeneration to greenspace: Delivery of project objectives for social and environmental gain. Urban For Urban Green 13:586–594. https://doi.org/10.1016/j.ufug.2013.04.002
6. Beatley T (2000) Green urbanism: Learning from European cities. Island Press, Washington DC

7. Bryant R, Bailey S (1997) Third world political ecology: An introduction. Taylor & Francis, London
8. Clement S (2021) Governing the anthropocene: Novel ecosystems, transformation and environmental policy, Palgrave Macmillan, Basingstoke
9. Clement S, Mell IC (2022) Nature, democracy, and sustainable urban transformations. In: Sustainability Transformations, Social Transitions and Environmental Accountabilities. Palgrave MacMillan., London
10. Couch C, Fowles S (2019) Metropolitan planning and the phenomenon of reurbanisation: The example of liverpool. Plan Pract Res 34:184–205. https://doi.org/10.1080/02697459.2018.1548237
11. Croeser T, Garrard GE, Thomas FM, Tran TD, Mell I, Clement S, Sánchez R, Bekessy S (2021) Diagnosing delivery capabilities on a large international nature-based solutions project. Npj Urban Sustain 1:32. https://doi.org/10.1038/s42949-021-00036-8
12. Dumitru A, Frantzeskaki N, Collier M (2020) Identifying principles for the design of robust impact evaluation frameworks for nature-based solutions in cities. Environ Sci Policy 112:107–116. https://doi.org/10.1016/j.envsci.2020.05.024
13. Eggermont H, Balian E, Azevedo JMN, Beumer V, Brodin T, Claudet J, Fady B, Grube M, Keune H, Lamarque P, Reuter K, Smith M, van Ham C, Weisser WW, Le Roux X (2015) Nature-Based Solutions: New influence for environmental management and research in Europe. GAIA Ecol Perspect Sci Soc 24:243–248
14. European Commission (2015) Towards an EU research and innovation policy agenda for Nature-Based Solutions & re-naturing cities. Final Report of the Horizon 2020 Expert Group on "Nature-Based Solutions and Re-Naturing Cities." Directorate-General for Research and Innovation Climate Action Environment Resource Efficiency and Raw Materials: European Commission, Brussels
15. Frantzeskaki N, Vandergert P, Connop S, Schipper K, Zwierzchowska I, Collier M, Lodder M (2020) Examining the policy needs for implementing nature-based solutions in cities: Findings from city-wide transdisciplinary experiences in Glasgow (UK), Genk (Belgium) and Poznań (Poland). Land Use Policy 96:104688. https://doi.org/10.1016/j.landusepol.2020.104688
16. Green Infrastructure North West, Mersey Forest (2010) Liverpool green infrastructure strategy. Green Infrastructure North West, Liverpool
17. Healey P (2006) Collaborative planning: Shaping places in fragmented societies, 2nd edn. Palgrave Macmillan, London
18. Kabisch N, Frantzeskaki N, Hansen R (2022) Principles for urban nature-based solutions. Ambio
19. Keeler BL, Hamel P, McPhearson T, Hamann MH, Donahue ML, Meza Prado KA, Arkema KK, Bratman GN, Brauman KA, Finlay JC, Guerry AD, Hobbie SE, Johnson JA, MacDonald GK, McDonald RI, Neverisky N, Wood SA (2019) Social-ecological and technological factors moderate the value of urban nature. Nat Sustain 21(2): 29–38. https://doi.org/10.1038/s41893-018-0202-1
20. Liverpool City Council (2016) Strategic green and open spaces review board: Final report. Liverpool City Council, Liverpool
21. Maes MJA, Jones KE, Toledano MB, Milligan B (2019) Mapping synergies and trade-offs between urban ecosystems and the sustainable development goals. Environ Sci Policy 93:181–188. https://doi.org/10.1016/j.envsci.2018.12.010
22. Markolf SA, Chester MV, Eisenberg DA, Iwaniec DM, Davidson CI, Zimmerman R, Miller TR, Ruddell BL, Chang H (2018) Interdependent infrastructure as linked social, ecological, and technological systems (SETSs) to address lock-in and enhance resilience. Earths Future 6:1638–1659. https://doi.org/10.1029/2018EF000926
23. Martin M, Deas I, Hincks S (2019) The role of temporary use in urban regeneration: Ordinary and extraordinary approaches in Bristol and Liverpool. Plan Pract Res 34:537–557. https://doi.org/10.1080/02697459.2019.1679429
24. Mell I (2020a) The impact of austerity on funding green infrastructure: A DPSIR evaluation of the Liverpool Green & Open Space Review (LG&OSR), UK. Land Use Policy 91. https://doi.org/10.1016/j.landusepol.2019.104284

25. Mell I, Clement S, O'Sullivan F (2022) Engineering nature-based solutions: examining the barriers to effective intervention. Proc Inst Civ Eng—Eng Sustain: 1–12. https://doi.org/10.1680/jensu.21.00033
26. North P, Nurse A, Barker T (2017) The neoliberalisation of climate? Progressing climate policy under austerity urbanism. Environ Plan Econ Space 49:1797–1815. https://doi.org/10.1177/0308518X16686353
27. Nurse A, North P (2020) A place for climate in a time of capitalist crisis? A case study of low-carbon urban policy making in Liverpool. England Town Plan Rev 91:155–179. https://doi.org/10.3828/tpr.2020.9
28. O'Sullivan F, Mell I, Clement S (2020) Novel solutions or rebranded approaches: Evaluating the use of Nature-Based Solutions (NBS) in Europe. Front Sustain Cities 2:572527. https://doi.org/10.3389/frsc.2020.572527
29. Rangelov M, Dylla H, Davies J, Sivaneswaran N (2020) Integration of life cycle assessment into planning and project delivery for pavements in the USA. Int J Life Cycle Assess 25:1605–1619. https://doi.org/10.1007/s11367-020-01777-x
30. Raymond CM, Frantzeskaki N, Kabisch N, Berry P, Breil M, Nita MR, Geneletti D, Calfapietra C (2017) A framework for assessing and implementing the co-benefits of nature-based solutions in urban areas. Environ Sci Policy 77:15–24. https://doi.org/10.1016/J.ENVSCI.2017.07.008
31. Seddon N, Chausson A, Berry P, Girardin CAJ, Smith A, Turner B (2020) Understanding the value and limits of nature-based solutions to climate change and other global challenges. Philos Trans R Soc B Biol Sci 375: 20190120. https://doi.org/10.1098/rstb.2019.0120
32. Sowińska-Świerkosz B, García J (2021) A new evaluation framework for nature-based solutions (NBS) projects based on the application of performance questions and indicators approach. Sci Total Environ 787:147615. https://doi.org/10.1016/j.scitotenv.2021.147615
33. Tait M (2011) Trust and the public interest in the micropolitics of planning practice. J Plan Educ Res 31:157–171. https://doi.org/10.1177/0739456X11402628
34. Wamsler C, Wickenberg B, Hanson H, Alkan Olsson J, Stålhammar S, Björn H, Falck H, Gerell D, Oskarsson T, Simonsson E, Torffvit F, Zelmerlow F (2020) Environmental and climate policy integration: Targeted strategies for overcoming barriers to nature-based solutions and climate change adaptation. J Clean Prod 247:119154. https://doi.org/10.1016/j.jclepro.2019.119154

Integrating Indigenous Lifestyle in Net-Zero Energy Buildings. A Case Study of Energy Retrofitting of a Heritage Building in the Southwest of Western Australia

Vittoria Strazzeri and Reena Tiwari

Abstract Net-zero energy buildings (NZEBs) are defined as highly performing buildings with minimal energy demand, which is often balanced by on-site renewable energy production. NZEBs are integrated within many national energy policies worldwide as an effective contribution toward the decarbonisation of the construction sector. The Council of Australian Government endorses the concept in the Trajectory for low energy buildings (2019) which sets a framework for improving the energy performance of new and existing buildings. Energy retrofitting against the zero-energy target offers energy-saving potential in terms of operational and embodied energy. Besides, heritage buildings retrofitting contributes to heritage management. However, the current approach to heritage buildings retrofitting aims almost exclusively to protect the physical structure and presents specific technical challenges concerning technical intervention on the building fabric and preserving the cultural heritage values. The study presents a case study of energy retrofitting of a State Heritage registered, in the south-west of WA, Carrolup Mission, which has significant cultural value for Aboriginal people. We targeted a single building assessed to be highly prioritised for the local community. We tailored the retrofitting strategy by integrating architectural design and architectural technology. We assessed the effectiveness of the retrofitting strategies against the net-zero energy target, and their impact on the buildings and the landscape. The analysis is based on building energy performance simulation (BEPS), and employs primary energy indicators to assess the building's energy performance. The outcome is a methodological path and best practice for energy retrofitting heritage buildings against net-zero energy targets in the south-west of WA. This study highlights the relevance of including architectural consideration in energy retrofitting heritage building against energy neutrality, with a specific focus on space quality and lifestyle. We found that integrating Indigenous

V. Strazzeri · R. Tiwari (✉)
School of Design and Built Environment, Curtin University, Perth, Australia
e-mail: r.tiwari@curtin.edu.au

V. Strazzeri
e-mail: vittoria.strazzeri@curtin.edu.au

Z. Allam (ed.), *Sustainable Urban Transitions*, Urban Sustainability,
https://doi.org/10.1007/978-981-99-2695-4_24

lifestyle and culture within the retrofitting strategy became considerably advantageous in enhancing the building space quality and provide environmental-specific sets of retrofitting strategies aiming to optimise the use of resources and the building system's efficiency. This study critically reviews the current Australian national policies concerning building energy efficiency, NZEBs and heritage retrofitting.

Keywords NZEBs · Energy efficiency · Building energy retrofitting · Building heritage · Australian heritage

1 Introduction

The construction sector's energy consumption rate is about 40% of the total national figure in Europe and the USA, contributing about 30% of the associated Green House Gas (GHG) emission [6]. In Australia, the construction sector's GHG emissions have been assessed to be about 25% of the total national annual emission [44], with buildings' embodied and operational energy primarily affecting this figure [63]. Globally, national economies have committed to common targets to reduce emissions and tackle climate change, as expressed by the Paris convention signed in 2015. Based on the correlation between GHG emissions and global warming, different policies and regulatory measures have been put into action by national economies worldwide to introduce more efficient, and sustainable energy uses, reducing GHG emissions across multiple sectors [27]. The urgency for reducing GHG emissions has led to a new approach to the building industry: designing sustainable new constructions and retrofitting the existing building stock to achieve highly-efficiency buildings [44].

Net-zero energy buildings (NZEBs) have been gaining relevance within many national energy efficiency strategies worldwide as a practical contribution toward the decarbonisation of the construction sector [13]. A widely accepted, non-specific definition of NZEBs nominates them as high-energy performance buildings, having minimal energy demand, balanced by on-site renewable energy production [47, 52]. This definition involves reducing the building energy demand by maximising the building energy efficiency as a core concept of NZEBS [44]. However, the literature [22, 61] converges on the fact that there is no standardised definition and methodological approach concerning NZEBs, which challenges comparing NZEBs outcomes. Besides, the definition of NZEBs is intrinsically context-related: climate conditions affect the effectiveness of design strategies and the energy performance of the buildings, as buildings in cooling-dominated climates may be facilitated in achieving energy neutrality [52].

The NZEBs concept also applies to energy retrofitting [8, 48], as retrofitting the existing building stock allows for minimising energy consumption associated to the construction of new buildings [7]. Europe, in 2010, introduced long-term strategies aiming to decreasing GHG by 80% concerning the 1990 figure by 2050. Concerning the construction sector, the introduction of net-zero energy buildings (NZEBs) plays a critical role [22]. The *Energy Performance Directive* 2018 mandates all new construction to be nearly zero-energy buildings and sets minimum energy requirements for energy retrofitting projects [29]. The USA adopted NZEBs-specific policies in 2007 with the *Energy Independence and Security Act* [61]. Australia has recently set mandatory requirements for enhancing energy efficiency of the building stock. NZEBs have been recently endorsed by several policies concerning energy production by renewable sources—the *Renewable Energy Target* [61]—and the *Trajectory for low energy buildings* (2019)—which may lead to a further upgrade of the current regulatory norms. The Trajectory is a national Plan supported by the Council of the Australian Government to realise energy-zero and carbon-ready new homes. The document includes the upcoming upgrade regarding energy efficiency requirements for residential buildings. The focus is indeed on the operational phase of the building [59], as it is the largest figure on the overall building energy consumption [63]. However, NZEBs remain marginally included in national policies as a tool for decarbonising the construction sector [61]. However, [61] observe that countries that include NZEBs in the national policy eventually develop technological advancements in making NZEBs a feasible option for the national industry. Besides, the potential of energy retrofitting against the zero-energy target is mainly excluded by the current Australian policy in the matter of energy efficiency, giving minimal target specifications.

Concerning historical/heritage buildings, energy retrofitting contribute to conserving the built heritage itself [32]. As a part of cultural heritage management, built energy retrofitting contributes to the local economy by optimising local resources while creating inclusive and resilient societies [55]. However, retrofitting historical and heritage buildings present specific technical challenges as heritage buildings might be subjected to partial or complete restrictions (e.g., restriction in modifying the facades), which limit the possibilities of intervention on the building fabric [55]. Such limitations constrain the retrofitting process, making energy renovation of existing buildings a new challenge from architectural design and architectural technology perspectives. Energy retrofitting of historical buildings takes considerable planning effort and involves a holistic multidisciplinary approach ensuring enhanced energy performance and preserving the building's historical and cultural values [55]. National construction schemes hardly ever include measures to support energy retrofitting intervention. However, substantial literature [7, 32, 41, 48] investigates the potential of buildings' energy retrofitting against the net-zero energy target, providing a solid methodological framework to address technical challenges for different building typologies and environmental contexts. A number of studies [9, 48, 55] set approaches for assessing the impact of retrofitting measures on building integrity and indicate tools to tune the retrofitting package to different building constraints. Stanojevic et al. [56], develop a methodological approach for heritage building energy retrofitting based on Analytic Hierarchy Process (AHP) to identify the most practical and feasible interventions for different levels of restrictions. The study identifies that designing retrofitting measures that simultaneously hit the NZEBs target and comply with high restriction levels is the main challenge in energy retrofitting against the zero energy target.

The literature illustrates many case studies located overseas, including residential and non-residential buildings [4, 12, 33]. These studies present a holistic approach to heritage building energy retrofitting, integrating different performance indicators and addressing specific technical aspects concerning the building type, the level of protection of the historical building and environmental consideration. Gremmelspacher et al. [32] illustrate the retrofitting process of a historical building in Sweden and evaluate the PV system's impact based on the LESO-QSV method. This approach is considered here. However, a limited number of studies investigate the effects of architectural design on builings' energy and thermal performance, and the potential of integrating traditional culture into the retrofitting project. Assimakopoulos et al. [12] emphasise the inclusion of daylight evaluation and thermal comfort as quality parameters in energy retrofitting projects. Alwetaishi [5] investigates the relevance of traditional architecture in achieving energy neutrality, presenting a case study of energy retrofitting of a traditional building in Saudi Arabia. All these studies are based on building dynamic simulations. To the best of these authors' knowledge, there is limited literature available reporting on case studies of energy retrofitting of heritage and historic buildings in Australia.

This study illustrates a case study of energy retrofitting against the NZEBs target of a historic building located in Carrolup in the south-west Western Australia. We tailored the retrofitting strategy to consider architectural design and space quality in the retrofitting package, and we assessed the impact of the interventions on the building's integrity and historical value. The retrofitting measures involve (i) re-designing the space layout of the building to enhance the indoor space quality responding to Aboriginal needs and culture, (ii) improving the building envelope to optimise the building thermal performance and minimise its energy demand (iii) selecting an HVAC system to meet the building energy demand (iv) integrate a PV system for to off-set the building energy demand with on-site energy production. The analysis is based on building energy performance simulation (BEPS) and comprises three main phases: (i) pre-retrofitting status to assess the current building performance and set the requirements to achieve energy neutrality [55]; (ii) retrofitting process to assess the effectiveness of the retrofitting measures; and (iii) analysis of the retrofitted building to ensure hitting the target. Primary energy indicators are employed to evaluate the building performance at different stages of the retrofitting process. The project's specific objectives are:

- Tailoring an environmental-specific retrofitting strategy of the local building heritage toward the NZEB target
- Investigating energy-saving implications of integrating architectural design in the retrofitting package.
- Providing a methodological approach for historic building energy retrofitting historic buildings against NZEBs in the south-west of Western Australia.

This study comprises three sections (besides this introduction). Section 2 contains a literature review on Australian and Western Australian policies concerning built heritage retrofitting. Section 3 presents the methodology, including the (i) definition of NZEBs and the theoretical approach on which this project is built upon, (ii) energy indicators used in the analysis, (iii) the case study, and (iv) the retrofitting measures. Section 4 presents the results regarding the building energy and thermal performance in the pre-and post-retrofitting status and the energy balance between building consumption and renewable production. Finally, Sect. 5 summarises the results and gives recommendations for further research.

2 State and National Policy on Building Retrofitting and Conservation of the Built Heritage in Western Australia

2.1 Sites of Cultural Heritage

Cultural heritage was defined by the International Council on Monuments and Sites (ICOMOS) 2002 as "*an expression of the ways of living developed by a community and passed on from generation to generation, including customs, practices, places, objects, artistic expression, and values.*" [38]. Cultural heritage assets are cultural resources that require protection in the same way as other non-renewable resources. Protection should aim at extending the useful life of buildings as much as feasibly possible, including recycling them for alternative uses when necessary. Aplin [6] highlighted a range of general heritage conservation and management principles from the Burra Charter. For example, he defined conservation as "*all the processes of looking after a place to retain its cultural significance […] including maintenance and may according to circumstances include preservation, restoration, reconstruction and adaptation […] commonly a combination of more than one of these*" [6]. Since conservation is a concept inclusive of protection, preservation, gentrification, renewal, and adaptive use, it is essential to define conservation in the context of adaptive re-use for the benefit of this paper (Hussein, 2019) [51].

Adaptive re-use/Adaptation which involves making changes to a building to accommodate a new use is often a means of enabling the continued usefulness of a historic building considering that the new use integrates with the original fabric. This is preferred instead of replacement to reduce climate change and global warming [53] taking into account the whole building's LC waste and cost and enhancing the building's functionality (Bullen & Love, 2011). Studies identify that the main driver for environmental benefits is "embodied energy" as the cumulative input energy required to construct the building initially [33]. However, re-using existing buildings may not ultimately reduce the need or desire for constructing new buildings [21]. Second, the adaptively re-used heritage buildings could fall short of today's expected standards, such as zero-emission building materials (Bullen & Love, 2011). Also, circular and adaptive re-use strategies can be more expensive than demolition and new construction regardless of the environmental and sustainability benefits [26]. Nevertheless, the literature stands for adaptive re-use of heritage buildings as a win for the environment despite these limitations [32]. The suggested approach in such cases is one that is conservative rather than being radical where—it "*involve[s] the least degree of intervention consistent with long term care* (ICOMOS, New Zealand Charter)."The question then is to identify ways of minimal intervention in heritage buildings for adaptive re-use through architectural design strategies that are culturally and socially responsive for the targeted community, in this case the Aboriginal community.

2.2 Cultural Beliefs and Practices Influencing Aboriginal Architecture

The main belief and practice that influences Aboriginal architecture, particularly housing, in the context of this case study, is the avoidance of relationships [50, p. 73]. The avoidance law is a set of rules that is practiced within Aboriginal culture which affects behavioural relationship between brothers and sisters, and mother-in-law and son-in-law. In such cases, eye contact is avoided, which influences the occupants' movement and interactions within the space. The suggestion is to decrease the likelihood of accidental meetings through maintaining visual privacy, separating bathrooms on different ends of the houses while providing access to these from outside, or creating thresholds between rooms. Breezeways within a single building have also been effectively used to provide partitions for privacy while at the same time improving natural ventilation. Designing houses with high ceilings and ceiling fans can help ventilate the space, keeping it cool during the day time which is a cheaper alternative to air conditioning systems [39, 45, 48] Breezeways can be designed to be narrower on one end and wider on the other to maximise ventilation or utilising the stack ventilation, where hot air can escape through operable skylights as it rises and is replaced with cooler air [42, p. 298]. These design strategies are particularly relevant to the case study discussed in this paper.

3 Methodology

3.1 Energy and Thermal Comfort Indicators

In line with the definition of NZEBs given by Gremmelspache et al. [32], our retrofitting plan involves minimising the building consumption and integrating a PV system to off-set the building's energy demand. The retrofitting plan comprises four steps: (i) re-designing the interior space layout to enhance daylighting, (ii) improving the building envelope to optimise the thermal performance and minimise energy consumption (iii) selecting an HVAC system to meet the building energy demand (iv) integrate a PV system for on-site energy production to balancing the building energy consumption. We conducted a set of simulations assuming that the building operates in mixed mode (HVAC + Natural ventilation). We conducted nine iterations starting from the building in its pre-retrofitting status (iteration 1), proceeding with the retrofitting measures changing a single variable per iteration, till obtaining the fully retrofitted building (iteration 9). We compared the building energy performance based on energy use intensity (EUI) [kWh/m^2/yr], annual electricity demand [kWh], and energy usage per unit area per year throughout the nine iterations. We observed the impact of each retrofitting measure on the overall building energy performance. For ease of comparison, we assumed the same HVAC system for all iterations, including iteration one. For iteration nine, we calculated the renewable energy generation [kWh], and evaluated the energy balance between the building energy demand and the on-site renewable energy generation [kWh].

Thermal comfort parameters found limited use to indicate NZEBs' performance in the literature. However, comfort parameters contribute to tailoring the strategies toward climate-specific responses [35]. In a study in a southern region in Belgium, Dartevelle et al. [25] evaluate the post-occupation summer thermal comfort in different NZEBs. They found that the buildings barely meet the CIBSE adaptive comfort criteria, showing a risk of overheating in buildings defined as NZEBs. Tang et al. [58] illustrate the risks of potentially negative impact on thermal comfort of energy-saving measures for building retrofitting against NZEBs target. Adaptive comfort models have been assessed to have the most significant energy-saving implications [52]. Sartori et al. [54], in their definition of NZEBs, transferred the correlation between building energy performance and the thermal comfort of occupants to the concept of NZEBs. The authors identify strategies that enhance the building's thermal performance concerning space use and local climate [54]. This approach facilitates the comparison of NZEBs outcomes in similar climates and for comparable designs.

3.2 Thermal Comfort Model

The Australian National Construction Code (NCC) divide the nation into eight major climate zones (Australian Building Code [15]. The building is located in Carrolup, in the south-west of Western Australia, near the town of Katanning, falling in zone 4—hot and dry summer and cool winter. According to the Koppen Climate classification, Katanning is classified as Csb—warm-summer Mediterranean climate. We employed the adaptive thermal comfort model set in the ANSI ASHRAE-Standard 55 (2020) [11] to set the heating and cooling HVAC thermostats. Adaptive comfort models establish quantitative relations between indoor comfort and climatic conditions, correlating the indoor thermal environment variations to outdoor conditions [38. The generic definition of adaptive thermal comfort models reads (1):

$$T_n = a + bT_o \tag{1}$$

where T_n is the neutral indoor temperature (or comfortable temperature), T_o is an estimate of the outdoor temperature, and a and b are two constants. Currently, the literature agrees on employing the exponentially weighted mean running outdoor temperature T_{pma} as the measure of the outdoor temperature. According to the ANSI-ASHRAE-55-2020 [11], T_{pma} for an infinite series reads (2):

$$T_{pma} = (1-\propto)[T_{o(d-1)} + \propto T_{o(d-2)_o} + \propto^2 T_{o(d-3)} + \ldots] \tag{2}$$

where $\propto$ is constant ranging between 0–1 and $T_o(d-i)$ is the temperature of day d-i. The ANSI-ASHRAE Standard 55 2020 recommends using an $\propto$-value ranging from 0.6 to 0.9 to model a slow and fast response of the running mean to the outdoor temperature variation [11]. The model defines the comfort bands on the statistical determination of acceptance levels, identifying an upper and lower limit for the comfort zone corresponding to the 80% and 90% of satisfied users, respectively (±3.5 °C and ±2.5 °C, respectively). It is assumed that an average expectancy of comfortable thermal conditions in residential buildings corresponds to 80% of acceptability [11]. The model's equation reads (3):

$$\mathrm{T_n} = 17.8 + 0.315\mathrm{T_{rmt}} \tag{3}$$

Despite being internationally adopted, the ASHRAE model has been assessed to be particularly suitable for C-classified climates (Koppen classification) [58]. We used Sefaira web application for the energy analysis, as the tool was explicitly developed for NZEBs. In Sefaira, the setpoints temperatures (read as dry-bulb temperatures) operates the HVAC system (when present) and the windows when natural ventilation is employed. This ensures that the mechanical system and natural ventilation target a convergent temperature range when the building operates in mixed mode. We conducted the first set of simulations assuming that the building operates in free-running mode. We defined the setpoint temperature values based on adaptive standard ASHRAE-55 (2020). To do so, we extrapolated the hourly-based outdoor temperature data from the weather data file used in the simulation and evaluated the prevailing running mean outdoor temperature according to Eq. (2), considering $\alpha =$ 0.65. We obtained a range of values between 6.5 and 24.4 °C. The corresponding range of comfortable temperature (Tn according to Eq. (1)) varies between 20 and 25.5 °C. Assuming a comfort range at 80% of acceptability, we set the setpoint temperature at 18 and 27.5 °C.

3.3 The Case Study: Carrolup Mission

Carrolup Marribank was founded in 1915 as a mission site under the name of Carrolup Native Settlement in response to the Aborigines Act of 1905 [33]. The site hosted a community of 125 people from the Noongar community and comprised thirteen primary buildings with different uses (including dormitories, a school building, and headquarters) and eight residences built between 1915 and 1973. The site worked discontinuously over time, underwent a number of changes, and eventually closed in 1995. Carrolup Marribank has been assessed as a significant cultural heritage for (i) the building's aesthetic value (ii) the building's architectural technology and traditional construction methods (iii) the economic role on the territory [33]. The site represents some distinctive cultural aspects and traditions of the Noongar people over generations and is associated with developing Southwestern Australia landscape art. The site has been listed in the Western Australia State Register of Heritage Places since 2007 and in the Aboriginal Heritage Sites Register (Heritage [36])

3.4 The Original Building

The selected building was built in 1973 and worked as a male dormitory. During the initial survey documented by [29], the building was assessed to be of high community priority. It is a timber-framed one storey-building and covered with a tiled timber-framed gable roof. The building is set on a timber floor covered with wooden floorb]oards, and it has a plasterboard ceiling. The walls have asbestos clad on the exterior and plasterboard on the interior. A 2.2 m deep under-roof veranda supported by steel posts runs along the two main sides of the building [30]. Openings comprise steel-framed sliding windows and solid-core wooden doors [30]. The total internal floor area is 239 m^2. The original layout includes seven bedrooms, four bathrooms, a lounge room, a kitchen, a living/common room, a clothing room, and a storage room. The building is currently abandoned. However, according to the restoration masterplan [59], the retrofitted building will host 12 people and work as a dormitory. The building will therefore fall in building class 1b according to the NCC (Australian Building Code [15]. The building components R-values (thermal resistance) and U-values (thermal conductance) employed in the analysis were either sourced directly from the Australian Building Code [15]or calculated following the procedure illustrated in [2].

As mentioned above, the external walls have asbestos-cement sheets (4.5 mm thick). Given that the asbestos sheets would be removed, for ease of comparison, we assumed the walls to be consistent with fibre-cement sheet clad timber framed illustrated in the NCC [1, p. 380]—total R-value 0.42. Given that the building is 50 years old, we only assumed R1 bulk insulation throughout the external walls. Concerning the roof, we considered the structure described above to be consistent with the typical detail illustrated in the NCC Volume 2 Tiled Pitched Roof with Flat Ceiling, unvented and with a downward direction of heat flow given the climate zone. We considered R1 bulk insulation throughout the ceiling. The building has a timber floor whose *R*-value is 0.39 for a thickness of 15 cm. The flooring consists of 10 mm thick hardwood sheets, except in wet areas. The sliding windows comprise aluminium-framed single clear glazing; hence we assumed $U = 6$ and SHGC = 0.81. Table 1 summarises the building technical parameters used in the analysis (Fig. 1).

Table 1 Technical parameters of the building envelope in the assumed current state

Building component	System	System *R*-value	Insulation	Total *R*-value [m^2K/W]	Total *U*-value [W/m^2K]
Wall	Timber frame, asbestos sheets exterior, and plasterboard interior	0.42	1	1.42	0.70
Roof	Pitched tiled roof with flat ceiling + R1 bulk insulation	0.56	1	1.56	0.64
Floor	Timber floor	0.21	–	0.21	4.76
Glazing	Single clear in aluminium frame	–	–	–	6
Glazing SHGC	0.7				

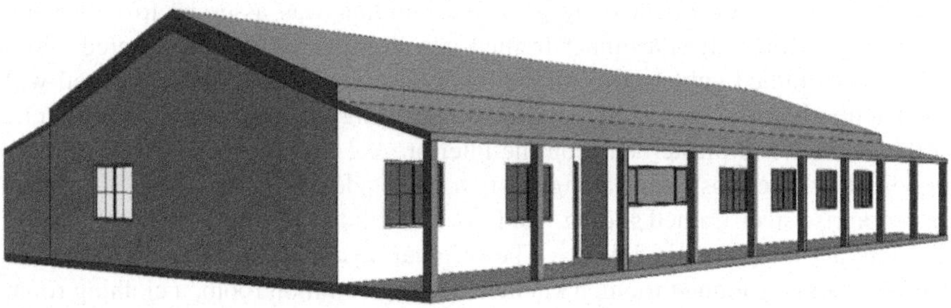

Fig. 1 The building (developed in Revit Autodesk)

The building is currently abandoned; however, for ease of comparison, concerning the building in its actual status, we assumed space occupancy, lighting and equipment consistent with the scheduling proposed for the retrofitted building, developing specific space use profiles. The lighting power density (LPD) was calculated following the *building area method* defined in the ASHRAE 90.2—2018 [10]. Assuming halogen lamps with 60W energy use [14], we obtained a value of 6.8 W/m^2. We consider 5 W/m^2 of equipment power density as specified in the Sefaira documentation. We considered 45% widows operability and 27% free area of openings according to the indications provided by Sefaira. We adopted a conservative approach considering all openings closed when the building is unoccupied and when it is windy outside based on the weather data file.

3.5 The Retrofitting Process

The cultural significance of the original building is bound up mainly with the attempt to replicate home life. We re-designed the interior spaces' configuration to enhance the indoor space's quality and provide the community with a contemporary usable space. Keeping the intact façade layout (except for substituting the asbestos-cement sheet) was not mandatory. However, it was assessed to be of high priority not to alter the setting of the building within the mission landscape. The retrofitting strategies were tailored to enhance the indoor space quality, daylight and thermal comfort, minimising heat losses through the building envelope and providing the building with an adequate conditioning system, artificial lighting and equipment. Our analysis comprises of the following iterations:

1. Current building
2. Modifying the space layout according to contemporary design standards and Aborignal culture and lifestyle
3. Upgrade windows and glazing
4. Enhance wall thermal resistance by increasing the insulation
5. Enhance floor thermal resistance by increasing the insulation
6. Enhance roof space and ceiling insulation
7. Increasing lighting efficiency
8. Increase equipment efficiency
9. Install an adequately dimensioned PV system.

We assumed the HVAC system was installed in the original building so that we could compare the building energy performance in the pre and post-retrofitting status. The selection of the HVAC system is a critical point in the development of NZEBs [17], and it depends on a few factors, including local climate, the building use and the building fabric. Given the building typology, use, and location, we integrated a user-demand-controlled VRF fan coil unit. In this circumstance, the main advantages of this system are (i) reduced dimension (ii) installation flexibility that minimises the system's impact on the building, and (iii) high efficiency. Besides, the system allows users to interact further with the environment. Thermostat temperatures were set according to Eq. (3), as explained above. The building geometry was modelled in Sketchup and then imported into the Sefaira web application, which uses the EnergyPlus solver.

3.6 Architectural Design

To the best of these authors' knowledge, limited literature investigates the effects of architectural design on the building energy performance and quality of the indoor environment based on quantitative data. Assimakopoulos et al. [12] present a qualitative-quantitative SWOT matrix to assess the impact of the retrofitting strategy.

The study includes the architectural design in the matrix strength quadrant since the high-quality design is positive for the occupants' well-being. We re-designed the internal space layout to accommodate the new building programme and meet contemporary aesthetic standards while being responsive to Aboriginal culture and lifestyle. Figure 2 and 3 illustrate the original and newly designed space layout. The new configuration includes larger bedrooms consistent with Aboriginal family living instead of individual occupancy. The clothing room, present in the original layout, was removed, and each bedroom was given built-in closets. The bathrooms maintain the original configuration, with minimal dimension modifications but with options for outdoor accessibility in line with Aboriginal lifestyle and avoidance needs. The original division of common rooms and lounge was converted into a shared kitchen/ dining area with a connected living/common area dedicated to common activities to favour socialisation. Two areas dedicated to socialisation open directly to the shaded verandah and connect with the outdoor thus aiding avoidance and privacy, if required. The new layout minimises circulation spaces, yet maintains the architectural principles of the original layout, with the line of the original corridor modified (partially integrated into the common spaces) but still visible. The kitchen/living and common areas have raking ceilings to allow the introduction of skylights to increase natural lighting, especially in the central area of the building. The central area (shaded green) also acts as a breezeway to partition off the building into two parts while giving maximum opportunity for natural ventilation. Re-designing the internal space layout implied also tuning the occupational schedules profiles in terms of occupation density.

3.7 Daylighting

We conducted a daylight analysis of the original building layout, assuming clear sky conditions outdoor and targeting 300 lx as illuminance value according to the EN17037. We observed that the building performed poorly—mostly underlit during the year. We included considerations derived from the deylight analysis in defining the building's new space layout. The new layout increases the well-lit time by 16% and reduces the under-lit time by 32%. Figure 6, 7, 8, and 9 summarise the daylight analysis of the original and modified design. The skylights' U-value and SHGC were optimised based on a sensitiveness analysis conducted on Sefaira, varying U from 0.4 to 6 Wm^2/K and SHGC from 0.05 to 0.95 against the EUI. We selected double-clear glazing.

3.8 The Building Envelope

We upgraded the thermal properties of the building's components. The tiled roof has a surface solar absorptance > 0.6. In zone 4 and 5, the NCC requires the roof s has

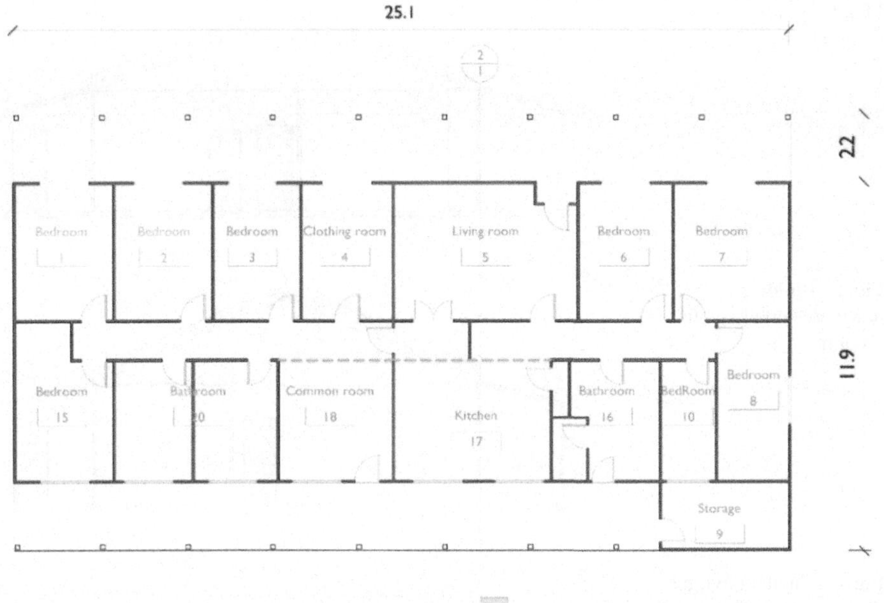

Fig. 2 Building original layout: plan

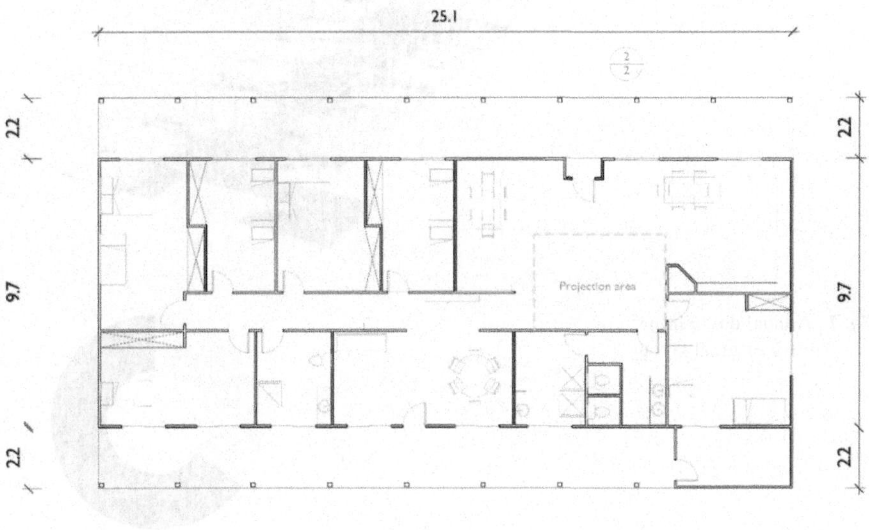

Fig. 3 Building newly designed internal layout, plan

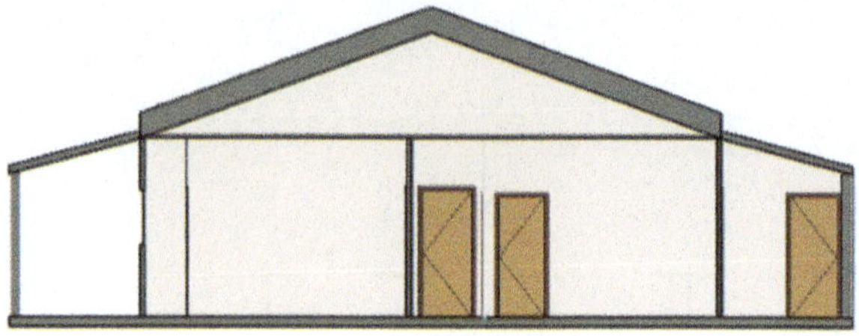

Fig. 4 Building original layout, section

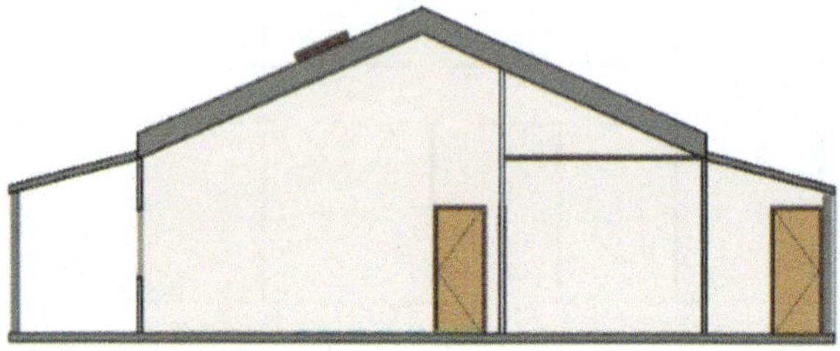

Fig. 5 Building newly-designed layout, section

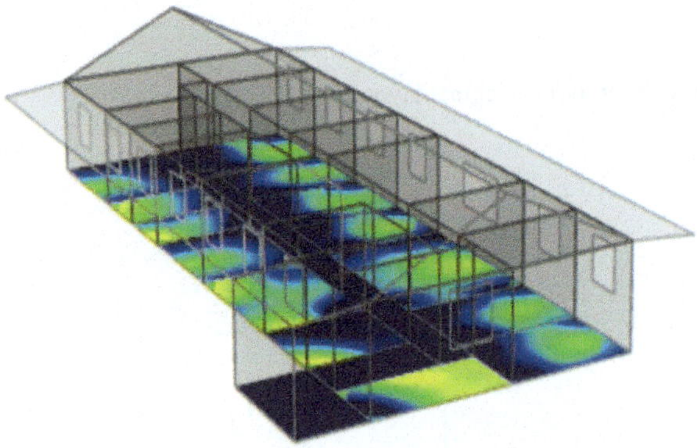

Fig. 6 Annual daylight analysis-original layout

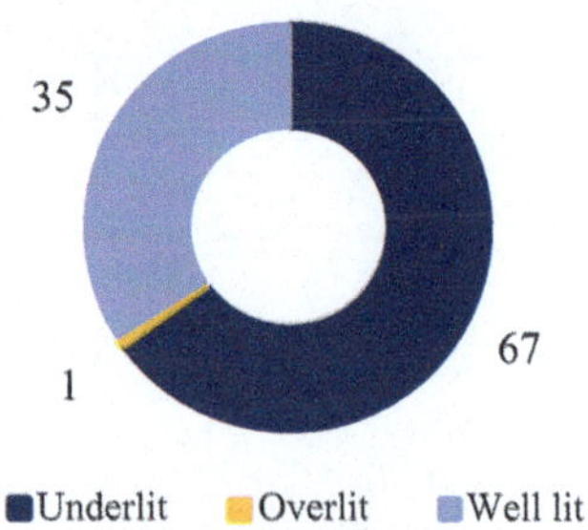

Fig. 7 Annual daylighting pie-chart for original layout

a total R-value of 5 m^2K/W. The bathroom and bedroom areas were designed with flat ceilings, insulated with R5 bulk insulation, and reflective sarking under the gable structure to prevent heat from radiating into the attic space. In the new layout, the

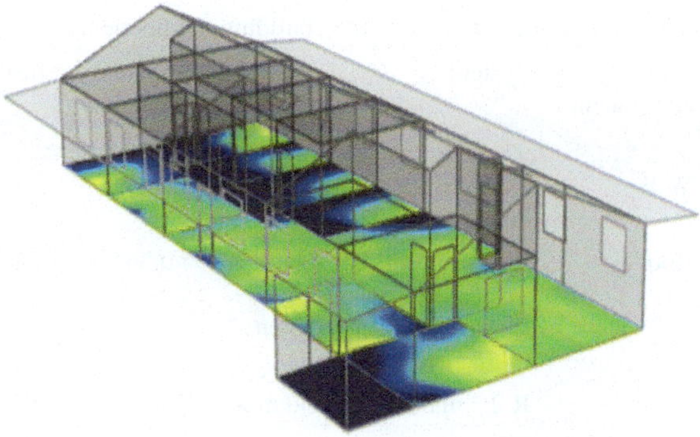

Fig. 8 Annual daylight analysis-new layout

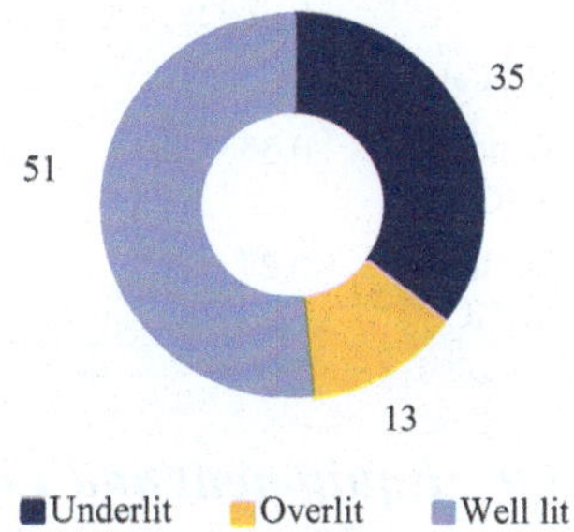

Fig. 9 Annual daylighting pie-chart for original layout

common/lounge area and the kitchen space have a raking ceiling with no roof space cavity. The raking ceiling is assumed to be insulated with R 5 high-density insulating batts and reflective sparking between the roof's rafters. The retrofitting of the external walls implied substituting the original asbestos-cement sheets. The aesthetically closest alternative was an external fibre cement sheet and internal plasterboard, with antiglare reflective wrap and R2.5 high-density bulk insulation placed into the stud cavity. This system slightly exceeds the minimum requirement of a 2.8 R-value specified in the NCC [1]. We assumed R2.3 underfloor insulation underneath the hardwood flooring. Windows were upgraded to thermally broken aluminium framed, air-filled, double glazing, characterised by low emissivity to minimise heat gain/loss by radiation throughout the glazing, and high SHGC, given the shading provided by the verandah. We did not include internal blinds or venetians because the windows are all shaded by the veranda, except for the window on the southeast-facing wall, which does not contribute substantially given the orientation and dimensions. Including vertical shading would, instead, impact the building facades. Table 2 summarises the *R*-values of the upgraded building components.

Table 2 *R*-values of the upgraded building 'components

Building component	System	System R-value	Insulation	Total R-value [m²K/W]	Total U-value [W/m²K]
Wall	Reflective wrap + R2.7 high density bulk insulation	0.42	2.7	3.12	0.32
Roof	Roof sarking + R 5 bulk insulation (high density throughout raking ceiling)	0.56	5.13	5.69	0.17
Floor	R 2.3 underfloor insulation	0.21	2.3	2.51	0.39
Windows	Thermally brocked air-filled double glazing, low-E, high SHGC	–	–	–	3.1
Skylight		–	–	–	2.2
Window SHGC	0.53				
Skylight SHGC	0.27				

3.9 Equipment and Lighting

Energy-efficient lighting and equipment were considered in the retrofitted building. We assumed to substitute the existing halogen lights with LED ranging between 10 and 13 W. For ease of comparison, we considered the same number of units (29) that we considered in the pre-retrofitting building model. The light power density was evaluated according to the ASHRAE-90.2 2018 [10], obtaining a value of 1.68 [W/m^2]. We accordingly adjusted the lighting power density values on the Space Use tab on Sefaira. Similarly, we tuned the equipment power density. Besides, it has been ensured that the artificial lighting of the single spaces meets the NCC requirement of lamp power density not exceeding 5 W/m^2 in class 1 buildings [1, p. 411].

3.9.1 Integrating the PV System

We used the building energy profiles (which coincide with the building electricity profiles) to dimension the building-integrated PV system. We adopted the LESO-QSV method (Munari [49] to evaluate the PV system's impact on the building. We aimed to identify the less impacting and most efficient area on the building roof to allocate the PV system. Hence, we employed the LESO-QSV methods and cross-read the results with the sensitiveness analysis for the system orientation, tilt and area conducted on Sefaira. The LESO-QSV method was originally developed for the urban context (Munari [49]. However, [32] used the method to evaluate the impact of integrating a PV system in a Sweden historic castle in a countryside area. The method develops a matrix that assesses the architectural integration of the PV field on the surface based on two main criteria (i) architectural sensitivity and (ii) visibility of the system (Munari [49]. The approach is fully explained by Munari [49].

Given the high cultural value and architectural quality of the site and the building, we considered a high level of *architectural sensitivity—heritage-protected or meaningful context building* (Munari [49], p. 675). The precinct is protected as a heritage enclosure and is located deep into the south-west forest. Installing a PV field will necessarily affect the materiality of the building and the integrity of the whole precinct. The LESO-QSV matrix is a valuable tool to assess and mitigate such impact. Hence, we designed the system considering the criteria described in the LESO-QSV method, to off-set the high level of architectural sensitivity—we reduce the system visibility to a moderate level, obtaining an overall moderate level of system criticality. We cross-read the LESO-QSV qualitative analysis results with the quantitative data concerning the renewable energy generation obtained by performing sensitivity analysis on Sefaira. We evaluated the response curves to optimise the panels' orientation and tilt on the base of the renewable energy generation indicator. The optimum orientation was 20° N-NW, with an optimum tilt 30°. This would mean locating it on the north side of the west roof pitch. However, the system would be highly visible (close visibility), according to Munari [49]. Thus, we decided to locate the system on the north side of the east roof pitch, reaching low visibility while remaining in the plateau of the optimisation response curve shown in Fig. 10. The panel's tilt was kept at 30°. We selected monocrystalline silicon panels with an efficiency of 0.18 and 250 W rated power (per panel). The roof cover (hosting surface) comprises of dark colour clay tiles. Hence, the PV system is considered partially coherent with the hosting surface in terms of materiality, surface texture and colour and not coherent in terms of the surface pattern (Munari [49]. According to the LESO-QSV approach, a homogenous PV surface better integrates with the hosting surface than a scattered surface (Munari [49]. Considering a strip of 99 cm wide by 157 cm height panels, we extend the PV field to the whole length of the roof gable for a total area of 78.5 m°.

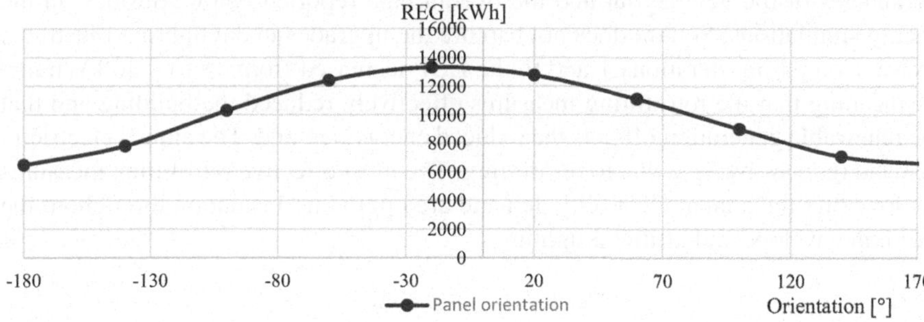

Fig. 10 Optimisation response curve for Panel system orientation

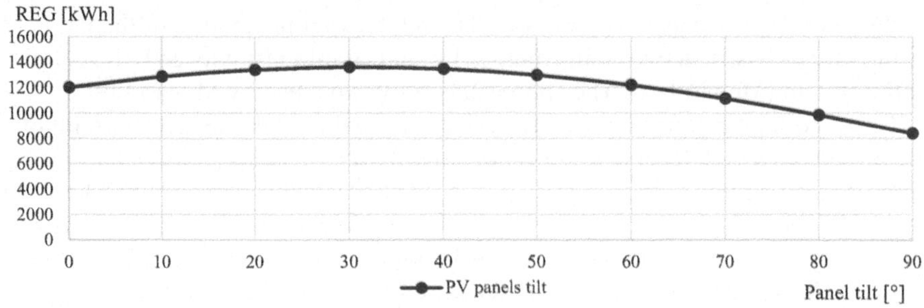

Fig. 11 Optimisation response curve for PV panels tilt system orientation

4 Results and Discussion

4.1 Energy Performance

This section reports and discusses the results of our analysis conducted on the building in the pre and post-retrofitting status and illustrates the impact of each retrofitting measure on the overall building energy performance. Table 3 shows the building EUI, the annual electricity demand, and the HVAC energy usage per unit area for each iteration. Figure 12 displays the building EUI per iteration, showing each retrofitting measure's impact. Figure 13 illustrates the annual electricity demand [kWh] and the HVAC energy use per unit area. Table 3 and Fig. 12 show that the EUI and the annual electricity demand decrease progressively from iteration 2 to iteration 9. However, the three parameters increase from iteration 1 to 2. Specifically, the increment in the HVAC energy usage per unit area is likely because the new building layout comprises of larger volume for the same building footprint compared with the original design. However, the annual electricity demand increases by only 2% compared with iteration 1; hence, the increment is marginal when considering the advantages of the new layout and the overall data reported below. Besides, in the energy simulations, Sefaira does not capture the upgrades in daylighting illustrated above. Comparing iterations 1 and 9, the EUI decreased from 48 to −46 kWh/m^2/yr, meaning that the retrofitting measures effectively reduced the building and that the renewable generation off-sets the reduced energy demand. The annual electricity demand reduces by 57% due to retrofitting. The most effective retrofitting measures for lowering the annual electricity demand are upgrading insulation throughout the building envelope and artificial lighting.

Table 3 Impact of the retrofitting measures on the energy (electricity) demand and EUI

Iteration	Retrofitting measures	EUI (kWh/m^2/yr)	Annual electricity demand (kWh)	HVAC Energy use per unit area (kWh/m^2/yr)
1	Initial status	48	11,392	16
2	New space layout	49	11,655	17
3	Thermally- broken Aluminium frame High SHGC, Low-E double glazing	48	11,550	17
4	Antiglare + R2.7 bulk insulation throughout external stud walls	48	11,398	16
5	R 2.3 underfloor insulation	41	9708	9
6	Reflective roof space + R5 bulk insulation throughout ceiling	38	8975	7
7	LED	23	5444	7
8	High efficiency equipments	21	4957	7
9	Roof PV system	−46	4948	7

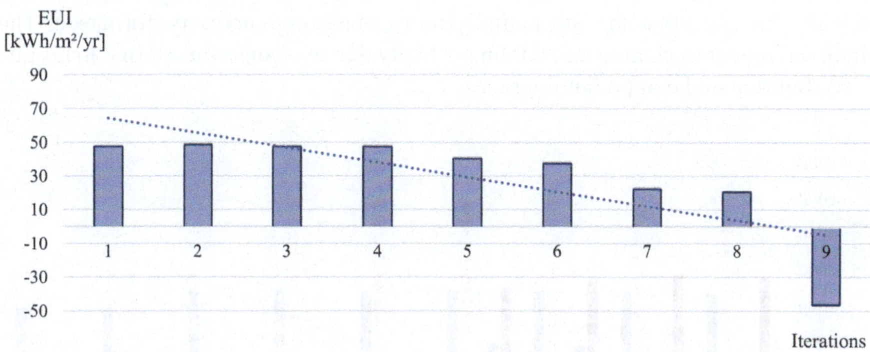

Fig. 12 Building EUI per each iteration

4.2 *Retrofitting the Building Envelope*

Iteration 3 concerns the upgrades of windows (frames and glazing). Despite the considerable technological advancements from aluminium framed single clear glazing to thermally broken low emissivity double glazing, this measure is observed to have a relative impact on the overall energy performance of the building. Likely

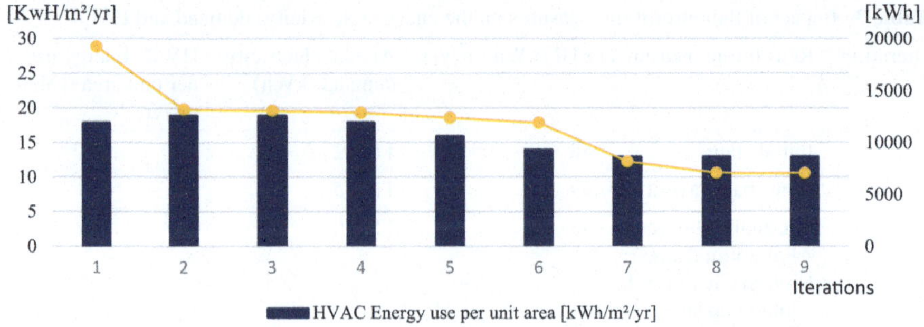

Fig. 13 Annual Electricity demand and HVAC energy use per unit area

the deep veranda on the building's front and rear mitigates the effectiveness of this measure, especially in summer. However, the local climate is characterised by large day-night temperature fluctuation, with a peak of −2 °C at night in winter. Low emissivity double glazing may reduce the heating demand, as shown in Fig. 14. Further research can optimise this measure by exploring different window upgrade options based on cost evaluation. As expected, insulating walls, roofs, and floors considerably impact the building's energy performance. Sefaira does not allow to fully model the insulation throughout the roof space (it does not allow to specify whether it is a reflective roof space, nor subdividing the overall R-value throughout roofing and ceiling). Hence, the effectiveness of the proposed upgrades of the roof space may be under-read in Sefaira. As expected, upgrading lighting and equipment to high-efficiency systems substantially impact building energy performance. The slight increment of heating and cooling is likely due to changes in internal loads due to the lighting and equipment upgrades.

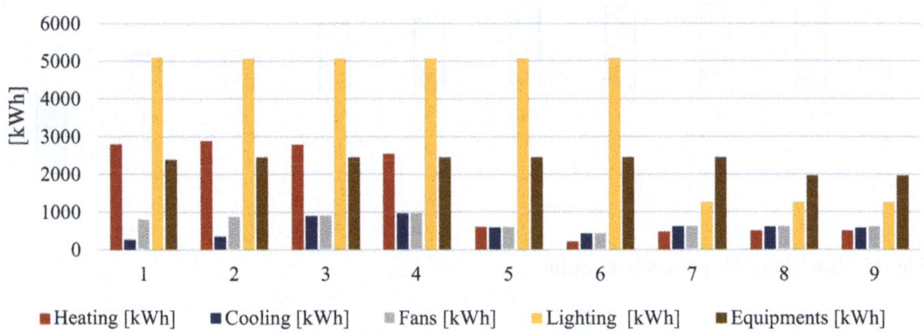

Fig. 14 Energy profile of the building for each iteration

4.3 Comparing the Pre- and Post- Retrofitting Energy Performance

We compared the energy profiles of the building concerning iterations 1 and 9 on an annual basis, as shown in Fig. 15 and Fig. 16. The two figures illustrate a substantial reduction in energy consumption associated with heating (−81%), lighting (−75%), and a considerable reduction associated with equipment. Figure 17 shows the building energy profile monthly after retrofitting. Tuning insulation throughout the roofing and the windows might decrease cooling loads.

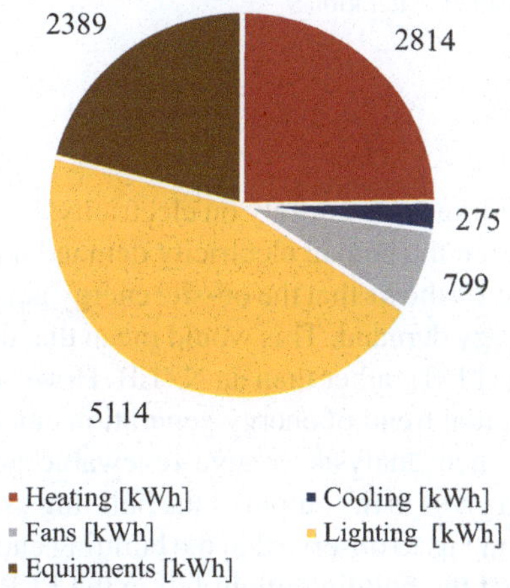

Fig. 15 Annual energy breakdown for iteration 1

Fig. 16 Annual energy breakdown for iteration 9

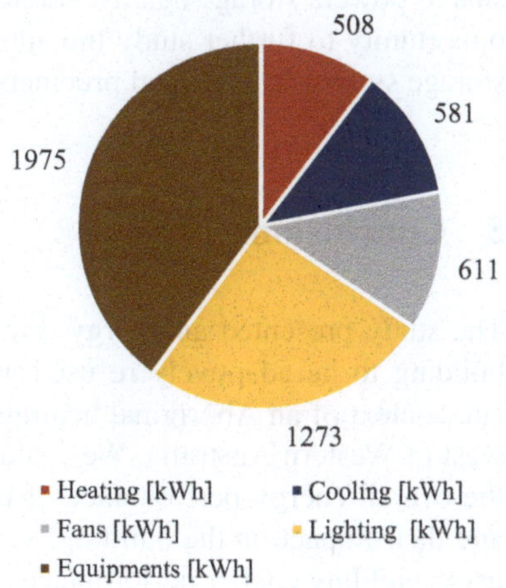

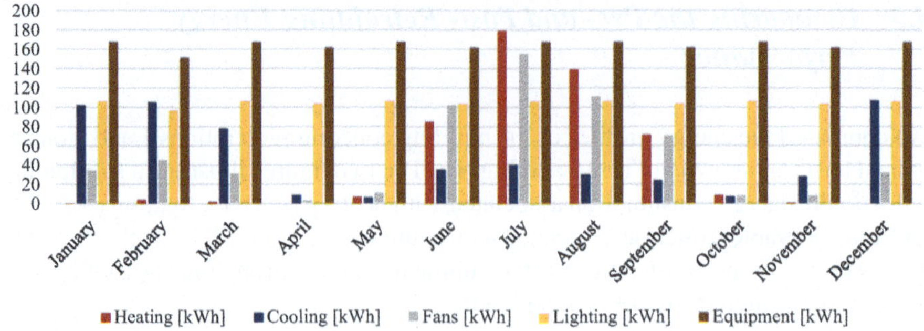

Fig. 17 Monthly energy profile for Iteration 9

4.4 Energy Balance

Considering that the building relay entirely on electricity, for iteration 9, we evaluate the energy balance between the annual electricity demand and the renewable energy generated on-site. Figure 18 shows that the on-site energy generation from PV largely exceeds the building energy demand. This would mean that the building tends to be a positive energy building (PEB) rather than an NZEB. However, only further analysis daily can illustrate the actual trend of energy generation concerning energy demand. Sefaira does not allow such analysis or give renewable generation per month in output. However, Sefaira gives in the output of the building usage of on-site generated energy per month, allowing us to observe that the building energy demands are off-set each month, meaning that the building might fall in the PEB category, Fig. 19. This means, that a power storage system is required. Recently) different pilot projects started Melbourne suburbs (Victoria) to compare the effectiveness of community shared powers storage against single dwelling private batteries. This site offers the opportunity to further study the applicability and effectiveness of the share power storage system in historical precincts.

5 Conclusions

The study presented an energy retrofitting strategy of a State Heritage registered building to be adaptively re used as an Aboriginal dormitory within the broader site context of an Aboriginal healing centre. The case study is located in the south-west of Western Australia. We evaluated the impact of the retrofitting measures on the overall energy performance of the building against the net-zero energy target, and their impact on the building' s cultural and historical values and the landscape using building energy performance simulation. The study highlighted the relevance of architectural design in energy retrofitting projects and provide quantitative data concerning the impact of design on the building's energy performance. Specifically,

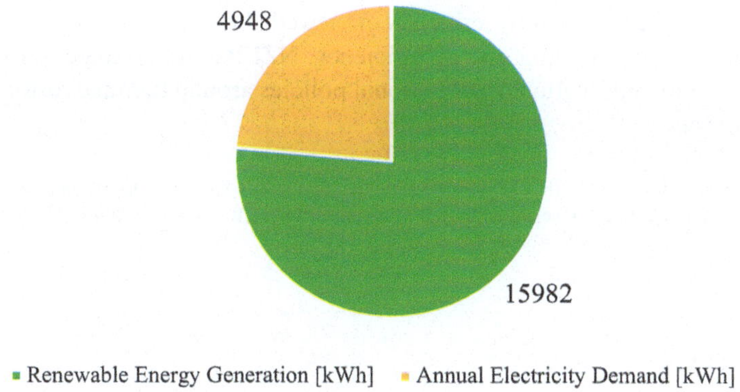

Fig. 18 Energy balance (retrofitted building)

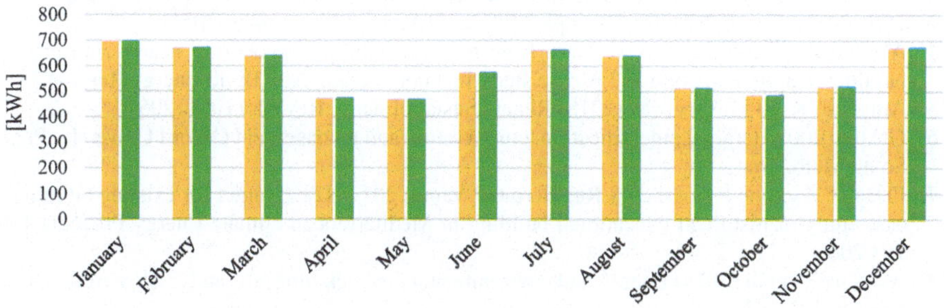

Fig. 19 Energy demand and renewable consumption per month

the study investigated the impact of integrating Aboriginal lifestyle on architectural design. The design strategy involved consultation with, and decisions made by the community stakeholders. We found that integrating Indigenous lifestyle and culture within the retrofitting strategy added positively in achieving the NZEBs target. The optimal use of natural ventilation with minimal intervention in the heritage structure and Aboriginal community living resulted in cultural and environmental-specific sets of outcomes increasing the building system's efficiency. Architectural technology plays a crucial role in energy retrofitting; upgrading the building elements substantially decreases the building energy demand, facilitating the hitting the energy neutrality target. Retrofitting historical heritage poses technical challenges to energy retrofitting. This study demonstrated that qualitative and quantitative data can be cross-red to develop design approaches that mitigate the impact of the retrofitting measures on the building fabric; this can optimise energy retrofitting of heritage buildings. Further studies is required to evaluate the cost effectiveness of the retrofitting measures presented here and their impact on the adaptive thermal comfort, which was partially integrated in the analysis. The outcome was a methodological path and best practice for energy retrofitting heritage buildings within a culturally responsive design framework against net-zero energy targets in the south-west of WA. Besides, the study investigate of study critically reviewed the current Australian national policies concerning building energy efficiency, NZEBs and heritage retrofitting. It also reviewed the international and national policies around heritage restoration and adaptive re use.

Acknowledgements We would like to thank Curtin University for their support towards this project. We thank Fatmaelzahraa Hussein for providing literature review on NetZero and Heritage.

References

1. ABCB (2019) National Construction Code—Volume 2, Canberra : ABCA
2. ABCB (2021) *Understanding the NCC – Assessment Methods.* [Online]
3. Available at: https://www.abcb.gov.au/sites/default/files/resources/2021/UTNCC_Thermal%20Bridging_Case%20Study_Example%202.pdf [Accessed November 2021]
4. Alwetaishi, M., 2020. Can we learn from heritage buildings to achieve nearly zero energy building and thermal comfort? A case study in a hot climate. Adv Build Energy Res 1–17
5. Amasyali K, El-Gohary NM (2018) Renew Sustain Energy Rev 81:1192–1205
6. G Aplin 2002 Heritage : identification, conservation and management Oxford University Press South Melbourne
7. FDMRF Ascione F Rossi de S Ruggiero GP Vanoli 2017 NZEB target for existing buildings: case study of historical educational building in Mediterranean climate Energy Procedia 140 194 206
8. Asdrubali F et al (2019) Energy and environmental payback times for an NZEB retrofit. Build Environ 147:461–472
9. Asdrubali F, Ballarini I, V. C., Evangelisti, L. & G. Grazieschi, a. C. G., (2019) Energy and environmental payback times for an NZEB retrofit. Build Environ 147:461–472
10. ASHRAE, 2018. Standard ASHRAE 90.2—Energy Efficient Design of Low-rise Residential Buildings, s.l.: ASHRAE
11. ASHRAE, 2020. ASHRAE 55-2020 Thermal Environmental Conditions for Human Occupancy (ANSI Approved), s.l.: ASHRAE
12. Assimakopoulos M-N et al (2020) Energy and Buildings 207(10):109630
13. S Attia 2018 Net Zero Energy Buildings (NZEB): concepts, frameworks and roadmap for project analysis and implementation Elsevier s.l.
14. Attia S, Hamdy M & Ezzeldin S (2017) Twenty-year tracking of lighting savings and power density in the residential sector. Energy Build 113–126
15. Australian Building Code Boards, 2020. National Construction Code, Volume Two, Canberra: Building Code of Australia
16. Beccali M et al (2018) Renew Sustain Energy Rev 82:1726–1736
17. Becchio C et al (2017) Toward NZEB by optimizing HVAC system configuration in different climates. Energy Proc 140:115–126
18. Bouden C, Ghrab N (2005) An adaptive thermal comfort model for the Tunisian context:a field study results. Energy Buildings 37:952–963
19. Buso T, Becchio C, Corgnati S (2017) NZEB, cost- and comfort-optimal retrofit solutions for an Italian Reference Hotel. Energy Proc 140:217–230
20. CEN 2005 EN 15251 Indoor environmental input parameters for design and assessment of energy performance of buildings addressing lighting and acoustics European Committee for Standardization Brussels
21. DR Cooper TG Gutowski 2017 The Environmental Impacts of Reuse: A Review J Ind Ecol 21 1 38 56

22. D'Agostino D, Mazzarella L (2019) What is a Nearly zero energy building? Overview, implementation and comparison of definition. Journal of Building Engineering 21:200–212
23. D'Agostino D, Tzeiranaki S, Zangheri P & Bertoldi P (2021) Assessing nearly zero energy buildings (nzebs) development in Europe. Energy Strategy Rev 100680
24. Daniel L, Soebarto V, Williamson T (2015) Energy and Buildings 88:34–44
25. Dartevelle OVMGMEaAS (2021) Long-term evaluation of residential summer thermal comfort: Measured vs. perceived thermal conditions in nZEB houses in Wallonia.. Build Environ 190:107531
26. Debacker W, Manshoven S & Denis F (2016) D1 Synthesis of the state-of-the-art: Key barriers and opportunities for Materials Passports and Reversible Building Design in the current system., s.l.: BAMB Horizon 2020
27. Den Elzen M et al. (2019) Are the G20 economies making enough progress to meet their NDC targets?. Energy policy pp. 238–250
28. Dong X, Soebarto V, Griffith M (2015) Energy and Building 86:852–863
29. European Commision, 2018. *Energy performance of buildings directive.* [Online] Available at: https://energy.ec.europa.eu/topics/energy-efficiency/energy-efficient-buildings/energy-performance-buildings-directive_en [Accessed 03 March 2022]
30. O Fenner 2017 Carrolup Marribank Mission building assessment report document Curtin University, School of Built Environment Perth
31. Foster G (2020) Resour Conserv Recycl 152:104507
32. Gremmelspacher JM et al (2021) Sol Energy 223:248–260
33. Grey, L. (n.d.) Carrolup Marribank conservation management plan, s.l.: s.n.
34. Guillen-Lambea S, Rodriguez-Soria B & Marin J (2017) Comfort settings and energy demand for residential nZEB in warm climates. Appl Energy 471–486
35. Hammond GP & Jones CI (2008) Embodied energy and carbon in construction materials. In: Proceedings of the Institution of Civil Engineers. s.l., Energy
36. Heritage Council (2020) *inHerit Carrolup Native Settlement.* [Online]. Available at: http://inherit.stateheritage.wa.gov.au/Public/Inventory/Details/2fd64e1e-d3a0-47aa-b96c-aa45ddbd04e6 [Accessed 16 Novemebr 2021]
37. Humphreys M, Rijal H, Nicol J (2013) Build Environ 63:40–55
38. ICOMOS (2002) ICOMOS World Report on Monuments and Sites (Heritage at Risk 2002/2003, Issue), s.l.: ICOMOS
39. ICOMOS (2010). Charter for the conservation of places of cultural herritage value. , s.l.: ICOMOS
40. ISO (2005). ISO 7730:2005 Ergonomics of the thermal environment — Analytical determination and interpretation of thermal comfort using calculation of the PMV and PPD indices and local thermal comfort criteria, Geneva: s.n
41. Klimowicz J (2018) Chosen case studies of nZeb retrofit buildings.. In: Design Solutions for nZEB Retrofit Buildings. s.l.: IGI Global, pp. 209–227
42. N Lechner 2014 Heating, Cooling, Lighting John Wiley & Sons Hoboken, New Jersey
43. Lovejoy FH (1971) Aust Q 43(1):79–90
44. Martek I et al (2019) J Clean Prod 211:281–292
45. Memmott P (2003) Housing Design in Indigenous Australia, Red Hill, ACT: The Royal Australian Institute of Architects
46. Mishra A, Ramgopal M (2013) Build Environ 64:94–106
47. Moazzen N, Karaguler M & Ashrafin T (2021) Comprehensive parameters for the definition of nearly zero energy and cost optimal levels considering the life cycle energy and thermal comfort of school buildings. Energy Build 253
48. Moran P, O'Connell J, Goggins J (2020) Sustainable energy efficiency retrofits as residenial buildings move towards nearly zero energy building (NZEB) standards. Energy and Buildings 211:109816
49. Munari Probst MCRC (2019) Criteria and policies to master the visual impact of solar systems in urban environments: The LESO-QSV method. Solar Energy 672–687

50. Omrany H, Soebarto V & Ghaffarianhoseini A (2021) Rethinking the concept of building energy rating system in Australia: a pathway to life-cycle net-zero energy building design. Arch Sci Rev 1–15
51. A Orbasli 2008 Architectural conservation : principles and practice Blackwell Science Oxford
52. Rabani M, Madessa HB & Natasa N (2017). A state-of-art review of retrofitting interventions in buildings toward nearly zero energy level. s.l., s.n., pp 317–326
53. Rodrigues C, . F. F., (2017) Adaptive reuse of buildings: Eco-efficiency assessment of retrofit strategies for alternative uses of an historic building. J Clean Prod 157:94–105
54. Sartori I, Napolitano A, Karsten V (2012) Energy and Buildings 48:220–232
55. Sharma V, Haque MH, Mahfuzul Aziz S (2019) Data Brief 25:104235
56. Stanojević AD et al. (2021) Developing multi-criteria model for the protection of built heritage from the aspect of energy retrofitting. Energy and Buildings, p. 111285
57. Strazzeri V, Karrech A (2022) Energy and Buildings 260:111901
58. Tang F, Chen J & Rodriguez D (2021) Energy saving actions toward NZEBs with multiple-criteria optimization in current residential buildings. Energy Report 3008–3022
59. Tiwari RSJHR (2019) Mission rehabilitation – A community centric approach to Aboriginal Healing in the Journal of Australian Aboriginal Studies. Aboriginal Studies Press
60. Wei S et al. (2011) Indoor thermal environment evaluations and parametric analyses in naturallyventilated buildings in dry season using afield survey and PMVe-PPDe mode. Building and Environment pp. 1275–1283
61. Wells L, Rismanchi B, Aye L (2018) Energy buildings 158:616–628
62. Williamson T, Daniel L (2020) A new adaptive thermal comfort modelfor homes in temperate climates of Australia. Energy Buildings 210:109728
63. Yu M, Wiedmann T, Crawford R, Tait C (2017) Proc Eng 180:211–220
64. Zuhaib S, Goggins J (2019) Energ Effi 12:1891–1920

The manufacturer's authorised representative in the EU is Springer Nature Customer Service Centre GmbH, Europaplatz 3, 69115 Heidelberg, Germany. If you have any concerns regarding our products, please contact ProductSafety@springernature.com

Printed and bound by CPI Group (UK) Ltd, Croydon, CR0 4YY
30/09/2026
02225318-0017